Remote Sensing: Analysis, Interpretation and Implementation

Remote Sensing: Analysis, Interpretation and Implementation

Edited by Jenson Gonzalez

SYRAWOOD
PUBLISHING HOUSE
New York

Published by Syrawood Publishing House,
750 Third Avenue, 9th Floor,
New York, NY 10017, USA
www.syrawoodpublishinghouse.com

Remote Sensing: Analysis, Interpretation and Implementation
Edited by Jenson Gonzalez

International Standard Book Number: 978-1-68286-866-9 (Hardback)

Cataloging-in-Publication Data

Remote sensing : analysis, interpretation and implementation / edited by Jenson Gonzalez.
p. cm.
Includes bibliographical references and index.
ISBN 978-1-68286-866-9
1. Remote sensing. 2. Remote sensing--Analysis. 3. Remote-sensing images. I. Gonzalez, Jenson.
G70.4 .R46 2020
621.367 8--dc23

Printed in China.

TABLE OF CONTENTS

PREFACE

Information can be gathered about an object, event or phenomenon without making direct physical contact with it. This is possible by applying the technology of remote sensing. Using satellite-based or aircraft-based sensor technology, objects on the Earth can be detected and classified. Data can be acquired from conventional radar, laser and radar altimeters, LIDAR, ultrasound and radar tide gauges, spectropolarimetric imaging, radiometers and photometers, etc. Remote sensing is useful in varied domains such as land surveying, hydrology, geography, ecology, glaciology, oceanography, etc. It also has commercial, economic, military, intelligence and humanitarian applications. This book unravels the recent studies in the field of remote sensing. Also included in this book is a detailed explanation of the different methods used for the analysis, interpretation and use of remote sensing data. A number of latest researches have been included to keep the readers up-to-date with the global concepts in this area of study.

This book is a comprehensive compilation of works of different researchers from varied parts of the world. It includes valuable experiences of the researchers with the sole objective of providing the readers (learners) with a proper knowledge of the concerned field. This book will be beneficial in evoking inspiration and enhancing the knowledge of the interested readers.

In the end, I would like to extend my heartiest thanks to the authors who worked with great determination on their chapters. I also appreciate the publisher's support in the course of the book. I would also like to deeply acknowledge my family who stood by me as a source of inspiration during the project.

Editor

Significant Remote Sensing Vegetation Indices: A Review of Developments and Applications

Jinru Xue and Baofeng Su

College of Mechanical and Electronic Engineering, Northwest A&F University, Yangling, Shaanxi 712100, China

Correspondence should be addressed to Baofeng Su; bfs@nwsuaf.edu.cn

Academic Editor: Chenzong Li

Vegetation Indices (VIs) obtained from remote sensing based canopies are quite simple and effective algorithms for quantitative and qualitative evaluations of vegetation cover, vigor, and growth dynamics, among other applications. These indices have been widely implemented within RS applications using different airborne and satellite platforms with recent advances using Unmanned Aerial Vehicles (UAV). Up to date, there is no unified mathematical expression that defines all VIs due to the complexity of different light spectra combinations, instrumentation, platforms, and resolutions used. Therefore, customized algorithms have been developed and tested against a variety of applications according to specific mathematical expressions that combine visible light radiation, mainly green spectra region, from vegetation, and nonvisible spectra to obtain proxy quantifications of the vegetation surface. In the real-world applications, optimization VIs are usually tailored to the specific application requirements coupled with appropriate validation tools and methodologies in the ground. The present study introduces the spectral characteristics of vegetation and summarizes the development of VIs and the advantages and disadvantages from different indices developed. This paper reviews more than 100 VIs, discussing their specific applicability and representativeness according to the vegetation of interest, environment, and implementation precision. Predictably, research, and development of VIs, which are based on hyperspectral and UAV platforms, would have a wide applicability in different areas.

1. Introduction

Remote sensed information of growth, vigor, and their dynamics from terrestrial vegetation can provide extremely useful insights for applications in environmental monitoring, biodiversity conservation, agriculture, forestry, urban green infrastructures, and other related fields. Specifically, these types of information applied to agriculture provide not only an objective basis (depending on resolution) for the macro- and micromanagement of agricultural production but also in many occasions the necessary information for yield estimation of crops [1]. These latter applications have been developed to be a well-known discipline category, precision agriculture, which could be tracked back to three decades ago [1]. However, the applicability of remote sensing and its different VIs extracted from these techniques usually relies heavily on the instruments and platforms to determine which solution is best to get a particular issue.

1.1. Remote Sensing Platform Considerations. In terms of platforms, the advantages of satellite based remote sensing include high spatial resolution, which makes possible the extraction of long time data series of consistent and comparable data, which can be cost effective [2]. Furthermore, some satellite platforms have free access to visible and multispectral data, such as Landsat 7-8. However, there are two main problems with these platforms for precision agriculture applications, which are related to the per pixel resolution (30 m^2 per pixel for Landsat and 500 m^2 for MODIS) and the orbit period (16 d for Landsat and 26 d for SPOT). More recently, pixel resolution has been increased by newer satellites, such as WorldView-2 and -3 (DigitalGlobe, Longmont, CO, USA). WorldView-2 was the first commercial high resolution satellite to provide eight spectral sensors in the visible to near infrared range. Along with the four typical multispectral bands: blue (450–510 nm), green (510–580 nm), red (630–690 nm), and near infrared (NIR) (770–895 nm), each sensor is narrowly focused on a particular range of the electromagnetic spectrum that is sensitive to a particular feature from the ground or a property of the atmosphere. However, images from this platform can be cost prohibitive for long time data series studies.

The second problem with satellite based remote sensing is the revisitation time, which is 16 days in average, which makes the agricultural applications difficult, specifically those related to water and nutrient management. Moreover, passive sensors cannot penetrate clouds; therefore, there is no usable data capture for overcast days.

To solve these two main problems, airborne and more recently UAV platforms can be used. The former can also be cost prohibitive due to the requirement of expensive aircrafts and pilots. The latter has become almost of ubiquitous use in the last five years with affordable aircrafts and camera payloads ranging from visible, near and thermal infrared, and 3D LIDAR, which has been referred to as Unmanned Aerial System (UAS). Among UAS platforms, there are mainly fixed wing and multirotor options available. There is a compromise using these UAS platforms in relation to payload weight versus flying time. In general, longer flying time achieved by fixed wing systems demands lighter weight payloads. For example, small high definition visible cameras weighting less than 300 grams as payload of a fixed wing UAS allow it to fly for around two hours using currently available battery-power [3]. On the contrary, battery-powered multirotor UAS with higher payload capacity have reduced fly time that at the moment is around 15- to 25-minute duration. Using these UAS, higher spatial and temporal data resolution can be achieved, which makes possible precision agriculture applications to the submeter resolution per pixel. This allows research and practical applications applied to growth and vigor dynamic assessment, plant water status sensing for irrigation scheduling applications, and evapotranspiration modelling, among others [4–9].

1.2. Remote Sensing and Vegetation Indices. Remote sensing of vegetation is mainly performed by obtaining the electromagnetic wave reflectance information from canopies using passive sensors. It is well known that the reflectance of light spectra from plants changes with plant type, water content within tissues, and other intrinsic factors [10]. The reflectance from vegetation to the electromagnetic spectrum (spectral reflectance or emission characteristics of vegetation) is determined by chemical and morphological characteristics of the surface of organs or leaves [3]. The main applications for remote sensing of vegetation are based on the following light spectra: (i) the ultraviolet region (UV), which goes from 10 to 380 nm; (ii) the visible spectra, which are composed of the blue (450–495 nm), green (495–570 nm), and red (620–750 nm) wavelength regions; and (iii) the near and mid infrared band (850–1700 nm) [11, 12]. The emissivity rate of the surface of leaves (equivalent to the absorptivity in the thermal waveband) of a fully grown green plant without any biotic or abiotic stress is generally in the range of 0.96–0.99 and is more often between 0.97 and 0.98 [13]. On the contrary, for dry plants, the emissivity rate generally has a larger range going from 0.88 to 0.94 [13]. Vegetation emissivity in the near and mid infrared regions has been widely studied within plant canopies. Indices extracted from this light spectra range can be attributed to a range of characteristics beyond growth and vigor quantification of plants related to water content, pigments, sugar and carbohydrate content, protein content, and aromatics, among others [2, 14]. Different applications are dependent on the reflectivity peaks or overtones for specific compounds within the visible and near/mid infrared regions of light spectra [14, 15]. Plant reflectivity in the thermal infrared spectral range (8–14 μm) follows the blackbody radiation law [16], which allows interpreting plant emission as directly related to plant temperature. Hence, indices obtained from this spectra range can be used as a proxy to assess stomata dynamics that regulates transpiration rate of plants. Therefore, the later indices can be used as indicator of plant water status [17–19] and abiotic/biotic stress levels [20, 21].

The latter considerations demonstrate that the quantitative interpretation of remote sensing information from vegetation is a complex task. Many studies have limited this interpretation by extracting vegetation information using individual light spectra bands or a group of single bands for data analysis. Thus, researchers often combine the data from near infrared (0.7–1.1 m) and red (0.6–0.7 m) bands in different ways according to their specific objectives [2]. These types of combinations present many disadvantages (e.g., lack of sensitivity) by using single or limited group of bands to detect, for example, vegetation biomass. These limitations are particularly evident when trying to apply these types of VI on heterogeneous canopies, such as horticultural tree plantations. A mixed combination of soils, weeds, cover crops in the interrow, and the plants of interest makes the discrimination regions of interest and extraction of simple VI very difficult, specifically, when the vegetation of interest has different VIs due to spatial variability, or VIs corresponding to other vegetation (weeds and cover crop), which can be similar to those of interest. The later will complicate imaging denoising and filtering processes. Several image analysis techniques and algorithms have been developed to go around these issues, which will be described later. Even though there are many considerations as described before, the construction of simple VI algorithm could many times render simple and effective tools to measure vegetation status on the surface of the Earth [6].

2. Vegetation Indices and Validation Process

With the use of high resolution spectral instrumentation, the number of bands obtained by remote sensing is increasing, and the bandwidth is getting narrower [7]. One of the most used and implemented indices calculated from multispectral information as normalized ratio between the red and near infrared bands is the Normalized Difference Vegetation Index (NDVI) [22]. A direct use of NDVI is to characterize canopy growth or vigor; hence, many studies have compared it with the Leaf Area Index (LAI) [23], where LAI is defined as the area of single sided leaves per area of soil [24].

Vegetation information from remote sensed images is mainly interpreted by differences and changes of the green leaves from plants and canopy spectral characteristics. The most common validation process is through direct or indirect correlations between VIs obtained and the vegetation characteristics of interest measured in situ, such as vegetation cover, LAI, biomass, growth, and vigor assessment. More established methods are used to assess VIs using direct and georeferenced methods by monitoring sentinel plants to be compared with VIs obtained from the same plants for calibration purposes.

The later process is known as allometric measurements and requires destructive methods to scan specific area of total leaves per plant or tree in the case of LAI [25]. Indirect validation methods are based on proximal instrumentation using the same or similar spectral instrumentation to assess georeferenced sentinel plants at the same angle as the aerial platforms. The latter method is useful to compare VIs obtained from satellite that are sensitive to atmospheric effects and serve as a mean to obtain correction factors. More recent indirect methods based on cover photography to estimate canopy cover, LAI, porosity, and clumping index have used automated analysis methods.

For this purpose, upward looking cover photogrammetry at zero zenith angle is taken with visible cameras to obtain canopy architecture parameters calculated using computer vision algorithms. An automated image acquisition and calculation method was proposed by Fuentes et al. 2008 applied to Eucalyptus trees [26] and it has been successfully applied for other crops such as grapevines compared to allometric measurements and to validate NDVI calculated from satellite information (WorldView-2) [27], apple trees with increased accuracy by using a variable light extinction coefficient (k) [28], and cherry trees improving the method by extracting nonleaf material such as branches for tall trees [29, 30]. In late 2015, a computer application (App) for smartphones and tablet PCs called VitiCanopy was released for free use to assess canopy architecture parameters using the cover photography automated algorithms, which can be applied to any other tree crop by changing to a specific k value [31, 32]. Other Apps using RGB photogrammetry to assess LAI have been later developed such as PoketLAI [33–35].

2.1. Basic Vegetation Indices. Jordan [36] proposed in 1969 one of the first VIs named Ratio Vegetation Index (RVI), which is based on the principle that leaves absorb relatively more red than infrared light; RVI can be expressed mathematically as

$$\mathrm{RVI} = \frac{R}{\mathrm{NIR}}, \tag{1}$$

where NIR is the near infrared band reflectance and R is red band reflectance. According to the spectral characteristics of vegetation, bushy plants have low reflectance on the red band and have shown a high correlation with LAI, Leaf Dry Biomass Matter (LDBM), and chlorophyll content of leaves [37]. The RVI is widely used for green biomass estimations and monitoring, specifically, at high density vegetation coverage, since this index is very sensitive to vegetation and has a good correlation with plant biomass. However, when the vegetation cover is sparse (less than 50% cover), RVI is sensitive to atmospheric effects, and their representation of biomass is weak.

The Difference Vegetation Index (DVI) was proposed later [38] and can be expressed as

$$\mathrm{DVI} = \mathrm{NIR} - R. \tag{2}$$

The DVI is very sensitive to changes in soil background; it can be applied to monitoring the vegetation ecological environment. Thus, DVI is also called Environmental Vegetation Index (EVI).

The Perpendicular Vegetation Index (PVI) [38] is a simulation of the Green Vegetation Index (GVI) in R, NIR 2D data. In the NIR – R coordinate system, the spectral response from soil is presented as a slash (soil brighten line). The latter effect can be explained as the soil presents a high spectral response in the NIR and R bands. The distance between the point of reflectivity (R, NIR) and the soil line has been defined as the Perpendicular VI, which can be expressed as follows:

$$\mathrm{PVI} = \sqrt{\left(\rho_{\mathrm{soil}} - \rho_{\mathrm{veg}}\right)_R^2 - \left(\rho_{\mathrm{soil}} - \rho_{\mathrm{veg}}\right)_{\mathrm{NIR}}^2}, \tag{3}$$

where ρ_{soil} is the soil reflectance; ρ_{veg} is the vegetation reflectivity; PVI characterizes the vegetation biomass in ρ_{red} the soil background; the greater the distance, the greater the biomass.

PVI can also be quantitatively expressed as

$$\mathrm{PVI} = \left(\mathrm{DN}_{\mathrm{NIR}} - b\right)\cos\theta - \mathrm{DN}_R * \sin\theta, \tag{4}$$

where $\mathrm{DN}_{\mathrm{NIR}}$ and DN_R are the radiation reflected luminance values from the NIR and R, respectively; b is the intercept of the soil baseline and the vertical axis of NIR reflectivity; and θ is the angle between the horizontal axis of R reflectivity and soil baseline. PVI filters out in this way the effects of soil background in an efficient manner; PVI also has less sensitivity to atmospheric effects and it is mainly used for the inversion of surface vegetation parameter (grass yield, chlorophyll content), the calculation of LAI, and vegetation identification and classification [39, 40]. However, PVI is sensitive to soil brightness and reflectivity, especially in the case of low vegetation coverage and needs to be adjusted for this effect [41].

As mentioned before, the Normalized Difference Vegetation Index (NDVI) is the most widely used as VI; it was proposed by Rouse Jr. et al. [42], which can be expressed as

$$\mathrm{NDVI} = \frac{\left(\rho_{\mathrm{NIR}} - \rho_R\right)}{\rho_{\mathrm{NIR}}} + \rho_R. \tag{5}$$

Since the index is calculated through a normalization procedure, the range of NDVI values is between 0 and 1, having a sensitive response to green vegetation even for low vegetation covered areas. This index is often used in research related to regional and global vegetation assessments and was shown to be related not only to canopy structure and LAI but also to canopy photosynthesis [43, 44]. However, NDVI is sensitive to the effects of soil brightness, soil color, atmosphere, cloud and cloud shadow, and leaf canopy shadow and requires remote sensing calibration.

2.2. Vegetation Indices considering Atmospheric Effects. Given the limitations of NDVI under atmospheric effects, Kaufman and Tanré [40] proposed the Atmospherically Resistant Vegetation Index (ARVI). This index is based on the knowledge that the atmosphere affects significantly R compared to the NIR. Thus, Kaufman and Tanré modified the radiation value of R by the difference between the blue (B) and R. Therefore, ARVI can effectively reduce the dependence of this VI to atmospheric effects, which can be expressed as

$$\text{ARVI} = \frac{(\text{NIR} - RB)}{(\text{NIR} + RB)}, \qquad \rho^*_{rb} = \rho^*_{\tau} - \gamma\left(\rho^*_b - \rho^*_{\tau}\right), \tag{6}$$

where RB is the difference between B and R, is the reflectivity related to the molecular scattering and gaseous absorption for ozone corrections, and represents the air conditioning parameters.

The ARVI is commonly used to eliminate the effects of atmospheric aerosols. The aerosols and ozone effects in the atmosphere still need to be eliminated by the 5S atmospheric transport model [45]. However, to implement the 5S atmospheric transmission model, actual atmospheric parameters must be considered, which are difficult to obtain. If the ARVI index is not calculated using the 5S model, this index is not expected to perform much better than NDVI considering atmospheric effects or large dust particles in the atmosphere. Thus, Zhang et al. [46] proposed a new atmospheric effect resistant vegetation index, namely, IAVI, that can eliminate atmospheric interference without the use of the 5S model.

$$\text{IAVI} = \frac{\{\rho_{\text{nir}} - [\rho_r - \gamma(\rho_b - \rho_r)]\}}{\{\rho_{\text{nir}} + [\rho_r - \gamma(\rho_b - \rho_r)]\}}, \tag{7}$$

where the range of γ values is between 0.65 and 1.12; a significant value of γ is close to 1 for ARVI. After testing, the error caused in IAVI by the atmosphere effect is between 0.4% and 3.7%, which is less than those found when using NDVI in the same conditions (14–31%).

2.3. Adjusted-Soil Vegetation Index. The distinction of vegetation from the soil background was originally proposed by Richardson and Wiegand [47] by analyzing the soil line, which can be considered as a linear relationship on the 2D plane of the soil spectral reflectance values between the NIR and R. Therefore, it can be considered as a comprehensive description of a large number of soil spectral information from a number of environments [48]. Many VIs that take into account the effect of soil background have been based on this principle. In addition to PVI ((3)-(4)), the Soil Line Atmospheric Resistance Index (SLRA) was developed based on the improvement of the soil line principle. The SLRA was then combined with the Transformed Soil-Adjusted Vegetation Index (TSAVI) to develop the Type Soil Atmospheric Impedance Vegetation Index (TSARVI) [49], which will be discussed later.

As shown before, NDVI (5) is very sensitive to background factors, such as the brightness and shade of the vegetation canopies and soil background brightness. Researches have shown that when the background brightness is increased, NDVI also increases systematically. Given the effect of soil background, R radiation increases significantly when the vegetation cover is sparse; conversely NIR radiation is reduced to make the relationship between vegetation and soil more sensible. Many VIs have been developed to adjust to the soil effect.

Since NDVI and PVI have some deficiencies in describing the spectral behavior of vegetation and soil background, Huete [50] established the Soil-Adjusted Vegetation Index (SAVI), which can be expressed as follows:

$$\text{SAVI} = \frac{(\rho_n - \rho_r)(1 + L)}{(\rho_n + \rho_r + L)}. \tag{8}$$

The above model of a soil vegetation system was established to improve the sensitivity of NDVI to soil backgrounds, where L is the soil conditioning index, which improves the sensitivity of NDVI to soil background. The range of L is from 0 to 1. In practical applications, the values of L are determined according to the specific environmental conditions. When the degree of vegetation coverage is high, L is close to 1, showing that the soil background has no effect on the extraction of vegetation information. This kind of ideal conditions is rarely found in natural environments and can be applicable only in the case of a large canopy density and coverage [40]. The value of L is around 0.5 under most common environmental conditions. When L is close to 0, the value of SAVI is equal to NDVI. However, L factor should vary inversely with the amount of vegetation present to obtain the optimal adjustment for the soil effect. Thus, a modified SAVI (MSAVI) replaces L factor in the SAVI equation (8) with a variable L function. In this way, MSAVI [51] reduces the influence of bare soil on SAVI, which can be expressed as follows:

$$\text{MSAVI} = 0.5 * \{2R_{800} + 1 - \text{SQRT}\left[(2R_{800} + 1)^2 - 8(R_{800} - R_{670})\right]\}. \tag{9}$$

The SAVI is much less sensitive than the RVI (1) to changes in the background caused by soil color or surface soil moisture content. Three new versions of SAVI (SAVI2, SAVI3, and SAVI4) were developed based on the theoretical considerations of the effects of wet and dry soils [41]. SAVI2, SAVI3, and SAVI4 reduce the effect of background soil brightness, by taking into account the effect of the variation of the solar incidence angle and changes in the soil physical structure.

Based on the implementation of the MSAVI, Richardson and Wiegand (1977) proposed a Modified Secondary Soil-Adjusted Vegetation Index (MSAVI2) [47], which can be expressed as

$$\text{MSAVI2} = 0.5 * \left[(2\text{NIR} + 1) - \sqrt{(2\text{NIR} + 1)^2 - 8(\text{NIR} - R)}\right]. \tag{10}$$

MSAVI2 does not rely on the soil line principle and has a simpler algorithm; it is mainly used in the analysis of plant growth, desertification research, grassland yield estimation, LAI assessment, analysis of soil organic matter, drought monitoring, and the analysis of soil erosion [39].

Baret et al. studied the sensitivity of five VIs (NDVI, SAVI, Transformed Soil-Adjusted Vegetation Index (TSAVI), Modified Soil-Adjusted Vegetation Index (MSAVI), and Global Environment Monitoring Index (GEMI)) to the soil background. They simulated the performance of the VIs for different soil textures, moisture levels, and roughness by using the Scattering from Arbitrarily Inclined Leaves (SAIL) model. They determined an optimum value of SAIL = 0.16 to reduce the effects of soil background and then proposed an Optimized Soil-Adjusted Vegetation Index (OSAVI) [48] that can be expressed as follows:

$$\text{OSAVI} = \frac{(\text{NIR} - R)}{(\text{NIR} + R + X)}, \tag{11}$$

where SAIL is 0.16 and OSAVI does not depend on the soil line and can eliminate the influence of the soil background effectively. However, the applications of OSAVI are not extensive; it is mainly used for the calculation of aboveground biomass, leaf nitrogen content, and chlorophyll content, among others [52].

2.4. Tasseled Cap Transformation of Greenness Vegetation Index (GVI, YVI, and SBI). Kauth and Thomas studied the spectral pattern of the vegetation growth process and called it the "spike cap" pattern, including the soil background reflectivity and brightness line. The Tasseled Cap Transformation is a conversion of the original bands of an image into a new set of bands with defined interpretations that are useful for vegetation mapping. A Tasseled Cap Transformation is performed by taking "linear combinations" of the original image bands, which is similar in concept to the multivariate data analysis technique called principal components analysis (PCA) [53].

The Tasseled Cap can convert Landsat MSS, Landsat TM, and Landsat 7 ETM data. For Landsat MSS data, furthermore, the Tasseled Cap performs orthogonal transformation on the original data, which converts it into a 4D space. This conversion includes the Soil Brightness Index (SBI) (14), degree of Green Vegetation Index (GVI) (12), and the degree of Yellow Vegetation Index (YVI) (13). It also includes Nonsuch Index (NSI) mainly for noise reduction. The NSI is closely related to atmospheric effects. For the Landsat 5 TM data, the Tasseled Cap results consist of three factors: brightness, greenness, and a third component related to soil. Among them, the brightness and the greenness are equivalent to SBI and GVI in the MSS Tasseled Cap. The third component is related to soil characteristics and humidity. For Landsat 7 ETM data, the Tasseled Cap Transformation generates six bands, namely, brightness, greenness, humidity, the fourth component (noise), a fifth component, and a sixth component.

$$\begin{aligned} \text{GVI} = &-0.290\text{MSS}_4 - 0.562\text{MSS}_5 + 0.600\text{MSS}_6 \\ &+ 0.491\text{MSS}_7, \end{aligned} \tag{12}$$

$$\begin{aligned} \text{YVI} = &-0.829\text{MSS}_4 - 0.522\text{MSS}_5 + 0.039\text{MSS}_6 \\ &+ 0.149\text{MSS}_7, \end{aligned} \tag{13}$$

$$\begin{aligned} \text{SBI} = &+0.433\text{MSS}_4 - 0.632\text{MSS}_5 + 0.586\text{MSS}_6 \\ &+ 0.264\text{MSS}_7. \end{aligned} \tag{14}$$

The GVI, YVI, and SBI ignore the interaction and effects of the atmosphere, soil, and vegetation. SBI and GVI can be used to evaluate the behavior of vegetation and bare soil [54]. The GVI has a strong correlation with different vegetation covers. Thus, GVI increases the processing of atmospheric effects. Jackson et al. (1980) proposed the Adjust Soil Brightness Index (ASBI) and Adjust Green Degree Vegetation Index (AGVI) [55], which can be expressed as follows:

$$\begin{aligned} \text{ASBI} &= 2.0\text{YVI}, \\ \text{AGVI} &= \text{GVI} - (1 + 0.018\text{GVI})\,\text{YVI} - \frac{\text{NSI}}{2}. \end{aligned} \tag{15}$$

Misra and Wheeler (1977) performed PCA of Landsat images and computed the multiple factors of these indexes. This analysis was the basis of the development of the Misra Soil Brightness Index (MSBI), Misra Green Degree Vegetation Index (MGVI), and Misra Yellow Degree Vegetation Index (MYVI) [56], which can be expressed as follows:

$$\begin{aligned} \text{MSBI} = &\ 0.406\text{MSS}_4 + 0.60\text{MSS}_5 + 0.645\text{MSS}_6 \\ &+ 0.243\text{MSS}_7, \\ \text{MGVI} = &-0.386\text{MSS}_4 - 0.53\text{MSS}_5 + 0.535\text{MSS}_6 \\ &+ 0.532\text{MSS}_7, \\ \text{MYVI} = &\ 0.723\text{MSS}_4 - 0.597\text{MSS}_5 + 0.206\text{MSS}_6 \\ &- 0.278\text{MSS}_7. \end{aligned} \tag{16}$$

Since NDVI has been found to be affected only by soil brightness, it presents a negative correlation between NDVI and soil brightness. A positive correlation is found when only atmospheric effects affect NDVI. Under natural conditions, the soil and atmosphere influence NDVI in a complex manner, which interacts with the vegetation cover influence. Therefore, atmosphere and vegetation have a collective effect on NDVI based on the soil characteristics and exposure. Liu and Huete comprehensively analyzed multiple soil types and atmospheric enhanced VIs. They developed the Atmosphere Antivegetation Index (ARVI) and Soil-Adjusted Vegetation Index (SAVI) for a comprehensive analysis of vegetation in these conditions. They found that, as a result of the interaction between the soil and the atmosphere, reducing one of them may increase the other. They introduced a feedback mechanism by building a parameter to simultaneously correct soil and atmospheric effects. This parameter is the Enhanced Vegetation Index (EVI) [57] that can be expressed as follows:

$$\text{EVI} = \frac{(\text{TM}_4 - \text{TM}_3)(1 + L)}{\text{TM}_4 - C_1\text{TM}_3 + C_2\text{TM} + L},$$
$$\text{EVI} = 2.5 * \frac{\rho_n^* - \rho_r^*}{\rho_n^* + C_1\rho_r^* - C_2\rho_b^* + L}, \quad (17)$$

which includes the values of NIR, R, and B, which are corrected by the atmosphere; L represents soil adjustment parameters, and its value is equal to 1; and parameters correspond to constant values equivalent to 6 and 7.5, respectively.

2.5. Vegetation Indices Based on UAS Remote Sensing in the Visible Spectra Region. UAS remote sensing is a low altitude remote sensing technology (50–100 m), which is less affected by atmospheric factors during the data acquisition process. It has the advantages of affordability, simple operation, fast imaging speed, and high spatial and temporal resolutions, which is unparalleled compared with traditional [58] remote sensing technologies based on satellites. At present, the UAS remote sensing technology plays a crucial role in the field of aerial remote sensing with increased interest in applying these platforms on different studies of vegetation assessment [59]. The practical applications of UAS are mainly related to images acquisition in the visible bands (RGB) due to easy access of ubiquitous high resolution cameras at low price and weight. However, due to rapid advances in technology, multispectral and infrared thermal cameras are becoming increasingly cheaper and miniaturized.

As previously shown through different VIs, most of them are based on the mixture of visible light bands and NIR to generate algorithms and those based only on the visible light spectra are not common. However, weightless high definition cameras are appearing in the market that includes the NIR band, which will enhance the practical applicability of UAS in the near future. These types of reflectance are commonly measured using visible, multispectral, and hyperspectral cameras [5]. According to Gago et al. (2015), NDVI is one of the most employed indices for UAS applications and is defined specifically as

$$\text{NDVI} = \frac{(R_{800} - R_{680})}{(R_{800} + R_{680})}, \quad (18)$$

where R_{800} is the reflectance at 800 nm and R_{680} at 680 nm. Due to the high NIR reflectance of chlorophyll, this index is used to detect plants greenness [60]. Some studies described the use of UAS with multispectral cameras and high resolution multispectral satellites to estimate LAI (Leaf Area Index) through NDVI [27, 61].

The optimized index transformed chlorophyll absorption in reflectance Transformed Chlorophyll Absorption in Reflectance Index/Optimized Soil-Adjusted Vegetation Index (TCARI/OSAVI) was proposed as more sensitive VI to chlorophyll content. In this way, avoiding other factors that could affect the reflectance values such as canopy reflectance and soil reflectance among others [5]. Another index evaluated by Zarco-Tejada et al. (2013) was the PRInorm, which is an improvement of the Photochemical Reflectance Index (PRI). This index considers xanthophyll changes related to water stress but also generates a normalization considering chlorophyll content and canopy leaf area reduction which is mainly affected by water stress [62]. However, by obtaining a quick and effective method to extract vegetation information based on UAS visible images, it will enhance and popularize the scope of application of UAS immediately [63]. In this sense, Wang et al. (2015) comprehensively considered the spectral characteristics of healthy green vegetation and the spectral characteristics of typical features of UAS imagery [63]. They use green (G) band instead of the NIR band to calculate NDVI, ($\rho_{\text{red}} + \rho_{\text{blue}}$) compared to R for NDVI, and the G band multiplied by 2 for ($\rho_{\text{red}} + \rho_{\text{blue}}$). Thus, a Visible-Band Difference Vegetation Index (VDVI) is created based on the three bands of visible light, which can be expressed as follows:

$$\text{VDVI} = \frac{\left(2 * \rho_{\text{green}} - \rho_{\text{red}} - \rho_{\text{blue}}\right)}{\left(2 * \rho_{\text{green}} + \rho_{\text{red}} + \rho_{\text{blue}}\right)}. \quad (19)$$

The values of VDVI are within [−1, 1] and the accuracy of the vegetation extraction based on VDVI is higher than other visible light band-based VIs and G band. Furthermore, the accuracy of VDVI has been reported to be over 90% [63].

2.6. Vegetation Indices Related to Vegetation Status. The NDVI, as shown previously, enhances the contrast of the reflectivity of the NIR and R channels (5). Therefore, it is a nonlinear extension of NIR and R ratios, resulting in the enhancement of the lower part of these values (higher values are suppressed). Hence, NDVI reaches saturation in this way more easily. Thus, Gitelson [64] proposed the Wide Dynamic Range Vegetation Index (WDRVI), which can be expressed as follows:

$$\text{WDRVI} = \frac{(\alpha\rho_{\text{nir}} - \rho_{\text{red}})}{(\alpha\rho_{\text{nir}} + \rho_{\text{red}})}. \quad (20)$$

WDRVI enhances the dynamic range of NDVI by applying a weighting parameter to the NIR reflectance. If α equals 1, WDRVI is equivalent to NDVI. If α is equal to the ratio ($\rho_{\text{red}}/\rho_{\text{NIR}}$), WDRVI is zero. After validation procedures, a coefficient value of 0.20 for appears to be generally effective for the WDRVI calculations. According to Gitelson (2004) [64], WDRVI offers a simple way to enhance the dynamic range for high biomass environments (LAI > 2). However, when biomass is low (LAI < 1), NDVI is still the best choice for the plant classification.

According to the spectral reflectance of plant leaves (between 550 nm and 700 nm) it can be considered constant even if the chlorophyll content of leaves is variable. Based on this relationship, Kim et al. (1994) measured the level of absorption at 670 nm and linked the reflection peak at 700 nm and 550 nm; the Chlorophyll Absorption Ratio Index (CARI) was then developed [65] and can be expressed as

$$\text{CAR} * \left(\frac{R_{700}}{R_{670}}\right),$$
$$\text{CAR} = \frac{\left|(a * 670 + R_{670} + b)\right|}{\left(a^2 + 1\right)^{0.5}}, \qquad (21)$$
$$a = \frac{\left(R_{700} - R_{500}\right)}{150},$$
$$b = R_{550} - (a * 550).$$

Later, Daughtry et al. improved the CARI by proposing a modified CARI (MCARI) [66], which can be expressed as

$$\text{MCARI} = \frac{1.5 * \left[2.5\left(R_{800} - R_{670}\right) - 1.3\left(R_{800} - R_{550}\right)\right]}{\sqrt{\left(2R_{800} + 1\right)^2 - \left(6R_{800} - 5R_{670}\right) - 0.5}}. \qquad (22)$$

The MCARI is more sensitive to leaf chlorophyll concentrations. Daughtry et al. (2000) found that LAI, chlorophyll, and the chlorophyll-LAI interaction accounted for 60, 27, and 13% of the MCARI variation. Even though the MCARI formula is not related to the NIR bands, good predictions were still found.

In agriculture, crop growth is directly linked to water supply and plant water status. When the soil water supply is insufficient, plants will be under water stress, which leads to reduced crop yield and even crop failures under extreme drought conditions. So it is very important to evaluate the crop water status in a timely and accurate manner, which has direct implications on crop growth, yield, and quality of produce [67]. In recent years, the development of infrared thermal remote sensing technology made it possible to measure canopy temperature changes and dynamics from crop populations. These changes are related to the transpiration rate of plants and stomatal conductance. Hence, crop leaf and canopy temperature have been used for the determination of crop water status [68]. In order to make the canopy temperature measurements consistent, Idso et al. (1981) established the Crop Water Stress Index (CWSI) to monitor crop water status [69].

$$\text{CWSI} = \frac{\left(T_{\text{canopy}} - T_{\text{nws}}\right)}{\left(T_{\text{dry}} - T_{\text{nws}}\right)}, \qquad (23)$$

where T_{canopy} is the temperature of fully sunlit canopy leaves (°C), T_{nws} is the temperature of fully sunlit canopy leaves (°C) when the crop is non-water-stressed (well-watered); T_{dry} is the temperature of fully sunlit canopy leaves (°C) when the crop is severely water stressed due to low soil water availability. T_{nws} and T_{dry} are the lower and upper baselines used to normalize CWSI for the effects of environmental conditions (air temperature, relative humidity, solar radiation, and wind speed) on T_{canopy}. The CWSI has two models, an empirical model and a theoretical model; however, the theoretical model involves too many parameters, and these parameters are not easy to obtain. Therefore, the theoretical model is only used for research purposes [70–72]. The empirical model can be obtained only by using crop canopy temperature, air temperature, and air saturation difference, so the empirical model has been further studied and used in many crop applications [73].

Besides the use of infrared thermal radiation to detect plant water stress detection, the visible part of the spectrum has also been useful for early water stress detection. This involves using indices focused on bands at specific wavelengths where photosynthetic pigments are affected by water stress conditions such as chlorophyll. The Photochemical Reflectance Index (PRI) has been used as a stress index of stress based on this principle, with initial developments to be applied to disease symptoms detection, which can be expressed as

$$\text{PRI} = \frac{\left(R_{531} - R_{570}\right)}{\left(R_{531} - R_{570}\right)}. \qquad (24)$$

It has been shown that the Light Use Efficiency (LUE) is a key variable to estimate Net Primary Productivity (NPP) [74, 75]. When obtaining reliable accuracy in LUE measurements, it is possible to study the distribution of energy and global climate change. The PRI is a normalized difference VI of reflectivity at 531 nm and 570 nm and the reflectance of these two bands is affected by the xanthophyll cycle and is closely related to LUE of leaves. Therefore, PRI provides a good estimation of leaf LUE.

2.7. Summary of Vegetation Indices. A summary of the main VIs discussed in this paper can be found in Table 1 with their respective citations.

3. Conclusions

Simple VIs combining visible and NIR bands have significantly improved the sensitivity of the detection of green vegetation. Different environments have their own variable and complex characteristics, which needs to be accounted when using different VIs. Therefore, each VI has its specific expression of green vegetation, its own suitability for specific uses, and some limiting factors. Therefore, for practical applications, the choice of a specific VI needs to be made with caution by comprehensively considering and analyzing the advantages and limitations of existing VIs and then combine them to be applied in a specific environment. In this way, the usage of VIs can be tailored to specific applications, instrumentation used, and platforms. With the development of hyperspectral and multispectral remote sensing technology, new VIs can be developed, which will broaden research areas. It is envisioned that these new developments will be readily applied and adopted by UAS platforms and will become one of the most important research areas in aerospace remote sensing in the near future.

TABLE 1: Summary of vegetation index expression.

Index	Definition	Reference
AGVI	$\mathrm{GVI} - (1 + 0.018\mathrm{GVI}) * \mathrm{YVI} - \frac{\mathrm{NSI}}{2}$	[76]
ARI	$\left(\frac{1}{R_{550}}\right) - \left(\frac{1}{R_{700}}\right)$	[76]
ARI2	$R_{800}\left[\left(\frac{1}{R_{550}}\right) - \left(\frac{1}{R_{700}}\right)\right]$	[76]
ARVI	$\frac{(\mathrm{NIR} - RB)}{(\mathrm{NIR} + RB)}$	[40]
ASBI	$0.2\mathrm{YVI}$	[77]
ATSAVI	$\frac{[a(\mathrm{NIR} - a\mathrm{Red} - b)]}{[a\mathrm{NIR} + \mathrm{Red} - ab + X(1 + a^2)]}$	[78]
AVI	$2.0\mathrm{MSS}_7 - \mathrm{MSS}_5$	[79]
AVI	$\tan^{-1}\left\{\left[\frac{(\lambda_3 - \lambda_2)}{\lambda_2}\right](\mathrm{NIR} - R)^{-1}\right\} + \tan^{-1}\left\{\left[\frac{(\lambda_2 - \lambda_1)}{\lambda_2}\right](G - R)^{-1}\right\}$	[80]
BGI1	$\frac{R_{400}}{R_{550}}$	[81]
BGI2	$\frac{R_{450}}{R_{550}}$	[81]
BRI1	$\frac{R_{400}}{R_{690}}$	[60]
BRI2	$\frac{R_{450}}{R_{690}}$	[60]
CAI	$0.5(R_{2000} + R_{2200}) - R_{2100}$	[82]
CARI	$\mathrm{CAR} * \left(\frac{R_{700}}{R_{670}}\right)$	[65]
CCCI	$\frac{(\mathrm{NDRE} - \mathrm{NDRE}_{\min})}{(\mathrm{NDRE}_{\max} - \mathrm{NDRE}_{\min})}$	[83]
CRCWD	$1 - \rho^r_{\min}$	[84]
CRI500	$\frac{(1/R_{515})}{(1/R_{550})}$	[85]
CRI700	$\frac{(1/R_{515})}{(1/R_{700})}$	[85]
CWSI	$\frac{((T_c - T_a) - (T_c - T_a)_{ll})}{((T_c - T_a)_{ul} - (T_c - T_a)_{ll})}$	[69]
DI1	$R_{800} - R_{550}$	[86]
DVI	$\int_{\lambda_1}^{\lambda_1}\left(\frac{d_\rho}{d_\lambda}\right)d\lambda$	[87]
DVI	$2.4\mathrm{MSS}_7 - \mathrm{MSS}_5$	[38]
EVI	$\frac{[(\mathrm{TM}_4 - \mathrm{TM}_3)(1 + L)]}{(\mathrm{TM}_4 - \mathrm{C}_1\mathrm{TM}_3 + \mathrm{C}_2\mathrm{TM} + L)}$	[88]
EXG	$2 * \rho_{\mathrm{green}} - \rho_{\mathrm{red}} - \rho_{\mathrm{blue}}$	[89, 90]
GARI	$\frac{\{\mathrm{NIR} - [\mathrm{Green} - \gamma(\mathrm{Blue} - \mathrm{Red})]\}}{\{\mathrm{NIR} + [\mathrm{Green} - \gamma(\mathrm{Blue} - \mathrm{Red})]\}}$	[24]
GDVI	$\mathrm{NIR} - \mathrm{Green}$	[23]

Table 1: Continued.

Index	Definition	Reference
GEMI	$\eta(1-0.25\eta)-\frac{(R-0.125)}{(1-R)}$ $\eta=\frac{[2(\mathrm{NIR}^2-R^2)+1.5\mathrm{NIR}+0.5R]}{(\mathrm{NIR}+R+0.5)}$	[90]
GLI	$\frac{(2R_g-R_r-R_b)}{(2R_g+R_r+R_b)}$	[91]
GM1	$\frac{R_{750}}{R_{550}}$	[92]
GM2	$\frac{R_{750}}{R_{700}}$	[92]
GNDVI	$\frac{(\rho_{\mathrm{NIR}}-\rho_G)}{(\rho_{\mathrm{NIR}}+\rho_G)}$	[91]
GRABS	$\mathrm{GVI}-0.09178\mathrm{SBI}+5.58959$	[93]
GRVI	$\frac{\mathrm{NIR}}{\mathrm{Green}}$	[23]
Greenness index (G)	$\frac{R_{554}}{R_{677}}$	[94]
GVI	$(-0.283\mathrm{MSS}_4-0.66\mathrm{MSS}_5+0.577\mathrm{MSS}_6+0.388\mathrm{MSS}_7)$	[53]
GVSB	$\frac{\mathrm{GVI}}{\mathrm{SBI}}$	[95]
LIC3	$\frac{R_{400}}{R_{740}}$	[96]
HJVI	$\frac{[2(\rho_{\mathrm{nir}}-\rho_{\mathrm{red}})]}{(\rho_{\mathrm{nir}}+6\rho_{\mathrm{red}}-7.5\rho_{\mathrm{blue}}+1)}$	[97]
HI	$\frac{(R_{534}-R_{698})}{(R_{534}+R_{698})}-0.5R_{704}$	[98]
IAVI	$\frac{\{\rho_{\mathrm{nir}}-[\rho_r-\gamma(\rho_b-\rho_r)]\}}{\{\rho_{\mathrm{nir}}+[\rho_r-\gamma(\rho_b-\rho_r)]\}}$	[46]
II	$\frac{\mathrm{TM}_5}{\mathrm{TM}_7}$	[99]
IPVI	$\frac{\mathrm{TM}_4}{(\mathrm{TM}_4+\mathrm{TM}_3)}$	[100]
MCARI	$[(R_{700}-R_{670})-0.2(R_{700}-R_{550})]\left(\frac{R_{700}}{R_{670}}\right)$	[66]
MCARI	$\frac{1.5*[2.5(R_{800}-R_{670})-1.3(R_{800}-R_{550})]}{\sqrt{(2R_{800}+1)^2-(6R_{800}-5R_{670})-0.5}}$	[101]
MGVI	$(-0.386\mathrm{MSS}_4-0.53\mathrm{MSS}_5+0.535\mathrm{MSS}_6+0.532\mathrm{MSS}_7)$	[102]
MNLI	$\frac{[(\mathrm{NIR}^2-\mathrm{Red})(1+L)]}{(\mathrm{NIR}^2+\mathrm{Red}+L)}$	[103]
MNSI	$(0.404\mathrm{MSS}_4-0.039\mathrm{MSS}_5-0.505\mathrm{MSS}_6+0.762\mathrm{MSS}_7)$	[102]
MRENDVI	$\frac{(\rho_{750}-\rho_{705})}{(\rho_{750}+\rho_{705}-2*\rho_{445})}$	[104, 105]

Table 1: Continued.

Index	Definition	Reference
MRESR	$\frac{(\rho_{750}-\rho_{445})}{(\rho_{705}-\rho_{445})}$	[104, 105]
MSAVI2	$0.5*\left[(2\text{NIR}+1)-\sqrt{(2\text{NIR}+1)^2-8(\text{NIR}-R)}\right]$	[106]
MSBI	$(0.406\text{MSS}_4+0.60\text{MSS}_5+0.645\text{MSS}_6+0.243\text{MSS}_7)$	[102]
MSAVI	$0.5*\left\{2R_{800}+1-\text{SQRT}\left[(2R_{800}+1)^2-8(R_{800}-R_{670})\right]\right\}$	[51]
MSR	$\frac{[(R_{800}/R_{670})-1]}{[\text{SQRT}(R_{800}/R_{670}+1)]}$	[106]
MSI	$\frac{\rho_{1599}}{\rho_{819}}$	[107]
MTVI	$1.2*[1.2(R_{800}-R_{550})-2.5(R_{670}-R_{550})]$	[101]
MTVI2	$\frac{1.5*[1.2(R_{800}-R_{550})-2.5(R_{670}-R_{550})]}{\sqrt{(2*R_{800}+1)^2-(6*R_{800}-5*\sqrt{R_{670}})-0.5}}$	[101]
MYVI	$(0.723\text{MSS}_4-0.597\text{MSS}_5+0.206\text{MSS}_6-0.278\text{MSS}_7)$	[102]
NDGI	$\frac{(G-R)}{(G+R)}$	[78]
NDI	$\frac{(\text{NIR}-\text{MIR})}{(\text{NIR}+\text{MIR})}$	[108]
NDI1	$\frac{(R_{780}-R_{710})}{(R_{780}-R_{680})}$	[109]
NDI2	$\frac{(R_{850}-R_{710})}{(R_{850}-R_{680})}$	[109]
NDI3	$\frac{(R_{734}-R_{747})}{(R_{715}-R_{726})}$	[110]
NDNI	$\frac{[\log(1/\rho_{1510})-\log(1/\rho_{1680})]}{[\log(1/\rho_{1510})+\log(1/\rho_{1680})]}$	[111]
NDLI	$\frac{[\log(1/\rho_{1754})-\log(1/\rho_{1680})]}{[\log(1/\rho_{1754})+\log(1/\rho_{1680})]}$	[111]
NDVI	$\frac{(R_{800}-R_{680})}{(R_{800}+R_{680})}$	[96]
NDVI	$\frac{(\rho_{\text{NIR}}-\rho_R)}{(\rho_{\text{NIR}}+\rho_R)}$	[42]
NDWI	$\frac{(\text{Green}-\text{NIR})}{(\text{Green}+\text{NIR})}$	[112]
NGBDI	$\frac{(G-R)}{(G+B)}$	[113]
NGRDI	$\frac{(G-R)}{(G+R)}$	[114]
NMDI	$\frac{[\rho_{860}-(\rho_{1640}-\rho_{2130})]}{[\rho_{860}+(\rho_{1640}-\rho_{2130})]}$	[115]
NLI	$\frac{(\text{NIR}^2-\text{Red})}{(\text{NIR}^2+\text{Red})}$	[116]

Table 1: Continued.

Index	Definition	Reference
OSAVI	$\frac{(1+0.16)(R_{800}-R_{670})}{(R_{800}+R_{670}+0.61)}$	[117]
PRI	$\frac{(R_{531}-R_{570})}{(R_{531}+R_{570})}$	[118]
PSRI	$\frac{(R_{680}-R_{500})}{R_{750}}$	[119]
PSNDc	$\frac{(R_{800}-R_{470})}{(R_{800}+R_{470})}$	[120]
PSSRa	$\frac{R_{800}}{R_{680}}$	[120]
PSSRb	$\frac{R_{800}}{R_{635}}$	[120]
PSSRc	$\frac{R_{800}}{R_{470}}$	[120]
PVI	$\sqrt{(\rho_{\text{soil}}-\rho_{\text{veg}})_R^2-(\rho_{\text{soil}}-\rho_{\text{veg}})_{\text{NIR}}^2}$	[38]
PVI	$\frac{(\text{NIR}-aR-b)}{\sqrt{a^2+1}}$	[121]
RARS	$\frac{R_{746}}{R_{513}}$	[85]
RDVI	$\frac{(R_{800}-R_{670})}{[\text{SQRT}(R_{800}+R_{670})]}$	[121]
RDVI	$\sqrt{\text{NDIVI}\cdot\text{DVI}}$	[121]
RENDVI	$\frac{(R_{750}-R_{705})}{(R_{750}+R_{705})}$	[122]
RGRI	$\frac{(\sum_{I=600}^{690} R_I)}{(\sum_{I=500}^{599} R_J)}$	[123]
RI	$\frac{(R-G)}{(R+G)}$	[124]
RVI	$\frac{R}{\text{NIR}}$	[125]
SAVI	$\frac{(\rho_{\text{NIR}}-\rho_G)}{(\rho_{\text{NIR}}+\rho_G+L)}+(1+L)$	[50]
SIPI	$\frac{(R_{800}-R_{445})}{(R_{800}+R_{680})}$	[126]
SBI	$(-0.283\text{MSS}_4-0.66\text{MSS}_5+0.577\text{MSS}_6+0.388\text{MSS}_7)$	[53]
SBL	$\text{MSS}_7-2.4\text{MSS}_5$	[38]
SDr	$\sum_N \rho'(\lambda_i)$	[127]
SGI	$\frac{\text{NIR}}{\text{Red}}$	[128]
SR	$\frac{R_{800}}{R_{670}}$	[36]
SR2	$\frac{R_{800}}{R_{550}}$	[86]

Table 1: Continued.

Index	Definition	Reference
SR3	$\frac{R_{700}}{R_{670}}$	[129]
SR4	$\frac{R_{740}}{R_{720}}$	[110]
SR5	$\frac{R_{675}}{(R_{700}R_{650})}$	[130]
SR6	$\frac{R_{672}}{(R_{550}R_{708})}$	[126]
SR7	$\frac{R_{860}}{(R_{550}R_{708})}$	[131]
TCARI	$3 * \left[(R_{700} - R_{670}) - 0.2 * (R_{700} - R_{550})\left(\frac{R_{700}}{R_{670}}\right)\right]$	[126]
TDVI	$\sqrt{0.5 + \left[\frac{(\mathrm{NIR} - \mathrm{Red})}{(\mathrm{NIR} + \mathrm{Red})}\right]}$	[132]
TSARVI	$\frac{[a_{rb}(\mathrm{NIR} - a_{rb}RB - b_{rb})]}{[RB + a_{rb}\mathrm{NIR} - a_{rb}B_{rb} + X(1 + a_{rb}{}^{2})]}$	[49]
TSAVI	$\frac{[a(\mathrm{NIR} - aR - B)]}{[R + a\mathrm{NIR} - ab]}$	[133]
TVI	$\sqrt{\mathrm{NDVI} + 0.5L}$	[134]
VARI	$(G - R)(G + R - B)$	[135]
VCI	$\frac{(\mathrm{NDVI}_i - \mathrm{NDVI}_{\min})}{(\mathrm{NDVI}_{\max} - \mathrm{NDVI}_{\min})}$	[136]
VDVI	$\frac{(2 * \rho_{\mathrm{green}} - \rho_{\mathrm{red}} - \rho_{\mathrm{blue}})}{(2 * \rho_{\mathrm{green}} + \rho_{\mathrm{red}} + \rho_{\mathrm{blue}})}$	[63]
VHI	$a * \mathrm{VCI} + (1 - a) * \mathrm{TCI}$	[136]
VREI1	$\frac{R_{740}}{R_{720}}$	[110]
VREI2	$\frac{(R_{734} - R_{747})}{(R_{715} + R_{726})}$	[110]
YVI	$(-0.283\mathrm{MSS}_4 - 0.66\mathrm{MSS}_5 + 0.577\mathrm{MSS}_6 + 0.388\mathrm{MSS}_7)$	[53]
WBI	$\frac{R_{970}}{R_{900}}$	[115]
WDRVI	$\frac{(\alpha\rho_{\mathrm{nir}} - \rho_{\mathrm{red}})}{(\alpha\rho_{\mathrm{nir}} + \rho_{\mathrm{red}})}$	[64]
WV-VI	$\frac{(\mathrm{NIR2} - \mathrm{Red})}{(\mathrm{NIR2} + \mathrm{Red})}$	[137]
ZM	$\frac{R_{750}}{R_{710}}$	[81]
1DZ -DGVI	$\sum_{\lambda_1}^{\lambda_n} \lvert\rho'(\lambda_i)\rvert \Delta\lambda_i$	[138]
2DZ -DGVI	$\sum_{\lambda_1}^{\lambda_n} \lvert\rho''(\lambda_i)\rvert \Delta\lambda_i$	[138]

Conflicts of Interest

The authors declare that they have no conflicts of interest.

Acknowledgments

This study was supported by the National Natural Science Foundation of China (no. 41401391); Key Science and Technology Program of Shaanxi Province, China (no. S2016YFNY0066); the Scientific Research Foundation for the Returned Overseas Chinese Scholars, State Education Ministry; the National Key Research and Development Program of China (no. 2016YFD0200601); the Fundamental Research Funds for the Central Universities of China (no. 2014YB071); and the Exclusive Talent Funds of Northwest A&F University of China (no. 2013BSJJ017).

References

[1] D. J. Mulla, "Twenty five years of remote sensing in precision agriculture: key advances and remaining knowledge gaps," *Biosystems Engineering*, vol. 114, no. 4, pp. 358–371, 2013.

[2] W. J. Foley, A. McIlwee, I. Lawler, L. Aragones, A. P. Woolnough, and N. Berding, "Ecological applications of near infrared reflectance spectroscopy - A tool for rapid, cost-effective prediction of the composition of plant and animal tissues and aspects of animal performance," *Oecologia*, vol. 116, no. 3, pp. 293–305, 1998.

[3] C. Zhang and J. M. Kovacs, "The application of small unmanned aerial systems for precision agriculture: a review," *Precision Agriculture*, vol. 13, no. 6, pp. 693–712, 2012.

[4] R. Ballesteros, J. F. Ortega, D. Hernández, and M. Moreno, "Characterization of vitis vinifera l. Canopy using unmanned aerial vehicle-based remote sensing and photogrammetry techniques," *American Journal of Enology and Viticulture*, vol. 66, no. 2, pp. 120–129, 2015.

[5] J. Gago, C. Douthe, R. E. Coopman et al., "UAVs challenge to assess water stress for sustainable agriculture," *Agricultural Water Management*, vol. 153, pp. 9–19, 2015.

[6] H. Hoffmann, H. Nieto, R. Jensen, R. Guzinski, P. J. Zarco-Tejada, and T. Friborg, "Estimating evapotranspiration with thermal UAV data and two source energy balance models," *Hydrology and Earth System Sciences Discussions*, vol. 12, no. 8, pp. 7469–7502, 2015.

[7] E. Honkavaara, H. Saari, J. Kaivosoja et al., "Processing and assessment of spectrometric, stereoscopic imagery collected using a lightweight UAV spectral camera for precision agriculture," *Remote Sensing*, vol. 5, no. 10, pp. 5006–5039, 2013.

[8] T. Xia, W. P. Kustas, M. C. Anderson et al., "Mapping evapotranspiration with high-resolution aircraft imagery over vineyards using one-and two-source modeling schemes," *Hydrology and Earth System Sciences*, vol. 20, no. 4, pp. 1523–1545, 2016.

[9] S. Ortega-Farías, S. Ortega-Salazar, T. Poblete et al., "Estimation of energy balance components over a drip-irrigated olive orchard using thermal and multispectral cameras placed on a helicopter-based unmanned aerial vehicle (UAV)," *Remote Sensing*, vol. 8, no. 8, article no. 638, 2016.

[10] L. Chang, S. Peng-Sen, and Liu Shi-Rong, "A review of plant spectral reflectance response to water physiological changes," *Chinese Journal of Plant Ecology*, vol. 40, no. 1, pp. 80–91, 2016.

[11] H. R. Bin Abdul Rahim, M. Q. Bin Lokman, S. W. Harun et al., "Applied light-side coupling with optimized spiral-patterned zinc oxide nanorod coatings for multiple optical channel alcohol vapor sensing," *Journal of Nanophotonics*, vol. 10, no. 3, Article ID 036009, 2016.

[12] B. A. Cruden, D. Prabhu, and R. Martinez, "Absolute radiation measurement in venus and mars entry conditions," *Journal of Spacecraft and Rockets*, vol. 49, no. 6, pp. 1069–1079, 2012.

[13] J. Hatfield, J. Baker, and T. J. Arkebauer, "Leaf radiative properties and the leaf energy budget," in *Micrometeorology in Agricultural Systems*, Agronomy Monograph, American Society of Agronomy, Crop Science Society of America, and Soil Science Society of America, Madison, Wis, USA, 2005.

[14] G. D. Batten, "Plant analysis using near infrared reflectance spectroscopy: The potential and the limitations," *Australian Journal of Experimental Agriculture*, vol. 38, no. 7, pp. 697–706, 1998.

[15] D. A. Burns and E. W. Ciurczak, *Handbook of Near-Infrared Analysis*, CRC Press, 2007.

[16] R. Karwa, "Laws of thermal radiation," in *Heat and Mass Transfer*, pp. 665–696, Springer, 2017.

[17] A. Prashar and H. G. Jones, "Assessing drought responses using thermal infrared imaging," *Methods in Molecular Biology*, vol. 1398, pp. 209–219, 2016.

[18] A. Martynenko, K. Shotton, T. Astatkie et al., "Thermal imaging of soybean response to drought stress: the effect of ascophyllum nodosum seaweed extract," *SpringerPlus*, vol. 5, no. 1, no. 1393, 2016.

[19] S. Fuentes, R. de Bei, J. Pech, and S. Tyerman, "Computational water stress indices obtained from thermal image analysis of grapevine canopies," *Irrigation Science*, vol. 30, no. 6, pp. 523–536, 2012.

[20] A.-K. Mahlein, E.-C. Oerke, U. Steiner, and H.-W. Dehne, "Recent advances in sensing plant diseases for precision crop protection," *European Journal of Plant Pathology*, vol. 133, no. 1, pp. 197–209, 2012.

[21] E.-C. Oerke, A.-K. Mahlein, and U. Steiner, "Proximal sensing of plant diseases," in *Detection And Diagnostics of Plant Pathogens*, pp. 55–68, Springer, 2014.

[22] A. Karnieli, N. Agam, R. T. Pinker et al., "Use of NDVI and land surface temperature for drought assessment: merits and limitations," *Journal of Climate*, vol. 23, no. 3, pp. 618–633, 2010.

[23] R. P. Sripada, R. W. Heiniger, J. G. White, and R. Weisz, "Aerial color infrared photography for determining late-season nitrogen requirements in corn," *Agronomy Journal*, vol. 97, no. 5, pp. 1443–1451, 2005.

[24] B. Zhang, D. Wu, L. Zhang, Q. Jiao, and Q. Li, "Application of hyperspectral remote sensing for environment monitoring in mining areas," *Environmental Earth Sciences*, vol. 65, no. 3, pp. 649–658, 2012.

[25] A. N. Ganeshamurthy, V. Ravindra, R. Venugopalan, M. Mathiazhagan, and R. M. Bhat, "Biomass distribution and development of allometric equations for non-destructive estimation of carbon sequestration in grafted mango trees," *Journal of Agricultural Science*, vol. 8, no. 8, 201 pages, 2016.

[26] S. Fuentes, A. R. Palmer, D. Taylor, M. Zeppel, R. Whitley, and D. Eamus, "An automated procedure for estimating the leaf area index (LAI) of woodland ecosystems using digital imagery, MATLAB programming and its application to an examination of the relationship between remotely sensed and field measurements of LAI," *Functional Plant Biology*, vol. 35, no. 10, pp. 1070–1079, 2008.

[27] S. Fuentes, C. Poblete-Echeverría, S. Ortega-Farias, S. Tyerman, and R. de Bei, "Automated estimation of leaf area index from grapevine canopies using cover photography, video and computational analysis methods," *Australian Journal of Grape and Wine Research*, vol. 20, no. 3, pp. 465–473, 2014.

[28] C. Poblete-Echeverría, S. Fuentes, S. Ortega-Farias, J. Gonzalez-Talice, and J. A. Yuri, "Digital cover photography for estimating Leaf area index (LAI) in apple trees using a variable light extinction coefficient," *Sensors (Switzerland)*, vol. 15, no. 2, pp. 2860–2872, 2015.

[29] M. Carrasco-Benavides, S. Ortega-Farías, L. O. Lagos et al., "Crop coefficients and actual evapotranspiration of a drip-irrigated Merlot vineyard using multispectral satellite images," *Irrigation Science*, vol. 30, no. 6, pp. 485–497, 2012.

[30] M. Mora, F. Avila, M. Carrasco-Benavides, G. Maldonado, J. Olguín-Cáceres, and S. Fuentes, "Automated computation of leaf area index from fruit trees using improved image processing algorithms applied to canopy cover digital photograpies," *Computers and Electronics in Agriculture*, vol. 123, pp. 195–202, 2016.

[31] R. De Bei, S. Fuentes, M. Gilliham et al., "Viticanopy: A free computer app to estimate canopy vigor and porosity for grapevine," *Sensors (Switzerland)*, vol. 16, no. 4, article no. 585, 2016.

[32] S. Fuentes, R. De Bei, C. Pozo, and S. Tyerman, "Development of a smartphone application to characterise temporal and spatial canopy architecture and leaf area index for grapevines," *Wine & Viticulture Journal*, vol. 20, no. 6, pp. 56–60, 2012.

[33] C. Francone, V. Pagani, M. Foi, G. Cappelli, and R. Confalonieri, "Comparison of leaf area index estimates by ceptometer and pocketlai smart app in canopies with different structures," *Field Crops Research*, vol. 155, pp. 38–41, 2014.

[34] F. Orlando, E. Movedi, L. Paleari et al., "Estimating leaf area index in tree species using the PocketLAI smart app," *Applied Vegetation Science*, vol. 18, no. 4, pp. 716–723, 2015.

[35] M. Campos-Taberner, F. J. García-Haro, R. Confalonieri et al., "Multitemporal monitoring of plant area index in the valencia rice district with PocketLAI," *Remote Sensing*, vol. 8, no. 3, article no. 202, 2016.

[36] C. F. Jordan, "Derivation of leaf-area index from quality of light on the forest floor," *Ecology*, vol. 50, no. 4, pp. 663–666, 1969.

[37] Z. Quan, Z. Xianfeng, and J. Miao, "Eco-environment variable estimation from remote sensed data and eco-environment assessment: models and system," *Acta Botanica Sinica*, vol. 47, pp. 1073–1080, 2011.

[38] A. J. Richardson and C. Weigand, "Distinguishing vegetation from soil background information," *Photogrammetric Engineering and Remote Sensing*, p. 43, 1977.

[39] X. D. L. Wenlong, "Vegetation index controlling the influence of soil reflection," 2009, http://www.paper.edu.cn/releasepaper/content/200906-376.

[40] Y. J. Kaufman and D. Tanré, "Atmospherically Resistant Vegetation Index (ARVI) for EOS-MODIS," *IEEE Transactions on Geoscience and Remote Sensing*, vol. 30, no. 2, pp. 261–270, 1992.

[41] D. J. Major, F. Baret, and G. Guyot, "A ratio vegetation index adjusted for soil brightness," *International Journal of Remote Sensing*, vol. 11, no. 5, pp. 727–740, 1990.

[42] J. W. Rouse Jr., R. Haas, J. Schell, and D. Deering, "Monitoring vegetation systems in the great plains with erts," NASA Special Publication 351, 309, 1974.

[43] J. A. Gamon, C. B. Field, M. L. Goulden et al., "Relationships between NDVI, canopy structure, and photosynthesis in three Californian vegetation types," *Ecological Applications*, vol. 5, no. 1, pp. 28 41, 1995.

[44] J. Grace, C. Nichol, M. Disney, P. Lewis, T. Quaife, and P. Bowyer, "Can we measure terrestrial photosynthesis from space directly, using spectral reflectance and fluorescence?" *Global Change Biology*, vol. 13, no. 7, pp. 1484–1497, 2007.

[45] D. Tanré, C. Deroo, P. Duhaut et al., "Description of a computer code to simulate the satellite signal in the solar spectrum: the 5S code," *International Journal of Remote Sensing*, vol. 11, no. 4, pp. 659–668, 1990.

[46] R.-H. Zhang, N. X. Rao, and K. N. Liao, "Approach for a vegetation index resistant to atmospheric effect," *Acta Botanica Sinica*, vol. 38, no. 1, pp. 53–62, 1996.

[47] A. J. Richardson and C. Wiegand, "Distinguishing vegetation from soil background information," *Photogrammetric Engineering & Remote Sensing*, vol. 43, no. 12, pp. 1541–1552, 1977.

[48] F. Baret, S. Jacquemoud, and J. F. Hanocq, "The soil line concept in remote sensing," *Remote Sensing Reviews*, vol. 7, no. 1, pp. 65–82, 1993.

[49] A. Bannari, D. Morin, and D. He, "High spatial and spectral resolution remote sensing for the management of the urban environment," in *Proceedings of the 1st International Airborne Remote Sensing Conference and Exhibition*, pp. 12–15, Strasbourg, France, 1994.

[50] A. R. Huete, "A soil-adjusted vegetation index (SAVI)," *Remote Sensing of Environment*, vol. 25, no. 3, pp. 295–309, 1988.

[51] J. Qi, A. Chehbouni, A. R. Huete, Y. H. Kerr, and S. Sorooshian, "A modified soil adjusted vegetation index," *Remote Sensing of Environment*, vol. 48, no. 2, pp. 119–126, 1994.

[52] X. Dandan and L. Wenlong, *Vegetation Index Controlling The Influence of Soil Reflection*, Department of Pastrol Agricultural Science and Technology, 2009.

[53] R. J. Kauth and G. Thomas, "The tasselled cap-a graphic description of the spectral-temporal development of agricultural crops as seen by landsat," in *Proceedings of the LARS Symposia*, 159 pages, 1976.

[54] T. Qingjiu and M. Xiangjun, "Advances in study on vegetation indices," *Department of Pastrol Agricultural Science and Technology*, vol. 13, pp. 327–333, 1998.

[55] R. D. Jackson, P. Pinter Jr., R. J. Reginato, and S. B. Idso, "Hand-held radiometry: a set of notes developed for use at the workshop of hand-held radiometry," in *Proceedings of the Early Warning and Crop Condition Assessment*, Phoenix, Ariz, USA, February 1980.

[56] P. N. Misra and S. G. Wheeler, "Landsat data from agricultural sites: crop signature analysis," pp. 1473–1482, 1977.

[57] H. Q. Liu and A. Huete, "Feedback based modification of the NDVI to minimize canopy background and atmospheric noise," *IEEE Transactions on Geoscience and Remote Sensing*, vol. 33, no. 2, pp. 457–465, 1995.

[58] L. Wang, J. Liu, L. Yang, Z. Chen, X. Wang, and B. Ouyang, "Applications of unmanned aerial vehicle images on agricultural remote sensing monitoring," *Nongye Gongcheng Xuebao/Transactions of the Chinese Society of Agricultural Engineering*, vol. 29, no. 18, pp. 136–145, 2013.

[59] B. Li, R. Liu, S. Liu, Q. Liu, F. Liu, and G. Zhou, "Monitoring vegetation coverage variation of winter wheat by low-altitude UAV remote sensing system," *Nongye Gongcheng Xuebao/Transactions of the Chinese Society of Agricultural Engineering*, vol. 28, no. 13, pp. 160–165, 2012.

[60] P. J. Zarco-Tejada, V. González-Dugo, and J. A. J. Berni, "Fluorescence, temperature and narrow-band indices acquired from a UAV platform for water stress detection using a micro-hyperspectral imager and a thermal camera," *Remote Sensing of Environment*, vol. 117, pp. 322–337, 2012.

[61] J. A. J. Berni, P. J. Zarco-Tejada, G. Sepulcre-Cantó, E. Fereres, and F. Villalobos, "Mapping canopy conductance and CWSI in olive orchards using high resolution thermal remote sensing imagery," *Remote Sensing of Environment*, vol. 113, no. 11, pp. 2380–2388, 2009.

[62] P. J. Zarco-Tejada, V. González-Dugo, L. E. Williams et al., "A PRI-based water stress index combining structural and chlorophyll effects: Assessment using diurnal narrow-band airborne imagery and the CWSI thermal index," *Remote Sensing of Environment*, vol. 138, pp. 38–50, 2013.

[63] X. Wang, M. Wang, S. Wang, and Y. Wu, "Extraction of vegetation information from visible unmanned aerial vehicle images," *Nongye Gongcheng Xuebao/Transactions of the Chinese Society of Agricultural Engineering*, vol. 31, no. 5, pp. 152–159, 2015.

[64] A. A. Gitelson, "Wide dynamic range vegetation index for remote quantification of biophysical characteristics of vegetation," *Journal of Plant Physiology*, vol. 161, no. 2, pp. 165–173, 2004.

[65] M. S. Kim, C. Daughtry, E. Chappelle, J. McMurtrey, and C. Walthall, "The use of high spectral resolution bands for estimating absorbed photosynthetically active radiation (a par)," in *Proceedings of the 6th International Symposium on Physical Measurements and Signatures in Remote Sensing*, CNES, Phoenix, Ariz, USA, January 1994.

[66] C. S. T. Daughtry, C. L. Walthall, M. S. Kim, E. B. De Colstoun, and J. E. McMurtrey III, "Estimating corn leaf chlorophyll concentration from leaf and canopy reflectance," *Remote Sensing of Environment*, vol. 74, no. 2, pp. 229–239, 2000.

[67] Z. Funian, Z. Hong, C. Jiayu, W. Ruijun, and Y. Fuklin, "Preliminary investigation on difference of crop water stress index baseline for maize," *Chinese Agricultural Science Bulletin*, vol. 29, pp. 46–53, 2013.

[68] S. A. O'Shaughnessy, S. R. Evett, P. D. Colaizzi, and T. A. Howell, "Using radiation thermography and thermometry to evaluate crop water stress in soybean and cotton," *Agricultural Water Management*, vol. 98, no. 10, pp. 1523–1535, 2011.

[69] S. B. Idso, R. D. Jackson, P. J. Pinter Jr., R. J. Reginato, and J. L. Hatfield, "Normalizing the stress-degree-day parameter for environmental variability," *Agricultural Meteorology*, vol. 24, pp. 45–55, 1981.

[70] B. B. D. Silva, J. A. Ferreira, and T. V. R. Rao, "Crop water stress index and water-use efficiency for melon (*Cucumis melo l.*) on different irrigation regimes," *Agricultural Journal*, 2007.

[71] G. Kar and A. Kumar, "Surface energy fluxes and crop water stress index in groundnut under irrigated ecosystem," *Agricultural and Forest Meteorology*, vol. 146, no. 1-2, pp. 94–106, 2007.

[72] V. Lebourgeois, J.-L. Chopart, A. Bégué, and L. Le Mézo, "Towards using a thermal infrared index combined with water balance modelling to monitor sugarcane irrigation in a tropical environment," *Agricultural Water Management*, vol. 97, no. 1, pp. 75–82, 2010.

[73] A. Anda, "Irrigation timing in maize by using the crop water stress index (CWSI)," *Cereal Research Communications*, vol. 37, no. 4, pp. 603–610, 2009.

[74] A. Ruimy, L. Kergoat, and A. Bondeau, "Comparing global models of terrestrial net primary productivity (NPP): Analysis of differences in light absorption and light-use efficiency," *Global Change Biology*, vol. 5, no. 1, pp. 56–64, 1999.

[75] A. Haxeltine and I. C. Prentice, "A general model for the light-use efficiency of primary production," *Functional Ecology*, vol. 10, no. 5, pp. 551–561, 1996.

[76] A. A. Gitelson, M. N. Merzlyak, and O. B. Chivkunova, "Optical properties and nondestructive estimation of anthocyanin content in plant leaves," *Photochemistry and Photobiology*, vol. 74, no. 1, pp. 38–45, 2001.

[77] R. D. Jackson, "Spectral indices in N-Space," *Remote Sensing of Environment*, vol. 13, no. 5, pp. 409–421, 1983.

[78] F. Baret and G. Guyot, "Potentials and limits of vegetation indices for LAI and APAR assessment," *Remote Sensing of Environment*, vol. 35, no. 2-3, pp. 161–173, 1991.

[79] P. Ashburn, "The vegetative index number and crop identification," NASA, Proc. of Tech., vol. 1, Johnson Space Center, 1979.

[80] S. Plummer, P. North, and S. Briggs, "The angular vegetation index: An atmospherically resistant index for the second along track sacnning radiometer (atsr-2)," in *Proceedings of the Sixth International Symposium of Physical Measurements and Signatures in Remote Sensing*, Val D'Isere, France, 1994.

[81] P. J. Zarco-Tejada, A. Berjón, R. López-Lozano et al., "Assessing vineyard condition with hyperspectral indices: Leaf and canopy reflectance simulation in a row-structured discontinuous canopy," *Remote Sensing of Environment*, vol. 99, no. 3, pp. 271–287, 2005.

[82] C. S. T. Daughtry, "Agroclimatology: Discriminating crop residues from soil by shortwave infrared reflectance," *Agronomy Journal*, vol. 93, no. 1, pp. 125–131, 2001.

[83] F. Li, Y. Miao, G. Feng et al., "Improving estimation of summer maize nitrogen status with red edge-based spectral vegetation indices," *Field Crops Research*, vol. 157, pp. 111–123, 2014.

[84] N. H. Broge and E. Leblanc, "Comparing prediction power and stability of broadband and hyperspectral vegetation indices for estimation of green leaf area index and canopy chlorophyll density," *Remote Sensing of Environment*, vol. 76, no. 2, pp. 156–172, 2001.

[85] A. A. Gitelson, G. P. Keydan, and M. N. Merzlyak, "Three-band model for noninvasive estimation of chlorophyll, carotenoids, and anthocyanin contents in higher plant leaves," *Geophysical Research Letters*, vol. 33, no. 11, Article ID L11402, 2006.

[86] C. Buschmann and E. Nagel, "In vivo spectroscopy and internal optics of leaves as basis for remote sensing of vegetation," *International Journal of Remote Sensing*, vol. 14, no. 4, pp. 711–722, 1993.

[87] T. H. Demetriades-Shah, M. D. Steven, and J. A. Clark, "High resolution derivative spectra in remote sensing," *Remote Sensing of Environment*, vol. 33, no. 1, pp. 55–64, 1990.

[88] A. Huete, K. Didan, T. Miura, E. P. Rodriguez, X. Gao, and L. G. Ferreira, "Overview of the radiometric and biophysical performance of the MODIS vegetation indices," *Remote Sensing of Environment*, vol. 83, no. 1-2, pp. 195–213, 2002.

[89] J. Torres-Sánchez, J. M. Peña, A. I. de Castro, and F. López-Granados, "Multi-temporal mapping of the vegetation fraction in early-season wheat fields using images from UAV," *Computers and Electronics in Agriculture*, vol. 103, pp. 104–113, 2014.

[90] B. Pinty and M. M. Verstraete, "GEMI: a non-linear index to monitor global vegetation from satellites," *Vegetatio*, vol. 101, no. 1, pp. 15–20, 1992.

[91] M. Louhaichi, M. M. Borman, and D. E. Johnson, "Spatially located platform and aerial photography for documentation of grazing impacts on wheat," *Geocarto International*, vol. 16, no. 1, pp. 65–70, 2001.

[92] A. A. Gitelson and M. N. Merzlyak, "Signature analysis of leaf reflectance spectra: Algorithm development for remote sensing of chlorophyll," *Journal of Plant Physiology*, vol. 148, no. 3-4, pp. 494–500, 1996.

[93] E. T. Kanemasu, J. L. Hellman, J. O. Bagley, and W. L. Powers, "Using Landsat data to estimate evapotranspiration of winter wheat," *Environmental Management*, vol. 1, no. 6, pp. 515–520, 1977.

[94] R. C. G. Smith, J. Adams, D. J. Stephens, and P. T. Hick, "Forecasting wheat yield in a Mediterranean-type environment from the NOAA satellite," *Australian Journal of Agricultural Research*, vol. 46, no. 1, pp. 113–125, 1995.

[95] G. Badhwar, "The use of parameters to separate and identify spring small grains," in *Proceedings of the Quarterly Technical Interchange Meeting NASA Johnson Space Flight Center*, Houston, Tex, USA, 1981.

[96] H. K. Lichtenthaler, M. Lang, M. Sowinska, F. Heisel, and J. A. Miehé, "Detection of vegetation stress via a new high resolution fluorescence imaging system," *Journal of Plant Physiology*, vol. 148, no. 5, pp. 599–612, 1996.

[97] Y. Zhang, Q.-Y. Meng, J.-L. Wu, and F. Zhao, "Study of environmental vegetation index based on environment satellite CCD data and LAI inversion," *Guang Pu Xue Yu Guang Pu Fen Xi/ Spectroscopy and Spectral Analysis*, vol. 31, no. 10, pp. 2789–2793, 2011.

[98] A.-K. Mahlein, T. Rumpf, P. Welke et al., "Development of spectral indices for detecting and identifying plant diseases," *Remote Sensing of Environment*, vol. 128, pp. 21–30, 2013.

[99] H. B. Musick and R. E. Pelletier, "Response to soil moisture of spectral indexes derived from bidirectional reflectance in thematic mapper wavebands," *Remote Sensing of Environment*, vol. 25, no. 2, pp. 167–184, 1988.

[100] R. E. Crippen, "Calculating the vegetation index faster," *Remote Sensing of Environment*, vol. 34, no. 1, pp. 71–73, 1990.

[101] D. Haboudane, J. R. Miller, E. Pattey, P. J. Zarco-Tejada, and I. B. Strachan, "Hyperspectral vegetation indices and novel algorithms for predicting green LAI of crop canopies: modeling and validation in the context of precision agriculture," *Remote Sensing of Environment*, vol. 90, no. 3, pp. 337–352, 2004.

[102] P. Misra and S. Wheeler, "Landsat data from agricultural sites-crop signature analysis," in *Proceedings of the International Symposium on Remote Sens*, pp. 1473–1482, Environ, 1977.

[103] Z. Yang, P. Willis, and R. Mueller, "Impact of band-ratio enhanced awifs image on crop classification accuracy," in *Proceedings of the 17th William Pecora Memorial Remote Sensing Symposium*, Denver, Colo, USA, 2008.

[104] B. Datt, "A new reflectance index for remote sensing of chlorophyll content in higher plants: Tests using Eucalyptus leaves," *Journal of Plant Physiology*, vol. 154, no. 1, pp. 30–36, 1999.

[105] D. A. Sims and J. A. Gamon, "Relationships between leaf pigment content and spectral reflectance across a wide range of species, leaf structures and developmental stages," *Remote Sensing of Environment*, vol. 81, no. 2-3, pp. 337–354, 2002.

[106] J. M. Chen, "Evaluation of vegetation indices and a modified simple ratio for boreal applications," *Canadian Journal of Remote Sensing*, vol. 22, no. 3, pp. 229–242, 1996.

[107] E. R. Hunt Jr. and B. N. Rock, "Detection of changes in leaf water content using Near- and Middle-Infrared reflectances," *Remote Sensing of Environment*, vol. 30, no. 1, pp. 43–54, 1989.

[108] H. Mcnairn and R. Protz, "Mapping corn residue cover on agricultural fields in oxford county, ontario, using thematic mapper," *Canadian Journal of Remote Sensing*, vol. 19, no. 2, pp. 152–159, 1993.

[109] B. Datt, "Visible/near infrared reflectance and chlorophyll content in Eucalyptus leaves," *International Journal of Remote Sensing*, vol. 20, no. 14, pp. 2741–2759, 1999.

[110] J. E. Vogelmann, B. N. Rock, and D. M. Moss, "Red edge spectral measurements from sugar maple leaves," *International Journal of Remote Sensing*, vol. 14, no. 8, pp. 1563–1575, 1993.

[111] L. Serrano, J. Peñuelas, and S. L. Ustin, "Remote sensing of nitrogen and lignin in Mediterranean vegetation from AVIRIS data: decomposing biochemical from structural signals," *Remote Sensing of Environment*, vol. 81, no. 2-3, pp. 355–364, 2002.

[112] S. K. McFeeters, "The use of the Normalized Difference Water Index (NDWI) in the delineation of open water features," *International Journal of Remote Sensing*, vol. 17, no. 7, pp. 1425–1432, 1996.

[113] J. Verrelst, M. E. Schaepman, B. Koetz, and M. Kneubühler, "Angular sensitivity analysis of vegetation indices derived from CHRIS/PROBA data," *Remote Sensing of Environment*, vol. 112, no. 5, pp. 2341–2353, 2008.

[114] C. J. Tucker, "Red and photographic infrared linear combinations for monitoring vegetation," *Remote Sensing of Environment*, vol. 8, no. 2, pp. 127–150, 1979.

[115] L. Wang and J. J. Qu, "NMDI: A normalized multi-band drought index for monitoring soil and vegetation moisture with satellite remote sensing," *Geophysical Research Letters*, vol. 34, no. 20, Article ID L20405, 2007.

[116] N. S. Goel and W. Qin, "Influences of canopy architecture on relationships between various vegetation indices and LAI and FPAR: a computer simulation," *Remote Sensing Reviews*, vol. 10, no. 4, pp. 309–347, 1994.

[117] G. Rondeaux, M. Steven, and F. Baret, "Optimization of soil-adjusted vegetation indices," *Remote Sensing of Environment*, vol. 55, no. 2, pp. 95–107, 1996.

[118] J. A. Gamon, J. Peñuelas, and C. B. Field, "A narrow-waveband spectral index that tracks diurnal changes in photosynthetic efficiency," *Remote Sensing of Environment*, vol. 41, no. 1, pp. 35–44, 1992.

[119] M. N. Merzlyak, A. A. Gitelson, O. B. Chivkunova, and V. Y. Rakitin, "Non-destructive optical detection of pigment changes during leaf senescence and fruit ripening," *Physiologia Plantarum*, vol. 106, no. 1, pp. 135–141, 1999.

[120] G. A. Blackburn, "Quantifying chlorophylls and carotenoids at leaf and canopy scales: An evaluation of some hyperspectral approaches," *Remote Sensing of Environment*, vol. 66, no. 3, pp. 273–285, 1998.

[121] J.-L. Roujean and F.-M. Breon, "Estimating PAR absorbed by vegetation from bidirectional reflectance measurements," *Remote Sensing of Environment*, vol. 51, no. 3, pp. 375–384, 1995.

[122] A. Gitelson and M. N. Merzlyak, "Spectral relfectance changes associated with autumn senescence of Aesculus hippocastanum L. and Acer platanoides L. leaves. Spectral features and relation to chlorophyll estimation," *Journal of Plant Physiology*, vol. 143, no. 3, pp. 286–292, 1994.

[123] J. A. Gamon and J. S. Surfus, "Assessing leaf pigment content and activity with a reflectometer," *New Phytologist*, vol. 143, no. 1, pp. 105–117, 1999.

[124] R. Escadafal and A. Huete, "Etude des propriétés spectrales des sols arides appliquée à l'amélioration des indices de végétation obtenus par télédétection," *Comptes Rendus de l'Académie des Sciences*, vol. 312, no. 2, pp. 1385–1391, 1991.

[125] R. L. Pearson and L. D. Miller, "Remote mapping of standing crop biomass for estimation of the productivity of the shortgrass prairie," *Remote Sensing of Environme*, vol. 8, p. 1355, 1972.

[126] D. Haboudane, J. R. Miller, N. Tremblay, P. J. Zarco-Tejada, and L. Dextraze, "Integrated narrow-band vegetation indices for

prediction of crop chlorophyll content for application to precision agriculture," *Remote Sensing of Environment*, vol. 81, no. 2-3, pp. 416–426, 2002.

[127] R. Pu and P. Gong, *Advances in Environmental Remote Sensing*, Higher Education, Beijing, 2011.

[128] G. S. Birth and G. R. McVey, "Measuring the color of growing turf with a reflectance spectrophotometer," *Agronomy Journal*, vol. 60, no. 6, p. 640, 1968.

[129] J. E. McMurtrey III, E. W. Chappelle, M. S. Kim, J. J. Meisinger, and L. A. Corp, "Distinguishing nitrogen fertilization levels in field corn (*Zea mays* L.) with actively induced fluorescence and passive reflectance measurements," *Remote Sensing of Environment*, vol. 47, no. 1, pp. 36–44, 1994.

[130] E. W. Chappelle, M. S. Kim, and J. E. McMurtrey, "Ratio analysis of reflectance spectra (RARS): an algorithm for the remote estimation of the concentrations of chlorophyll A, chlorophyll B, and carotenoids in soybean leaves," *Remote Sensing of Environment*, vol. 39, no. 3, pp. 239–247, 1992.

[131] B. Datt, "Remote sensing of chlorophyll a, chlorophyll b, chlorophyll a+b, and total carotenoid content in eucalyptus leaves," *Remote Sensing of Environment*, vol. 66, no. 2, pp. 111–121, 1998.

[132] A. Bannari, H. Asalhi, and P. M. Teillet, "Transformed difference vegetation index (TDVI) for vegetation cover mapping," in *Proceedings of the IEEE International Geoscience and Remote Sensing Symposium IGARSS '02*, pp. 3053–3055, Toronto, Canada, 2002.

[133] F. Baret, G. Guyot, and D. J. Major, "TSAVI: A vegetation index which minimizes soil brightness effects on LAI and APAR estimation," in *Proceedings of the 12th Canadian Symposium on Remote Sensing Geoscience and Remote Sensing Symposium*, pp. 1355–1358, IEEE, 1989.

[134] J. Rouse Jr., "Monitoring the vernal advancement and retrogradation (green wave effect) of natural vegetation," NASA Technical Reports Server, Tex, USA, 1974, https://ntrs.nasa.gov/search.jsp?R=19740022555.

[135] A. A. Gitelson, Y. J. Kaufman, R. Stark, and D. Rundquist, "Novel algorithms for remote estimation of vegetation fraction," *Remote Sensing of Environment*, vol. 80, no. 1, pp. 76–87, 2002.

[136] F. N. Kogan, "Application of vegetation index and brightness temperature for drought detection," *Advances in Space Research*, vol. 15, no. 11, pp. 91–100, 1995.

[137] A. F. Wolf, "Using WorldView-2 Vis-NIR multispectral imagery to support land mapping and feature extraction using normalized difference index ratios," in *Proceedings of the 18th Annual Conference on Algorithms and Technologies for Multispectral, Hyperspectral, and Ultraspectral Imagery*, Baltimore, Md, USA, 2012.

[138] C. D. Elvidge and Z. Chen, "Comparison of broad-band and narrow-band red and near-infrared vegetation indices," *Remote Sensing of Environment*, vol. 54, no. 1, pp. 38–48, 1995.

2

Landscape Pattern and Ecological Security Assessment and Prediction using Remote Sensing Approach

Pei Liu, [1,2] **Shoujun Jia**, [1,2] **Ruimei Han**, [1,2] **and Hanwei Zhang** [1,2]

[1]*Key Laboratory of Mine Spatial Information Technologies of SBSM, Henan Polytechnic University, Jiaozuo 454003, China*
[2]*School of Surveying and Mapping Land Information Engineering, Henan Polytechnic University, Jiaozuo 454003, China*

Correspondence should be addressed to Pei Liu; cumtlp@qq.com and Ruimei Han; hrm@hpu.edu.cn

Academic Editor: Carmine Granata

In this work, we present a processing chain for landscape pattern and ecological security status assessment and prediction based on cellular automata Markov (CA-Markov) and pressure status response pattern (PSRP) models using remotely sensed data (RSD) captured in 1986, 1996, 2006, 2016, and RSD simulated in 2026 over Zhengzhou city, Henan province, China. Three major findings can be withdrawn through the experiments. First, there is a significant changing of landscape type area, especially for building land. The area of building land is up to more than 5%, from 1986 to 2016. Secondly, the heterogeneity of landscape is increasing, and the diversity of landscape is becoming more and more diversifying and complex. Third, the changing trend of ecological security of Zhengzhou city shaped as decreasing and increasing gradually during the last 40 years. While the ecological security status, nowadays, appeared to a good trend by contrast of the previous stages. The predicted results with CA-Markov model show that the level of ecological security is still in moderate and has a trend of moving toward to the center in 2026.

1. Introduction

The change of land use and land cover (LULC) is the key point for urban sprawl research and is becoming an advantage and hot pot for research of global environmental change and sustainable development [1]. The changes of landscape, as a complex structure of land use, at different scales, shapes, and quantities, express inherence mechanism of land use and land cover change (LULCC) [2]. The research of interaction mechanism between landscape pattern and LUCC has become an essential scientific topic in earth surface environmental change [3]. Ecological security assessment (ESA) is the foundation of regional ecological environment governance and prevention, because the comprehensively ecological risk and vulnerability of natural, economic, and social levels are considered in the changing procedure of landscape pattern [4]. And the integrity of ecosystem and health of overall ecosystem status can be reflected by ESA.

At present, more and more attentions are paid to the changing trend, vulnerability, sensitivity, and heterogeneity analysis methods of landscape pattern, such as landscape pattern index, the particle size effect, and spatial statistics algorithm [2, 5–7]. The wildly accepted ecological security assessment system include the "ecological footprint" concept and model, the concept of pressure-state-response (PSR) framework model, and the concept of nature-economy-society framework model [8, 9].

The motivation of this paper is to develop a processing chain for landscape and ESA monitoring, analysis, and prediction with mathematical and geography models. The objective of this work was to study LULC change statues during 40 years and to research relationships among LULC and landscape pattern and ecological security.

The paper is organized as follows: a brief introduction about research area and preprocessing of datasets are given in Section 2. The proposed chain based on remotely sensed observation techniques and CA-Markov model, PSRP model, and analytic hierarchy process (AHP) method are applied to the test area in Section 3. The results and analysis are shown in Section 4, and the conclusions are drawn in Section 5.

2. Study Area and Datasets

2.1. Study Area. Zhengzhou city, capital of Henan province, China, which is situated at the transitional zone among piedmont, torrential plain, and the Yellow River alluvial plain, is selected as the study area. The geographic coordinates lie in 112°42′–114°14′E, 34°15′–34°57′N, shown in Figure 1. The city is bounded by Nanyang Road. West part of the road is the piedmont inclined plain with the height 100–130 m and surface lithology of Pleistocene series loess powder on the soil. Moreover, east part is the Yellow River alluvial plain with low height 85–100 m. This test area has climatic conditions of continental monsoon climate, with a total area of 7446 square kilometers and 11 million resident population. The annual average temperature is 14.5°C. The general terrain trend is tilt from southwest to northeast. Nowadays, the test area is the center of politics, economy, science and technology, culture, finance, transportation, and information in Henan province and undergoing Chinese urban agglomeration, and economic development and urban expansion are accelerating.

Figure 1: Geographical location of the study area.

2.2. Datasets and Preprocessing. The datasets collected in this research include multitemporal remote sensing (RS) datasets, LULC product data, simulated LULC data in future, vector data, and demographic data. The selected multitemporal remote sensing (RS) datasets acquired for four different times include Landsat TM/ETM+ (captured on 12 August 1986, 15 April 1996, 9 April 2006, row in 36, and path in 121), Landsat OLI (captured on 23 April 2016, row in 36, and path in 121), GlobeLand30 product obtained in 2010, simulated LULC map in 2026, the division vector data of administrative area with a scale of 1 : 30000, and population statistic data (captured in 1986, 1996, 2006, and 2016, respectively). More detailed description about datasets collected can be found in Table 1. The data availability statement: all datasets used in this paper including the primary remotely sensed data and boundary vector data, as well as process results data can be obtained from https://pan.baidu.com/s/1rThlB5U2RQExrKAlM26Zig. Researchers who are interested in this topic can download the data to replicate the analysis or conduct secondary analysis.

The preprocessing includes geometry correction, remotely sensed data resizing, and LULC types obtaining. In detail, (1) the geometry correction is applied to different satellite images based on the topography map over test area using the polynomial transformation equations. The degree is set to one, and the nearest neighbour resampling method is selected in the polynomial method, which is defaulted by ENVI® software. In this process, firstly, seventy-four ground control points, seventy-seven ground control points, and seventy-eight ground control points are selected between remotely sensed data 1986 and 2016, 1996 and 2016, and 2006 and 2016, respectively. Then remotely sensed data obtained on 1986, 1996, and 2006 are geocoding to the same geometry system based on remotely sensed data obtained on 2016. The root mean square error (RMSE) value for different scenes is less than 0.8 pixel (0.67, 0.71, and 0.56 for 1986, 1996, and 2006, resp.); (2) the 30-meter resolution remotely sensed data over test area in 1986, 1996, 2006, and 2016 are obtained with data resize method based on administrative boundary vector file. In this process, the spatial subset of remotely sensed data over research area was obtained through data resize method based on vector data. Firstly, the vector data was imported and overlapped to the whole remotely sensed data. Then the resize model with spatial subset in ENVI packages was performant.

3. Methodology

3.1. LULC Map Classification. LULC status are categorized into five classes using the support vector machine (SVM) classifier [10] according to the Classification Criteria for Land Use Status/GB-T21010–2015 and GlobeLand30 standard products [11]. The GlobelLand30 is an open-access 30 m resolution global land cover data product which comprises ten land cover types, including water bodies, wetlands, artificial surfaces, cultivated lands, forests, shrub lands, grasslands, and barren lands [12]. Considering the situation and comparing with GlobelLand30 product of the selected research area, the selected five LULC classes are agricultural land, vegetation, water body, bare land, and building land. The selected SVM classifier, which uses a various of kernel function to construct nonlinear decision boundaries in the original feature space and transfers the sample data into linear boundaries of a higher dimension, has been successfully applied for regression, classification and has been demonstrated as one of the most powerful and efficient tool for remotely sensed data classification [13]. Therefore, in this research, the SVM classifier which derived from statistical learning theory was selected to obtain LULC status. The classification scheme proposed in this research are as follows. Firstly, about 100 points of training samples for each class are selected by expert in remote sensing field; secondly, the radial basis function (RBF) kernel, where parameters are determined automatically to minimize the upper limit of the expected test error, and according to the results from previous research, it works well in most cases, was selected in this work. The mathematical representation of RBF is K(x_i,

TABLE 1: Datasets selected.

Data type	Date	Sensor	Band combinations	Resolution/Scale	Source of datasets
RS data	1986.8.12	Landsat TM	1,2,3,4,5,7	30 m	https://www.usgs.gov
RS data	1996.4.15	Landsat TM	1,2,3,4,5,7	30 m	https://www.usgs.gov
RS data	2006.4.9	Landsat TM	1,2,3,4,5,7	30 m	https://www.usgs.gov
RS data	2016.4.23	Landsat OLI/TIRS		30 m	https://www.usgs.gov
GlobeLand30	2010	HJ-1,BJ-1,Landsat TM/ETM+	—	30 m	http://www.globallandcover.com/GLC30Download/index.aspx
Vector data	2009	—	—	1 : 4,000,000	http://muchong.com/
Population	1986	—	—	4,915,000	http://www.stats.gov.cn/tjsj/
	1996	—	—	6,076,000	http://www.stats.gov.cn/tjsj/
	2006	—	—	7,243,000	http://www.stats.gov.cn/tjsj/
	2016	—	—	9,569,000	http://www.stats.gov.cn/tjsj/

$x_j) = \exp(-\gamma|x_i - x_j|^2)$, $\gamma > 0$, where γ is the bias term in kernel function for polynomial and sigmoid kernels; third, inputting feature vectors, such as band 1 to 5 and band 7 for Landsat TM/ETM+ data; and lastly, applying SVM classifier to feature combinations obtained in the third step and obtaining LULC results. Additionally, LULC map of 2026 was simulated based on LULC map of 2006 and 2016 using CA-Markov model. The thematic maps produced using SVM classifier in different years were accuracy evaluated for further procedure, such as change trend analysis, LULC trend prediction, landscape pattern and ecological security assessment, and so on. The commonly used accuracy evaluation indexes including overall accuracy (OA), user's accuracy (UA), producer's accuracy (PA), and kappa coefficients (KC) can be performed based on matrix statistics [31modelling]. In this research, the overall and comprehensive indicator include OA and KC, which can be calculated using expressions $\mathrm{OA} = 1/N\sum_{i=1}^{r} n_{ii}$ and $\mathrm{KC} = N\sum_{n}^{r} n_{ii} - \sum_{n}^{r}(n_{i\mathrm{col}} n_{i\mathrm{row}}/N^2) - \sum_{i=1}^{r} n_{i\mathrm{col}} n_{i\mathrm{row}}$, respectively, are selected. Where n_{ii} is number of pixels correctly classified, N is the total number of pixels in the confusion matrix, r is the number of rows, and $n_{i\mathrm{col}}$ and $n_{i\mathrm{row}}$ are the column and row total, respectively [14]. There are total of 15,538, 13,829, 15,991 and 13,937 pixels selected for training in remotely sensed data captured on 12 August 1986, 15 April 1996, 9 April 2006, and 23 April 2016, respectively. The number of selected test pixels (ground truth points) are 5643 points. The classification map results are shown in Figure 2. The classification accuracy shown in Table 2 demonstrates that LULC maps obtained with our method are good enough for further processing.

3.2. SDDLUC and IDDLUC Model. The changing speed and changing amplitude are analyzed by calculating landscape type transfer matrixes in order that changing trend can be characterized from spatial and temporal aspects. The variation of amplitude is expressed in the area change of landscape, while the changing speed is embodied in the single dynamic degree of LULC and integrated dynamic degree of land use change [15, 16]. The single dynamic degree of land use change (SDDLUC) describes the changing degree rate of a specific land use type of the area. SDDLUC can be described in the following equation [17]:

$$V = \frac{u_2 - u_1}{u_1} \times \frac{1}{\tau}, \quad (1)$$

where V is the SDDLUC index, u_1 and u_2 are areas of a specific land use type at the beginning and ending time, respectively, during the study period. τ is the length of the research period.

The integrated dynamic degree of land use change (IDDLUC) describe overall changing degree of the study area with a certain period. IDDLUC can be described in the following equation [18]:

$$K = \left[\frac{\sum_{i=1}^{n} \Delta u_{ij}}{2\sum_{i=1}^{n} u_i}\right] \times \frac{1}{T} \times 100\%, \quad (2)$$

where u_i is area of the ith land cover type in initial time, while Δu_{ij} is transfer area amount of the ith land cover type to other land cover types during the research period.

3.3. Simulation with CA-Markov Model. The CA-Markov is a combination model of Markov model which describes the probability of land cover changes between start and end period by developing a transition probability matrix between them and CA model which allows the transition probabilities of one pixel to be function of adjacent pixel [14]. The CA-Markov model has a capability to simulate and to predict LULC changing status in future with remotely sensed data [19]. It is helpful to the research area in establishing and optimizing urban development decisions at different spatial-temporal dimensions [14, 19–21], when simulating and prediction urban landscape changes comprehensively in quantity and spatial aspects. As given in [19] by Singh et al., the Markov chain process can be described as $F_X(X(t_{n+1}) \le x_{(n+1)} \mid X(t_n)) = x_n, X(t_{(n+1)}) = x_{(n-1)}, \dots, X(t_1) = x_1 = (X(t_{(n+1)}) \le x_{(n+1)} \mid X(t_n) = x_n)$, where $X(t)$ means Markov chain process for a particular time (t),t_{n-1}, t_n, and t_{n+1} means previous time, current time, and end time in future, respectively.

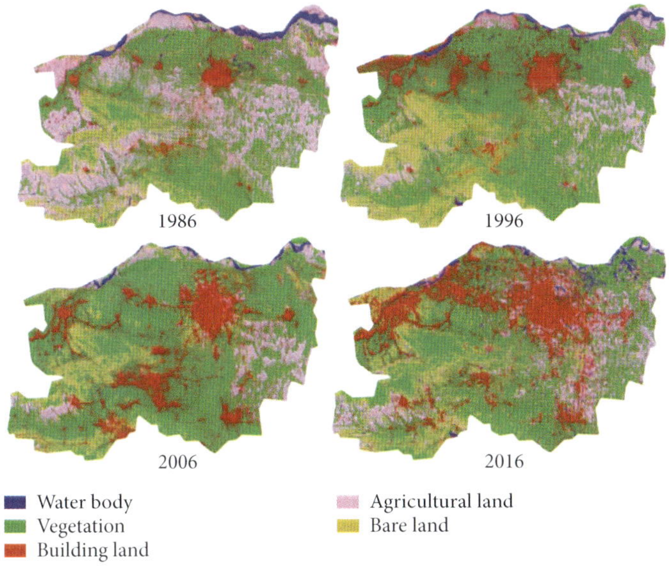

FIGURE 2: LULC classification maps over Zhengzhou city.

TABLE 2: Classification accuracy.

	1986	1996	2006	2016
OA	79.26%	84.01%	82.86%	85.13%
KC	0.78	0.82	0.81	0.83

In order to realize this objective, strategies can be divided into several procedures as follows: (1) setting LULC classification results obtained using classification method as initial variable; (2) calculating transition probability matrix and condition probability matrix of LULC types as conversion rules with GIS analysis method; (3) analyzing accuracy of simulating and predicting results based on the contrast of classification maps and predicted maps; and (4) changing trend analysis and ecological security assessment on the basis of classification results and simulation results. In this particular experiment, the four scenes LULC maps obtained using SVM classifier from remotely sensed data captured in 1986, 1996, 2006, and 2016 were set as initial variable and used to develop transition probability matrixes for LULC types between adjacent year; and then 10 times for CA iteration and 5*5 standard filter was selected to simulate LULC maps in 2006, 2016, and 2026, respectively. And simulation accuracy was calculated through contrasting and comparing the simulated LULC maps in 2006 and 2016 with classification maps obtained in the correspondent year. Then the changing trend and ecological security status between the periods of 1986 to 2026 are analyzed and assessed.

3.4. Landscape Pattern Index Analysis. The landscape pattern indexes are important indicators to quantitative describe and monitor the transformation characteristics of landscape structure. While it is a challenge to choose appropriate landscape indexes that would reflect the landscape properties and is not necessary to calculate all landscape indexes for a special landscape, since some of these indexes are strongly correlated. The choice of landscape pattern indexes can be based on both quantitative analysis and knowledge of the study area [19, 22, 23]. One of previous study suggests the combining use of Spearman's correlation coefficient (SCC) and principal component analysis (PCA) method to reduce redundancies [22]. The method proposed in [22] used SCC to reduce high correlation between two landscape indexes by checking SCC whether greater than 0.9 or less than −0.9 and used PCA [14, 22–24] to reduce the number of landscape indexes to a manageable level. In this research, the selection of appropriate landscape indexes was based on experience of previous research as well as the knowledge of the study area when the spatial distribution status and data quality of research area were considered. The slight correlation landscape pattern indexes at both patch level and landscape level are selected to analyze long-term series of landscape pattern change of research area from 1986 to 2026 [25]. The number of patches (NP), patch density (PD), largest patch index (LPI), edge density (ED), mean shape index (MSI), fractal dimension index (FDI), aggregation index (AI), and interspersion juxtaposition index (IJI) were selected when landscape pattern was analyzed in patch level. While number of patches (NP), patch density (PD), largest patch index (LPI), fractal dimension index (FDI), contagion index (CONTAG), Shannon's evenness index (SHDI), Shannon's diversity index (SHEI), fragmentation index (FI) were selected when landscape pattern was analyzed in landscape level. The integrated application of multiple landscape indexes can avoid the limitation in some extent when individual landscape index was selected and can reduce the adverse effect on result analysis caused by remotely sensed data resolution and human factors.

3.5. Establishment of ESA System. The ecological security assessment system established in this research is based on the concept of pressure-state-response (PSR) put forward by The UN Organization for Economic Cooperation and Development (OECD). The establishment of PSRP model is according to the principle of dynamism, hierarchy, and feasibility [9, 26–28], and according to the stability of the ecosystem structure, the resilience of the landscape, the service function of ecosystem, and the changing factors of landscape pattern. The established so-called pressure-state-response-pattern (PSRP) model system included target layer, criterion layer, index layer, and 12 single indexes [29, 30].

While there is no comparability between the various indexes because of different units of different indexes, for the sake of quantification and computation, the primitive value of indexes is standardized as between 0 to 10. The standardization model for positive indicators (the larger the values, the safer the ecology) can be described in the following equation:

$$P_i = 10 \times \frac{X_i - X_{min}}{X_{max} - X_{min}}. \tag{3}$$

The standardization model for negative indicators (the smaller the values, the safer the ecology) can be described in the following equation:

$$P_i = 10 \times \left(1 - \frac{X_i - X_{min}}{X_{max} - X_{min}}\right), \tag{4}$$

where P_i is standardization value, X_i is primitive value of indexes, X_{max} is the maximum value of a set of indicators, and X_{min} is the minimum value of a set of indicators.

Different indexes have significant different contribution for ecological security assessment. To overcome the diversity and to guarantee the relative importance of indexes in the ecological security assessment system, entropy method and AHP method were applied comprehensively to determine weight of indicators (results are shown in Table 3). Entropy method was used to determine the weight coefficient of indexes by using the diversity of the difference of indexes information, which can avoid the randomness caused by subjective decision. At the meantime, as a complement, AHP method was employed to verify the results of weight determined and to overcome the limitation of absolute objectivity in entropy method [31].

With the help of GIS spatial analysis mechanism, the matrix of transition probability and matrix of condition probability were calculated as conversion rules in evaluation level of ecological security assessment. The status of cellular automaton model is defined as the class level of pixels for ecological security assessment. Taking the data in 2006 and 2016 as an instance, ecological security assessment results in 2006 and 2016 were setting as initial status, with standard 5*5 filter and 10 times iteration parameters, the ecological security trend can be predicted [21].

To make sure the equitable using various indexes in assessment system and establish the importance of different indexes, a linear weighted synthetic evaluation model (LWSEM) was built. The LWSEM can be described in the following equation:

$$y = \sum_{j=1}^{m} w_j x_j, \tag{5}$$

where y is comprehensive evaluation value of ecological security; w_j is weight of index; x_j is standardized value of index. Domestic and foreign methods of classifying level of ecological security were referenced to determine criteria of ecological security assessment class [26]. The established criteria of ecological security assessment are shown in Table 4.

4. Results and Analysis

4.1. Changing Speed and Amplitude. To analysis landscape changing character, LULC changing speed and amplitude between 1986 and 1996, 1996 and 2006, and 2006 and 2016 need to be calculated in advance. The results were shown as Tables 5–7.

The results of LULC changing statistics shown that the transformation relationships of various landscape types were complex, and the variation range of the single dynamic degree was wide in this research area. While the integrated dynamic degree during three periods (1986–1996, 1996–2006, and 2006–2016) was 21.65%, 20.17%, and 19.21%, respectively. The integrated dynamic degree indicated that there was a dramatically landscape type change in this period. The characteristics of landscape changing were discussed as follows:

(1) There was a downward-downward-growth changing trend of water body. There was a sustained downward trend of water body from 1986 to 1996 and from 1996 to 2006. The changing amplitude of water was great, and especially during the period between 2006 and 2016 was the biggest among all LULC types. While, overall, the area of water was reduced by 59461.2 ha. The dominated changing direction of water body areas are vegetation and agricultural land. There were 124191.9 ha water body translated to water body and building area. The possible reason was because that reclaiming land and agricultural land from the river have reduced the area of the river that flows through the research area.

(2) There was a growth-downward-growth trend of vegetation. During the period from 1996 to 2006, vegetation had a tendency of slight increase. And single dynamic degree of vegetation was 0.89%. Overall, the area of vegetation has decreased by 16,920.9 ha in the last 30 years. The main flow direction of vegetation was building area and bare land. And the conversion ratio was 4.31% and 2.85% for building area and bare land, respectively. The primary land types that translate to vegetation were building land, agricultural land, and bare land. The main reason were that urban growth and development are influenced by the national sustainable development strategy and the concept of green harmonious development.

TABLE 3: Weight setting for synthetic assessment system of ecological security index.

Target layer	Criterion layer	Index layer	Weight
Comprehensive evaluation	Pressure 0.1196	Density of population	0.3828
		Urban expansion intensity	0.3392
		Building land index	0.2780
	State 0.4481	NDVI	0.3079
		Biological richness	0.3752
		Ecological resilience	0.3169
	Response 0.3075	Green index	0.3748
		Humidity index	0.3687
		Dryness index	0.2565
	Pattern 0.1248	FI	0.3588
		FDI	0.3049
		SHDI	0.3363

TABLE 4: Class level of ecological security evaluation criteria.

Evaluation value	(0,4)	[4,5)	[5,6)	[6,8)	[8,10]
Class	Morbidity	Danger	Criticality	Security	Stability

The reforestation activities make the area increasing. During the period from 2006 to 2016, the severe shortage of agricultural leading to vegetation transform into agricultural land.

(3) Building land appeared a sustainable increasing trend. Building area of the research city increased by 1016336.7 ha from 1986 to 2016. The mainly transformation is that vegetation and farmland contributed more to building area than other LULC types. The primary reasons were that the urbanization process in the research city has been accelerating, and the urban expand was increasing by industrial infrastructure development. The expansion rate of urban building land is larger than reduced rate of building in villages, thus the urban building area increased overall.

(4) There was a downward-growth-downward trend of farmland. Farmland mainly translated to vegetation and building land during this period. The conversion ratio of farmland was 14.57% from 1986 to 2016. And the main LULC types that translate to farmland were vegetation and bare land. That was because building land expansion occupied a large amount of farmland with the development of urbanization and industrialization. Furthermore, people in the village moved into the city, which cause some areas of farmland to be shelved and abandoned. The development of forestry reduced the area of farmland to a certain extent.

(5) Bare land appeared downward-growth-downward trend. There was a sharp reduce of bare land from 1996 to 2006. The reduced rate of bare land was 3.54% during the period and was slower than the previous period. The total area of bare land increased by 56415.6 ha. The bare land was mainly converted into vegetation and farmland. The primary reasons were that the contradiction between population and resource is protruding, and the demand of urban land is increasing. These result in the exploit of unused land. Simultaneously, with the development of the economy and the afforestation from the bare land, the whole area of the bare land tends to decrease.

4.2. Prediction of Changing Trend. Based on the result of comparison between simulated land cover map in 2016 and the classification map in 2016, the prediction accuracy of the CA-Markov model was analyzed and verified (shown in Table 8). Results demonstrate that all relative errors between classified and predicted LULC type area are less than 8%. So it is scientific and reasonable to predict landscape changes based on the CA-Markov model. Then the CA-Markov model was selected to simulate land cover map in 2026 based on LULC maps captured in 2006 and 2016. The results are shown as Table 9 and Figure 3.

The results in Table 9 indicate that the largest changing amplitude of landscape type with a decrease trend is vegetation, which has an amount of 38.07 percent of total LULC types during the period of 2016 and 2026. The secondary change in the biggest of all is agriculture land, which will increase by 191006.1 ha from 2016 to 2026. The area of vegetation, farmland, bare land, and water body all will be increased at different degree based on the simulated results. It demonstrates that the urbanization in the research area is accelerating with the development of the urban economy. Simultaneously, a large reduction of vegetation shows that the ecological balance in the research area might be destroyed at a certain extent.

4.3. Landscape Pattern Analysis at Different Levels

4.3.1. Analysis of Landscape Pattern at Patch Level. Based on primary statistics, in total, the results shown an interesting phenomenon that relative large number of patches (NP) was bare land, the largest interspersion juxtaposition

Table 5: LULC change of Zhengzhou city from 1986 to 1996.

Land type	Area (ha)		Changing amplitude	Changing speed
	1986	1996	(%)	(%)
Water body	213,439.5	167,060.7	−21.73%	−2.17%
Vegetation	3,387,718.8	4,073,481.9	20.24%	2.02%
Building area	626,895.9	822,069.9	31.13%	3.11%
Agricultural land	2,101,129.2	908,448.3	−56.76%	−5.68%
Bare land	1,170,585	1,528,706.7	30.59%	3.06%
Integrated dynamic degree			21.65%	

Table 6: LULC change of Zhengzhou city from 1996 to 2006.

Land type	Area (ha)		Changing amplitude	Changing speed
	1996	2006	(%)	(%)
Water body	167,060.7	74,314.8	−21.73%	−2.17%
Vegetation	4,073,481.9	4,436,432.1	20.24%	2.02%
Building area	822,069.9	1,348,050.6	31.13%	3.11%
Agricultural land	908,448.3	654,131.7	−56.76%	−5.68%
Bare land	1,528,706.7	986,783.4	30.59%	3.06%
Integrated dynamic degree			20.17%	

Table 7: LULC change of Zhengzhou city from 2006 to 2016.

Land type	Area (ha)		Changing amplitude	Changing speed
	2006	2016	(%)	(%)
Water body	74,314.8	153,978.3	107.20%	10.72%
Vegetation	4436,432.1	3,370,797.9	−24.02%	−2.40%
Building area	1,348,050.6	1,643,232.6	21.90%	2.19%
Agricultural land	654,131.7	1,104,622.2	68.87%	6.89%
Bare land	986,783.4	1,227,000.6	24.34%	2.43%
Integrated dynamic degree			19.21%	

Table 8: The prediction accuracy analysis of landscape evolution model (ha).

Land type	Reality in 2016	Prediction in 2016	Difference value	Relative error
Water body	153,978.3	157,022.6	3044.3	0.0198
Vegetation	3,370,797.9	3,357,674.7	−13,123.2	−0.0039
Building land	1,643,232.6	1,765,627.3	122,394.7	0.0745
Agricultural land	1,104,622.2	1,085,974.0	−18,648.2	−0.0169
Bare land	1,227,000.6	1,230,871.6	3,871.0	0.0032

index (IJI), and the least edge density (ED) was water body. And the landscape types in the research, which has relative large largest patch index (LPI) was farmland and building area, while relative large aggregation index (AI) was vegetation and water body. The landscape types with relative large mean shape index (MSI) was vegetation, and the relative large fractal dimension index (FDI) was agricultural land and vegetation. So the landscape of various LULC types has changed significantly since 1986 shown from Figures 4–11 for NP, LPI, PD, ED, IJI, MSI, and AI, respectively. And the detailed analysis of landscape change can be discussed as follows:

(1) The changing of farmland landscape. NP of farmland increased by 27,474, and patch density (PD) of farmland increased by 1.144 from 1986 to 2016. By prediction, largest patch index (LPI), aggregation index (AI), and interspersion juxtaposition index (IJI) will be reduced at different degrees by 2026. These results indicate that farmland is seriously disturbed by human activity in the process of urbanization and industrialization. Massive activities of economic construction and development occupied many farmlands. The increase of NP and the decrease of AI demonstrate that the patches were more fragmented, and the original patches were split up into discrete pieces of debris and distributed more widely. The transfer between farmland and other landscape types becomes more frequent. Dominance

TABLE 9: The prediction of Zhengzhou land use types in 2026.

Land type	Water body	Vegetation	Building land	Agricultural land	Bare land
Area (ha)	207,370.8	2,864,382.3	1,779,116.4	1,295,628.3	1,353,154.8
Percentage	2.76%	38.07%	23.65%	17.22%	17.99%
Variation (ha)	53,392.5	−506,415.6	135,883.8	191,006.1	126,154.2
Changing amplitude	0.71%	−6.73%	1.81%	2.54%	1.68%

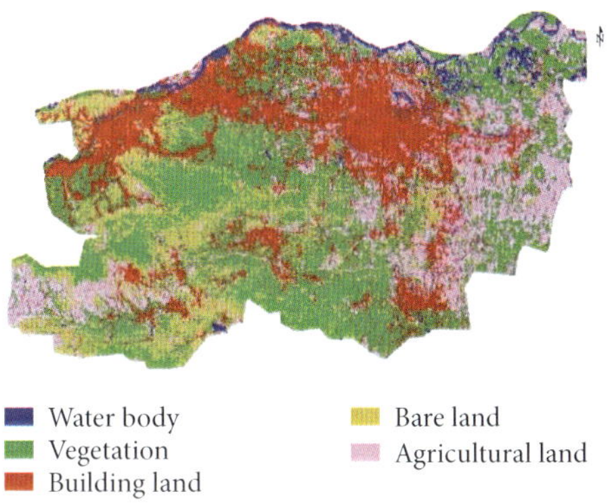

FIGURE 3: Predicted LULC map of Zhengzhou city in 2026.

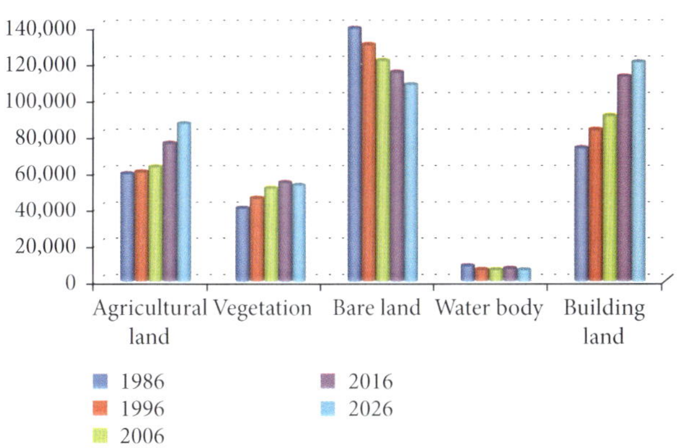

FIGURE 4: Changing trend of NP index.

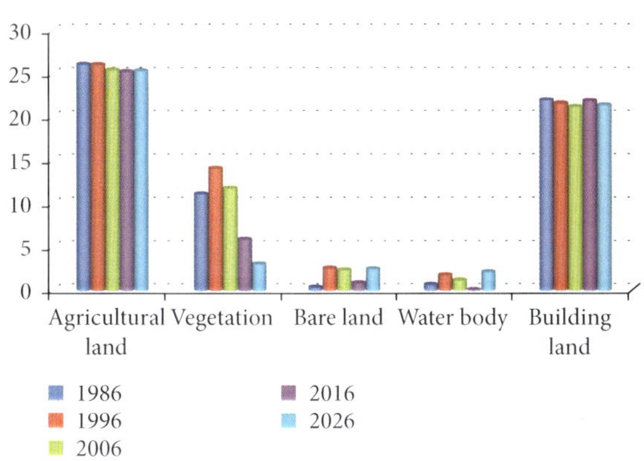

FIGURE 5: Changing trend of LPI index.

role of farmland is reducing, and landscape pattern of selected area was more and more broken. The degree of invading border is higher; land spatial structure will be more complicated.

(2) The transformation of vegetation landscape. The mean shape index (MSI), fractal dimension index (FDI) and aggregation index (AI) of vegetation were relatively large from 1986 to 2016. At the meantime, AI and LPI of vegetation were decreasing during the period. Due to the ecological construction activities in the research area, the NP of vegetation increased and LPI was decreased. These show that human activities and interference were becoming more and more frequent in recent years. Human being efforts aggravated fragmentation of vegetation and increased the connectivity intensify. The fact that the rising of NP, IJI, and MSI of vegetation indicates that distribution tend of vegetation was disperse. The complex shape, means the deep irregular shape, which demonstrate human activity of developing vegetation land has been enhanced. According to results of prediction, NP and IJI will decline, AI will rise, and the vegetation landscape tends to be concentrated by 2026. The number of adjacent patches will increase, and the control power of the superior patches will decrease by that time.

(3) The changing of bare land landscape. The bare land had the largest NP index and had the least LPI index during the period from 1986 to 2016. NP of bare land decreased by 24,402, ED of bare land reduced by 9.382, IJI declined, and AI increased. These indicated that the area of the bare land landscape was decreasing, and the distribution of bare land was more concentrated. According to the predicted results, NP of bare land will be further reduced and AI will rise significantly by 2026, which demonstrate that with the development of urbanization in the research area, the utilization of land will be become more efficienct and remarkable.

(4) The landscape pattern changes of water body. Water body had the relative high AI and the largest IJI, while it has the least LPI and ED, during the period from 1986 to 2016. There was a sustained downward trend in NP and PD of water body, but there was a continued growth tendency in AI and IJI of water body. These demonstrates that due to interference of human activities, rainfall, urban construction, and other factors, the patch shape of the water body

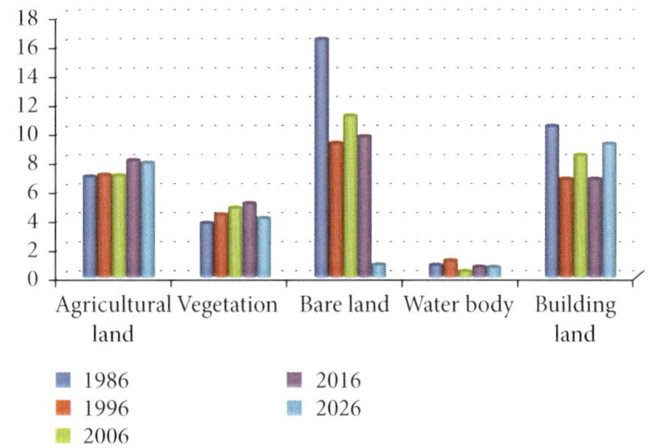

FIGURE 6: Changing trend of PD index.

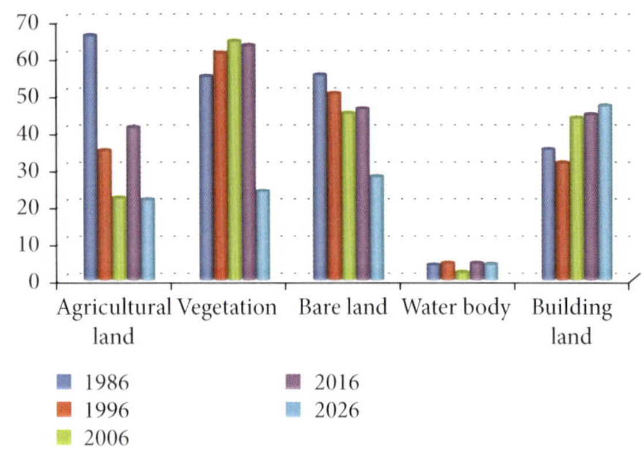

FIGURE 7: Changing trend of ED index.

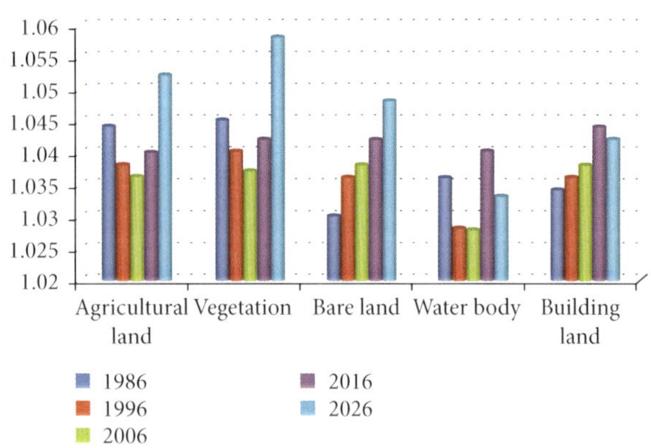

FIGURE 8: Changing trend of MPFD index.

landscape tends to be simplified, the separated patches are joined together, and the distribution tend to be concentrated. The predicted results shown that NP will rise and AI and IJI will decline by 2026. These phenomena indicate that the area of the water body will increase, and the distribution become dispersed.

(5) The changing of built-up landscape. In recent 30 years, the NP of building land increased by 39,155, and the FDI increase by 0.01. These numbers demonstrated that the area of building land is soar and

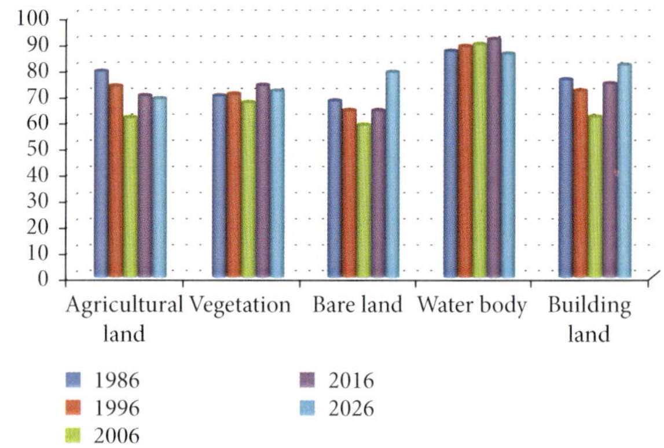

FIGURE 9: Changing trend of IJI index.

occupying a domination because the increase in population and the development of urbanization. The phenomenon that AI and LPI of building land rise demonstrated the increasing connectivity for expansion of urban and the continuous distribution of large-scale area rural building land. The rising MSI of building land rising shown that the patches were clustered, the boundaries were fragmented, the distribution is more concentrated, and the shape is more complex. Based on the results of prediction, there will be a sustained downward trend in NP of building land, but a continued growth trend tendency in FDI of building land by 2026. These predicted results demonstrate that there is lack of reasonable plans for infrastructure construction in the preliminary stage, with the development of economic and growth of population, which makes the structure of patches more complex. While in the late stage, there is more scientific and reasonable construction plan, which makes the structure of patches not so much complicated.

4.3.2. Landscape Pattern Analysis at Landscape Level. In terms of the landscape view of the entire research area (shown in Table 10), there has similar changing trend of the total NP and FI, shown as downward-growth-downward trend. FDI has undergone a process from large to small and small to large, which is contrary to contagion index (CONTAG). The statistics shown that there is a sustained downward trend of LPI. The mentioned results above demonstrate that the dominate landscape in the previous stage has been replaced by the small patches gradually, and the landscape of the research area tend to be more complex and diversified. During the period from 1986 to 2006, the area of vegetation landscape increased significantly, the dominance of regional landscape was increasing, and the types of patches formed satisficed connectivity. While the construction of the research area lacks reasonable plans, the structure of the patches is more complex with economic development and population growth. But with the help of scientific and reasonable plan by decision maker, the structure of patches become simplification during the latter period (from 2006

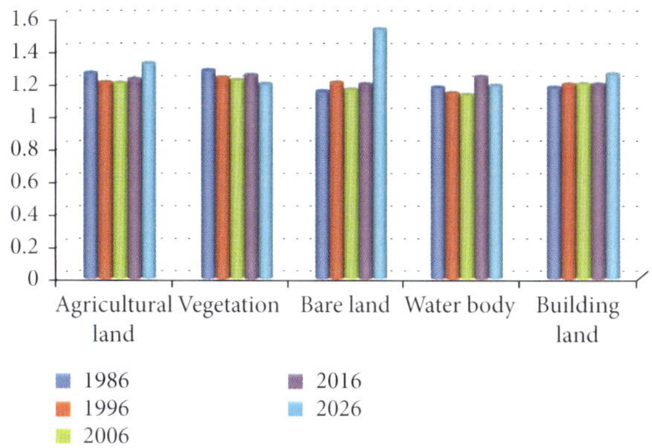

FIGURE 10: Changing trend of MSI index.

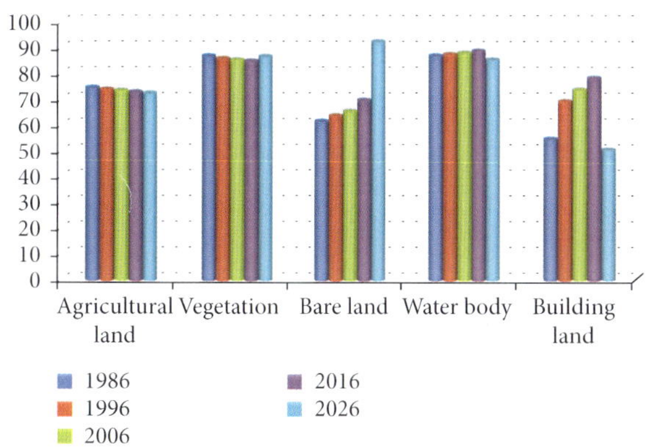

FIGURE 11: Changing trend of AI index.

to 2026). Simultaneously, with the rapid decrease of vegetation, the fragmentation degree of the landscape will increase, the landscape dominance will decrease, and the regional landscape will form a dense pattern with various elements.

During the first period from 1986 to 2006, there was a sustained downward trend in Shannon's evenness index (SHEI) and Shannon's diversity index (SHDI) shown that the heterogeneity of landscape decrease, the proportion of the dominant landscape rise, and the control function of the single component increase. The reason of the result is because that there is plain area domination in the selected research area, and that the policy of returning farmland to forest causes vegetation area increasing. The vegetation, as a dominant landscape type, whose proportion further improved and whose ability of landscape controlling has been enhanced significantly. During the second period (from 2006 to 2026), SHEI and SHDI rise observably. This indicate that interference intensity of human activities enhanced, the rare patches area increased, landscape diversity rises, the effect of controlling action of advantage components of landscape to overall landscape has decreased, the landscape stability decreased, and the degree of landscape heterogeneity is gradually increased.

The primary reasons were that to meet the objective requirements of urban expansion over the next decade, the economy has been developing rapidly over the past decade, and the greatly increasing building area occupied huge amount of vegetation and farmland that led to the increase of rare patches such as water body. On a whole, the proportion differences of different landscape were gradually narrowed, and the diversity of the landscape and the complexity of the structure increased significantly.

4.4. Analysis, Assessment, and Prediction of Ecological Security Status. With the help of the ecological security assessment system of above mentioned, maps of ecological security assessment classes of Zhengzhou city were obtained (shown as Figure 12).

Based on the primary results, the mean value of five phases of comprehensive security assessment from 1986 to 2026 were calculated (shown in Table 11), and the ecological security curve of Zhengzhou was drawn (Figure 13). The proportion of ecological security class of Zhengzhou city was statistical according to the assessment standard in Table 3; the changing situation is shown in Figure 14 and (shown in Table 12).

During the 40 years, the ecological security assessment value shows a "V" shape trend, decrease first and then increase. The ecological security assessment value of the research area was 7.01 in 1986, which means the ecological condition was at the "security" level. While the ecological security situation worsens significantly in 1996, and the ecological security was at "danger" level at the time. From 2006 to 2016, the ecological security evaluation value increased gradually, and the ecological security situation improved. According to the predicted results, the ecological security assessment value of the research area will further increase by 2026, but the ecological security situation will be still at the "criticality" level. The changing curve of ecological security shown that the ecological security value of the research area in 1986 is relatively gentle; the ecological security was more balanced. The two peaks of the ecological security value in the research area from 1996 to 2026 are approximately 5.5 and 8, which shown that the ecological security value of most areas is about 5.5 and 8. Based on the analysis of remotely sensed classification images and the maps of comprehensive assessment ecological security level, a conclusion can be drawn as follows. The area with low ecological security value concentrates mainly upon the water body and building land. The area whose ecological security value is 5.5 distributes in vegetation and farmland types, and the area whose ecological security value is around 8 concentrates in vegetation and bare land where human disturbance is weak.

The results of horizontal comparison of ecological security status shown that ecology of the research area is major in the "security" level in 1986, with a proportion of 38.81%. And the proportion of "morbidity" level of ecology is less than 1%. These conclusions indicated that a complete and stable ecosystem had established in the research area, but the ecological security areas with "morbidity" level are mainly focused in the northern of the research area due to the lack of reasonable management. In 1996, the "criticality" level ecology was dominant in the research area. And the percentage of "Morbidity" level ecology reached to 9.17%, exceeding the "security" level, which shown that the

TABLE 10: The overall landscape index changes in the research area.

Time	NP	PD	LPI	FDI	CONTAG	SHDI	SHEI	FI
1986	428,512	0.602	26.052	1.036	42.693	1.522	0.850	1.398
1996	306,914	0.614	26.015	1.036	45.654	1.471	0.821	1.309
2006	310,988	0.625	25.338	1.034	48.135	1.406	0.785	1.236
2016	319,784	0.628	25.281	1.038	42.299	1.547	0.863	1.428
2026	299,653	0.633	25.263	1.045	46.471	1.594	0.890	1.497

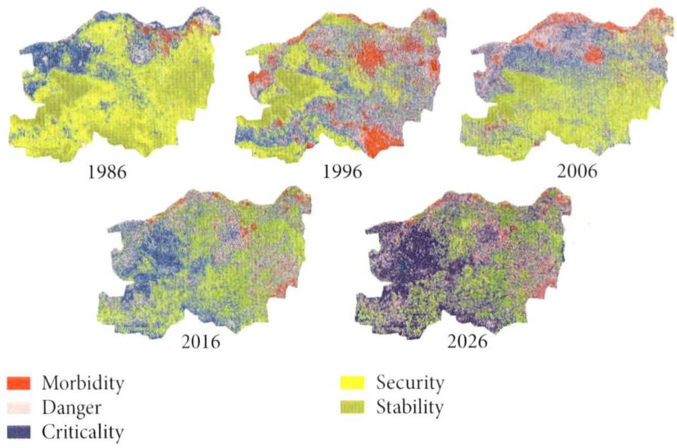

FIGURE 12: Ecological security assessment results from 1986 to 2026.

TABLE 11: The mean value of ecological security evaluation.

Time	1986	1996	2006	2016	2026
Mean value	7.01	4.97	5.07	5.42	5.87

ecosystem stability of the research area is affected seriously by human disturbance. In 2006, the area with "criticality" level of ecology was more than 60%, the area with "security" and "stability" level of ecology was more than 10%. In 2016, the area with "criticality" level of ecology remained dominant in the research area. According to the predicted results, the area with "criticality" level of ecology will occupy a greater advantage by 2026, with an obviously trend of aggregating to "criticality" level.

Combining with maps of comprehensive assessment of ecological security level, a 40-year long-term vertical comparison analysis of ecological security assessment over the research area was carried out. Due to the rapid development of urbanization and the weak awareness of human ecological protection, during the period from 1986 to 1996, the "danger" level of ecology in the central and northern research area converted to "morbidity" level. At the meantime, the "stability" level of ecology transformed to "criticality" and "security" level. The ecological vitality of the research area decreased obviously, and the ecological security of the research area was deteriorated significantly. The rapid increase of population may be caused by rapid development of urbanization, the boom of economy, and the extensive construction of infrastructure. The increase of population pressure and the decrease of vegetation coverage led to the decline of ecological security in the research area overall.

During the period from 1996 to 2006, with the implementation of the sustainable development strategy, the increasing of vegetation coverage, the "morbidity" level of ecology changed to the "danger" and "criticality" level gradually. So, the ecological security situation has been alleviated in some extend. The primary reason is that with the promoting of strategy of "central rise" and sustainable development by Chinese administrate, the economic development model has changed from traditional economy to intensive economy, which strengthen effort of ecological and environmental protection significantly. Meanwhile, with the overall implementation of "one family, one child" policy, the contradictions between population and resources has been relieved in some extent. So, the landscape ecological security situation was improved.

From 2006 to 2016, the area with "danger" level of ecology increased slightly, and the area with "security" level of ecology reduced slightly, which indicated that with the further strengthening environmental protection by government, the improvement of awareness of environmental protection by residents in the process of urban development, ecological security has been further improved. It is predicted that by 2026, the "morbidity" level area will further reduce. In the meantime, the tendency of gathering to middle level of ecological security is more obvious with the increasing of "criticality" level. Therefore, it is an urgent problem to be solved that protecting the current security landscape, improving the landscape which was in "danger" and "morbidity" level.

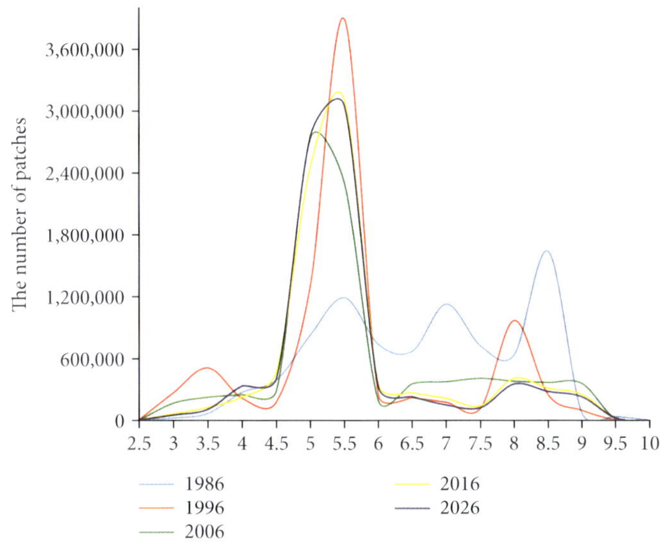

FIGURE 13: Changing curve of ecological security.

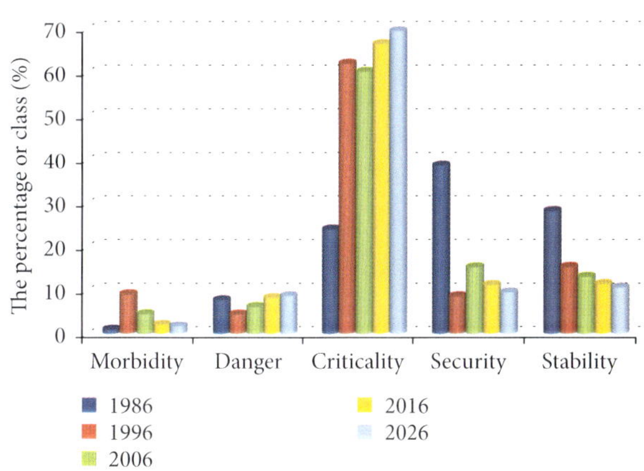

FIGURE 14: Class level of ecological security.

TABLE 12: The proportion of ecological security class in Zhengzhou city.

Time	Morbidity	Danger	Criticality	Security	Stability
1986	0.98%	7.85%	24.09%	38.81%	28.27%
1996	9.17%	4.55%	62.02%	8.77%	15.49%
2006	4.63%	6.39%	60.21%	15.51%	13.26%
2016	2.14%	8.27%	66.51%	11.43%	11.65%
2026	1.82%	8.78%	68.36%	12.67%	8.37%

5. Conclusions

In this work, the synthetically analysis of urban surface environment changes and landscape ecological indexes extends the application scope and applicable environment of the remotely sensed data. An ecological security evaluation system was established by collaborative processing of society, economy, nature, and environment factors. This study provides a first-time 40-year period, from past to future, comprehensive changing trend of typical LULC types over the selected area. The study proves that a growth-changing trend in building land and vegetation, a downward trend in water body, farmland, and bare land. During this LULC change process, the patch shape of the water body tends to be simplified, and the distribution of this landscape tends to be dispersed. The diversity of the landscape and the complexity of the structure had increased significantly overall. The ecological security evaluation value of Zhengzhou city in the last 40 years shows a "V" shape trend.

There are also some suggestions that can be drawn based on this research to realize the sustainable development of the research area. These suggestions include but not limit to the establishment of urban smart growth plan, strengthen of urban landscape ecological construction, controlling the total amount of urban land use scientifically, allocating the urban land area reasonably, following the conversion regulation of farmland to built-up land strictly, sustainably preventing and controlling soil erosion, optimizing the urban planning and design, and so on. The administrators are also recommended to promote the relationships between economic development and ecological balance by taking reasonable measures.

Conflicts of Interest

The authors declare no conflict of interest. The founding sponsors had no role in the design of the study; in the collection, analyses, or interpretation of data; in the writing of the manuscript; and in the decision to publish the results.

Authors' Contributions

Pei Liu led the research work, proposed the idea and structure of this manuscript, wrote the instruction and conclusion parts, and integrated the contributions from all authors. Ruimei Han and Shoujun Jia integrated the modules and codes, performed the experiments, wrote LULC map classification and CA-Markov simulation parts and tested. Ruimei Han who implemented landscape indexes selection and extraction, established ESA system, wrote the corresponding parts, and also collaborated with Pei Liu to conceive and design the experiments. Hanwei Zhang was the coleader of the group and involved in the algorithm implementations through the paper, performed changing detection, and wrote the corresponding parts.

Acknowledgments

This work was supported by the National Natural Science Foundation of China (Grant nos. 41601450, 41601392, 41401403), the Key Research Project Plan of Colleges and Universities in Henan Province (Grant no. 16A420004), the Key Technology R&D Program of Henan Province (Grant nos. 182102310860, 162102310089), Henan postdoctoral foundation and Ph.D. fund of Henan Polytechnic University (Grant nos. B2015-20, B2014-18).

References

[1] B. C. Pijanowski and K. D. Robinson, "Rates and patterns of land use change in the Upper Great Lakes States, USA: a framework for spatial temporal analysis," *Landscape and Urban Planning*, vol. 102, no. 2, pp. 102–116, 2011.

[2] C. R. Fichera, G. Modica, and M. Pollino, "Land cover classification and change-detection analysis using multi-temporal remote sensed imagery and landscape metrics," *European Journal of Remote Sensing*, vol. 45, no. 1, pp. 1–18, 2012.

[3] D. Wang and S. Zhang, "Landscape pattern change and its driving forces in Sanjiang plain from 1976 to 2000," in *2012 2nd international conference on remote sensing, environment and transportation Engineering*, pp. 1–4, Nanjing, China, June 2012.

[4] J. Tzanopoulos and I. N. Vogiatzakis, "Processes and patterns of landscape change on a small Aegean island: the case of Sifnos, Greece," *Landscape and Urban Planning*, vol. 99, no. 1, pp. 58–64, 2011.

[5] M. G. Turner, R. V. O'Neill, R. H. Gardner, and B. T. Milne, "Effects of changing spatial scale on the analysis of landscape pattern," *Landscape Ecology*, vol. 3, no. 3-4, pp. 153–162, 1989.

[6] H. Liu and Q. Weng, "Landscape metrics for analysing urbanization-induced land use and land cover changes," *Geocarto International*, vol. 28, no. 7, pp. 582–593, 2013.

[7] M. J. van Strien, C. T. J. Slager, B. de Vries, and A. Grêt-Regamey, "An improved neutral landscape model for recreating real landscapes and generating landscape series for spatial ecological simulations," *Ecology and Evolution*, vol. 6, no. 11, pp. 3808–3821, 2016.

[8] X. Zhou, S. Shekhar, and R. Y. Ali, "Spatiotemporal change footprint pattern discovery: an inter-disciplinary survey," *Wiley Interdisciplinary Reviews: Data Mining and Knowledge Discovery*, vol. 4, no. 1, pp. 1–23, 2014.

[9] D. Liu and S. Hao, "Ecosystem health assessment at county-scale using the pressure-state-response framework on the Loess Plateau, China," *International Journal of Environmental Research and Public Health*, vol. 14, no. 1, p. 2, 2017.

[10] X. Ma, X. Tong, S. Liu, X. Luo, H. Xie, and C. Li, "Optimized sample selection in SVM classification by combining with DMSP-OLS, Landsat NDVI and GlobeLand30 products for extracting urban built-up areas," *Remote Sensing*, vol. 9, no. 3, p. 236, 2017.

[11] J. Chen, X. Cao, S. Peng, and H. Ren, "Analysis and applications of GlobeLand30: a review," *ISPRS International Journal of Geo-Information*, vol. 6, no. 8, p. 230, 2017.

[12] J. Chen, J. Chen, A. Liao et al., "Global land cover mapping at 30m resolution: a POK-based operational approach," *ISPRS Journal of Photogrammetry and Remote Sensing*, vol. 103, pp. 7–27, 2015.

[13] S. K. Singh, P. K. Srivastava, M. Gupta, J. K. Thakur, and S. Mukherjee, "Appraisal of land use/land cover of mangrove forest ecosystem using support vector machine," *Environmental Earth Sciences*, vol. 71, no. 5, pp. 2245–2255, 2014.

[14] S. K. Singh, P. B. Laari, S. Mustak, P. K. Srivastava, and S. Szabó, "Modelling of land use land cover change using earth observation data-sets of Tons River Basin, Madhya Pradesh, India," *Geocarto International*, pp. 1–21, 2017.

[15] R. G. Pontius, J. Huang, W. Jiang et al., "Rules to write mathematics to clarify metrics such as the land use dynamic degrees," *Landscape Ecology*, vol. 32, no. 12, pp. 2249–2260, 2017.

[16] L. P. Basommi, Q. F. Guan, D. D. Cheng, and S. K. Singh, "Dynamics of land use change in a mining area: a case study of Nadowli District, Ghana," *Journal of Mountain Science*, vol. 13, no. 4, pp. 633–642, 2016.

[17] Y. Li, G. Liu, and C. Huang, "Dynamic changes analysis and hotspots detection of land use in the central core functional area of Jing-Jin-Ji from 2000 to 2015 based on remote sensing data," *Mathematical Problems in Engineering*, vol. 2017, Article ID 2183585, 16 pages, 2017.

[18] Y. Wang, X. Li, and J. Li, "Study on the response of ecological capacity to land-use/cover change in Wuhan city: a remote sensing and GIS based approach," *The Scientific World Journal*, vol. 2014, Article ID 794323, 11 pages, 2014.

[19] S. K. Singh, S. Mustak, P. K. Srivastava, S. Szabó, and T. Islam, "Predicting spatial and decadal LULC changes through cellular automata Markov chain models using earth observation datasets and geo-information," *Environmental Processes*, vol. 2, no. 1, pp. 61–78, 2015.

[20] E. F. Lambin, M. D. A. Rounsevell, and H. J. Geist, "Are agricultural land-use models able to predict changes in land-use intensity?," *Agriculture, Ecosystems & Environment*, vol. 82, no. 1–3, pp. 321–331, 2000.

[21] B. Abuelaish and M. T. C. Olmedo, "Scenario of land use and land cover change in the Gaza Strip using remote sensing and GIS models," *Arabian Journal of Geosciences*, vol. 9, p. 274, 2016.

[22] L. Pei and D. Peijun, "Analysis and simulation of land cover and thermal environment in mining area based on remote sensing data and CA_Markov model," *Journal of China Coal Society*, vol. 11, no. 37, pp. 1847–1853, 2012.

[23] S. Lamine, G. P. Petropoulos, S. K. Singh et al., "Quantifying land use/land cover spatio-temporal landscape pattern dynamics from Hyperion using SVMs classifier and FRAGSTATS®," *Geocarto International*, pp. 1–17, 2017.

[24] S. K. Singh, P. K. Srivastava, S. Szabó, G. P. Petropoulos, M. Gupta, and T. Islam, "Landscape transform and spatial metrics for mapping spatiotemporal land cover dynamics using earth observation data-sets," *Geocarto International*, vol. 32, pp. 1–15, 2017.

[25] D. Liu, S. Hao, X. Liu, B. Li, S. He, and D. N. Warrington, "Effects of land use classification on landscape metrics based on remote sensing and GIS," *Environmental Earth Sciences*, vol. 68, no. 8, pp. 2229–2237, 2013.

[26] P. Du, J. Xia, Q. Du, Y. Luo, and K. Tan, "Evaluation of the spatio-temporal pattern of urban ecological security using remote sensing and GIS," *International Journal of Remote Sensing*, vol. 34, no. 3, pp. 848–863, 2013.

[27] H. Jia, D. Pan, and W. Zhang, "Health assessment of wetland ecosystems in the Heilongjiang River Basin, China," *Wetlands*, vol. 35, no. 6, pp. 1185–1200, 2015.

[28] H. Pei, S. Fang, L. Lin, Z. Qin, and X. Wang, "Methods and applications for ecological vulnerability evaluation in a hyper-arid oasis: a case study of the Turpan Oasis, China," *Environmental Earth Sciences*, vol. 74, no. 2, pp. 1449–1461, 2015.

[29] J. W. Rouse, R. H. Haas, and J. A. Schell, *Monitoring vegetation systems in the Great Plains with ERTS. Third ERTS Symposium, NASA SP-351*, vol. 1, pp. 309–396, 1973.

[30] Y. Zha, S.-x. Ni, and S. Yang, "An effective approach to automatically extract urban land-use from TM imagery," *Journal of Remote Sensing-Beijing*, vol. 7, no. 1, pp. 37–40, 2003.

[31] H. Fang and L. Li, "Assessing urban eco-security by PSR model in fast developing regions: a case of Chongqing in China," *Disaster Advances*, vol. 5, no. 4, pp. 855–860, 2012.

A Rapid UAV Image Georeference Algorithm Developed for Emergency Response

Shumin Wang, Ling Ding, Zihan Chen, and Aixia Dou

Institute of Earthquake Forecasting, China Earthquake Administration, Beijing 100036, China

Correspondence should be addressed to Shumin Wang; wmcnu@163.com

Academic Editor: Vincenzo Marletta

The image collection system based on the unmanned aerial vehicle plays an important role in the postearthquake response and disaster investigation. In the postearthquake response period, for hundreds of image stitching or 3D model reconstruction, the traditional UAV image processing methods may take one or several hours, which need to be improved on the efficiency. To solve this problem, the UAV image rapid georeference method for postearthquake is proposed in this paper. Firstly, we discuss the rapid georeference model of UAV images and then adopt the world file designed and developed by ESRI to organize the georeferenced image data. Next, the direct georeference method based on the position and attitude data collected by the autopilot system is employed to compute the upper-left corner coordinates of the georeferenced images. For the differences of image rotation manners between the rapid georeference model and the world file, the rapid georeference error compensation model from the image rotation is considered in this paper. Finally, feature extraction and feature matching for UAV images and referenced image are used to improve the accuracy of the position parameters in the world file, which will reduce the systematic error of the georeferenced images. We use the UAV images collected from Danling County and Beichuan County, Sichuan Province, to implement the rapid georeference experiments employing different types of UAV. All the images are georeferenced within three minutes. The results show that the algorithm proposed in this paper satisfies the time and accuracy requirements of postearthquake response, which has an important application value.

1. Introduction

The high-resolution image collection system based on the unmanned aerial vehicle has the advantages of lightweight, low cost, and flexibility. It has to be a vital tool for rapid collecting and viewing of the disaster information in post-earthquake response and reaction and attracts more attention [1–6]. At present, the main processing methods of UAV images are image stitching [7–11] and 3D point cloud model reconstruction [12–15] based on the computer vision theory. The image stitching can get the panoramic image of the parts of a disaster area even the whole disaster area. It generally needs steps such as feature extraction, feature matching, image transformation, bundle adjustment, and image fusion, which usually consumes one or several hours to achieve the image stitching. The 3D point cloud reconstructed from the UAV images can acquire the disaster information from different sides of the concerned objects. It generally needs steps such as sparse match, dense match, adjustment, and point cloud calculation. We need much time to achieve these steps. All the methods mentioned above are implemented in the commercial software, such as PhotoScan (http://www.agisoft.com/), an image stitching software, and Smart 3D (http://www.acute3d.com/), a three-dimensional model reconstruction software. The image map and three-dimensional model output from the mentioned software could be provided to the postearthquake commanders as the next level results. However, in the postearthquake response period, one to several hours is quite long for the emergency response commanders; they want to know the disaster information as soon as possible and make the correct decision to muster rescue force and distribute necessary supplies. To improve the image processing speed, [16–19] adopt

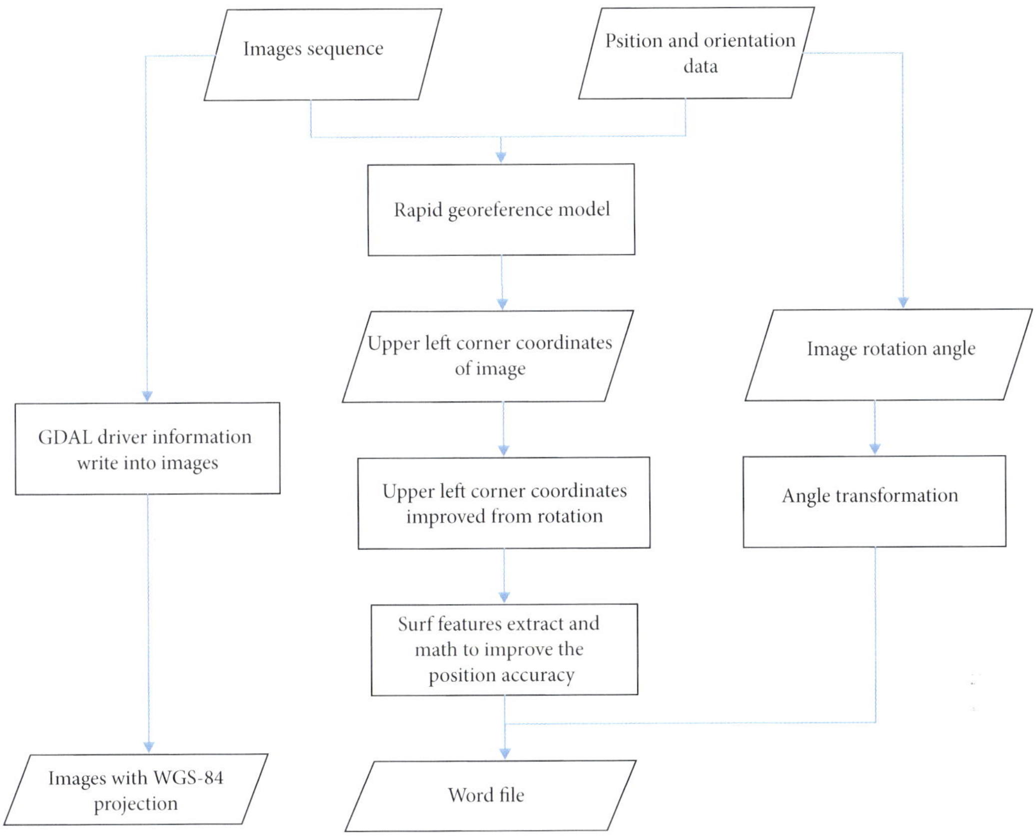

FIGURE 1: The UAV image rapid georeference for postearthquake response.

the SLAM (simultaneous localization and mapping) framework to create a panorama image, which uses the incremental manner to mosaic the UAV images. For large-scale image mosaics which contain thousands of images, it becomes computationally infeasible to use the filtering approach since the size of the state vector grows very large. The commercial solutions such as DroneDeploy Live Map produces mosaics instantly during the flight, which is designed to create and view a low-resolution map on the mobile device and, if you want to get the high-resolution map, you have to upload the images to the cloud through the network and download the result after image stitching. Processing the high-quality data can take up to a few hours for a very large job with high-resolution imagery (https://support.dronedeploy.com/docs/live-map). It depends on the image network speed, and it is hard to achieve the real-time georeference. To satisfy the postearthquake response application requirements, the rapid georeference method of UAV images for postearthquake response is proposed in this paper, the technique flow (see Figure 1). Firstly, the rapid georeference model is established combined with the position and attitude data collected from the autopilot system. And then, we deduce the image rotation transformation method, which converts the image rotation around the image center to the image rotation around the upper-left corner. Considering the limitation of GPS accuracy, which does not have the differential function, the feature extraction and matching method are employed to reduce the systematic error and improve the georeference accuracy. Finally, we validate the algorithm proposed in this paper through the rapid georeference experiments, which collected the UAV images from Danling County and Beichuan County, Sichuan Province, using different types of UAV. The results indicate that the algorithm could achieve rapid georeference within three minutes for up to hundreds of UAV images. It saves much more time and provides important technical support in the early postearthquake response period.

2. The Rapid Georeference of UAV Images

2.1. The Rapid Georeference Model of UAV Images. The disaster information collecting system based on the UAV is usually equipped with a high-resolution digital camera and an autopilot system, which could record the time, position, and attitude of the camera exposure. The UAV images could achieve the georeference by synchronizing the images and POS (positioning and orientation system) data. Because of the digital camera and the autopilot system which is installed along with the main optical axis of the camera are installed tightly, the lever arm and boresight angle will be ignored at the situation of emergency response. It satisfies the accuracy requirement. We can acquire the position and attitude data of the camera theoretically, which output from the autopilot system. In order to avoid the distortion of georeferenced images in the process of spatial transformation, which [illegible] image interpretation, we adopt the image rigid body transformation comprised of scale, rotation, and translation.

$$\begin{bmatrix} x' \\ y' \end{bmatrix} = \lambda R \begin{bmatrix} x \\ y \end{bmatrix} + T, \qquad (1)$$

$$\begin{bmatrix} x' \\ y' \end{bmatrix} = \lambda \begin{bmatrix} \cos\theta & \sin\theta \\ -\sin\theta & \cos\theta \end{bmatrix} \begin{bmatrix} x \\ y \end{bmatrix} + \begin{bmatrix} T_x \\ T_y \end{bmatrix}. \qquad (2)$$

In the above formulas (1) and (2), $[x'y']^{\mathrm{T}}$ is the georeferenced image pixel coordinate. $[xy]^{\mathrm{T}}$ is the image pixel coordinate before georeference. λ is the scale factor, which indicates the image spatial resolution. θ is the camera rotation angle around the optical axis. T is the translation matrix, T_x is the translation amount along with the x-axis, and T_y is the translation amount along with the y-axis.

2.2. The Data Organization Form of Georeferenced UAV Images. Considering the processing efficiency of the UAV image rapid georeference, we do not implement the pixel resampling in the process of the rapid georeference but adopt the world file to organize the georeference parameters. The world file is a data organization format, designed and developed by ESRI, a Geographic Information System company. It comprises six parameters, which are organized in a file with the specified suffix name and text encoding. By this data organizing manner, the Geographic Information System software could automatically identify the georeferenced images. All of these parameters consist of the affine transformation matrix as flows.

$$\begin{bmatrix} x' \\ y' \end{bmatrix} = \begin{bmatrix} A & B & C \\ D & E & F \end{bmatrix} \begin{bmatrix} x \\ y \\ 1 \end{bmatrix}. \qquad (3)$$

In formula (3), $[x'y']^{\mathrm{T}}$ is the georeferenced image pixel coordinate. $[xy]^{\mathrm{T}}$ is the image pixel coordinate before georeference. A is the scale factor on the x-axis. B and D are the rotation parameters. E is the scale factor on the y-axis, and it is negative. C and F are the longitude and latitude coordinates of the georeferenced image upper-left corner. All the rotation and scale factors could be deduced from formula (2).

Each pixel will acquire the geographic coordinates, when the Geographic Information System software loads the georeferenced image data as Figure 2. When x equals zero and y equals zero, the georeferenced pixel x' equals C and y' equals F. The key problem of UAV image rapid georeference is how to quickly calculate parameters C and F. Since the autopilot system collects the angle data, which indicates that the image rotates around its symmetry center but the world file adopts the upper-left as its rotation center, we have to change the center rotation to the upper-left rotation and reduce the error caused by rotation.

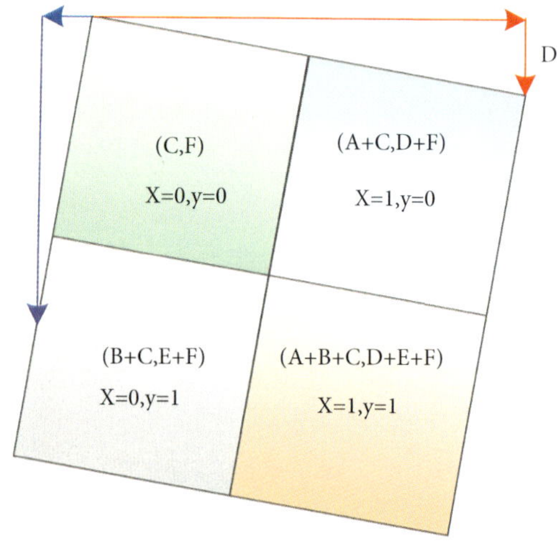

Figure 2: The imaged georeference based on the world file.

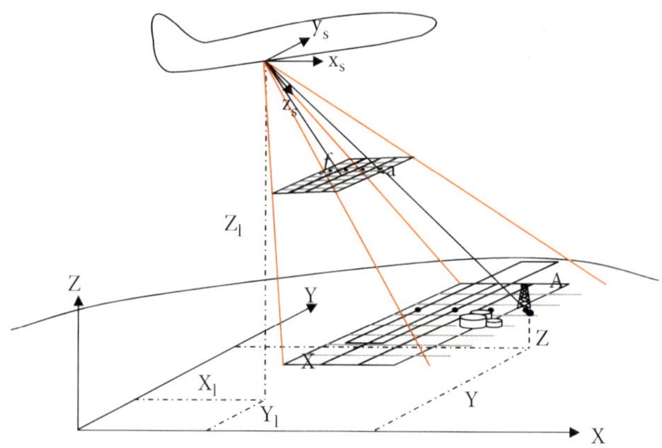

Figure 3: The image rapid georeference frame.

3. The Parameters of Rapid Georeference Determination

3.1. The Position Parameter Calculation. The image upper-left corner coordinates are key parameters in the UAV image rapid geographic reference based on the world file. A direct and simple way to obtain the coordinate parameters is to calculate the numbers of pixels between the center and the upper-left corner, then multiply the image resolution, and finally plus the center coordinates. The calculated results are the coordinates of the upper-left corner. But it does not take the error brought from the position and the attitude data change into consideration. To solve this problem, the method based on the collinear equation is employed to calculate the image upper-left corner coordinates.

Ground point A is given and its geographic coordinates are (X, Y, Z) as shown in Figure 3, and the coordinates of correspondence pixel a in the camera frame are $(x, y, -f)$; the transformation from pixel a to A is shown in formula (4). This transformation process includes such coordinate systems.

$$\begin{bmatrix} X \\ Y \\ Z \end{bmatrix}^e = \begin{bmatrix} X_S \\ Y_S \\ Z_S \end{bmatrix}^e + kR_g^e R_b^g R_c^b \begin{bmatrix} x \\ y \\ -f \end{bmatrix}. \quad (4)$$

(1) Camera frame: the z-axis is pointing up (aligned with gravity), the y-axis is defined by how the user has mounted the IMU (inertial measurement unit), and the x-axis is defined by how the user has mounted the IMU

(2) IMU body frame: the z-axis points up through the roof of the vehicle perpendicular to the ground, the y-axis points out the front of the vehicle in the direction of travel, and the x-axis completes the right-handed system (out the right-hand side of th vehicle when facing forward)

(3) Navigation frame: it is a local frame, which is tangent to the reference ellipsoid. Its origin is at the center of the reference ellipsoid, and the x-axis points to the equator intersection of the Greenwich meridian; the y-axis points to the equator with a 90° meridian intersection and the z-axis through the North Pole.

(4) Earth-centered earth-fixed frame: a coordinate system (X, Y, Z) is selected which is earth-centered earth-fixed (ECEF); with the X-Y plane containing the equator, X intersects, in the positive direction, the Greenwich Meridian (from where the longitude is measured; longitude equals 0 degrees at Y equals zero); Z is parallel to the Earth's rotation axis and has a positive direction toward the North Pole, and Y is in the equatorial plane and perpendicular to X and completes a right-handed coordinate system; i.e., the cross-product of X and Y is a vector in the direction of Z.

In formula (4), R_c^b is the transformation matrix from the camera frame to the IMU body frame. R_b^g is the transformation matrix from the IMU body frame to the navigation frame. R_g^e is the transformation matrix from the navigation frame to the earth-centered earth-fixed frame.

$$R_b^g = \begin{bmatrix} a_1 & a_2 & a_3 \\ b_1 & b_2 & b_3 \\ c_1 & c_2 & c_3 \end{bmatrix} = \begin{bmatrix} \cos R & 0 & -\sin R \\ 0 & 1 & 0 \\ \sin R & 0 & \cos R \end{bmatrix} \begin{bmatrix} 1 & 0 & 0 \\ 0 & \cos P & -\sin P \\ 0 & \sin P & \cos P \end{bmatrix} \begin{bmatrix} \cos H & -\sin H & 0 \\ \sin H & \cos H & 0 \\ 0 & 0 & 1 \end{bmatrix}$$
$$= \begin{bmatrix} \cos H \cos R + \sin H \sin P \sin R & \sin H \cos P & \cos H \sin R - \sin H \sin P \cos R \\ -\sin H \cos R + \cos H \sin P \sin R & \cos H \cos P & -\sin H \sin R - \cos H \sin P \cos R \\ -\cos P \sin R & \sin P & \cos P \cos R \end{bmatrix}. \quad (5)$$

In formula (5), H, P, and R indicate the heading pitch and roll angles, gained from the autopilot system.

$$R_g^e = \begin{bmatrix} -\sin L & \cos L & 0 \\ -\sin B \cos L & -\sin B \sin L & \cos B \\ \cos B \cos L & \cos B \sin L & \sin B \end{bmatrix}. \quad (6)$$

In formula (6), L and B, respectively, indicate the longitude and latitude, which are outputs from the autopilot system. The rotation transformation matrix could be deduced according to the Euler transformation referred in [20–22]. When $[x\, y\, -f]^T$ equals $[-(w/2) \quad h/2 \quad -f]^T$, the image upper-left corner coordinates could be obtained through formula (4).

When the UAV collects the images in the postearthquake areas, it flies about hundreds of meters above the ground and there is no signal blocking from the GPS receiver to the satellite. If the signal discontinuity problem happens, we could adopt the linear interpolation method to calculate the lost position and the attitude information within a short time interval. The UAV collects image 1 at time t_1, and the autopilot records the position and the attitude information $p_1(x_1, y_1, z_1, \text{roll}_1, \text{pitch}_1, \text{yaw}_1)$. It collects image 2 at time t_2, and the autopilot records the position and the attitude information $p_2(x_2, y_2, z_2, \text{roll}_2, \text{pitch}_2, \text{yaw}_2)$. However, it collects image i without recording the position and attitude information at time t_i between t_1 and t_2; we can calculate the position and attitude information p_i from the flowing formula:

$$A_i = \frac{t_i - t_1}{t_2 - t_1} \Delta A = \frac{t_i - t_1}{t_2 - t_1} (A_2 - A_1). \quad (7)$$

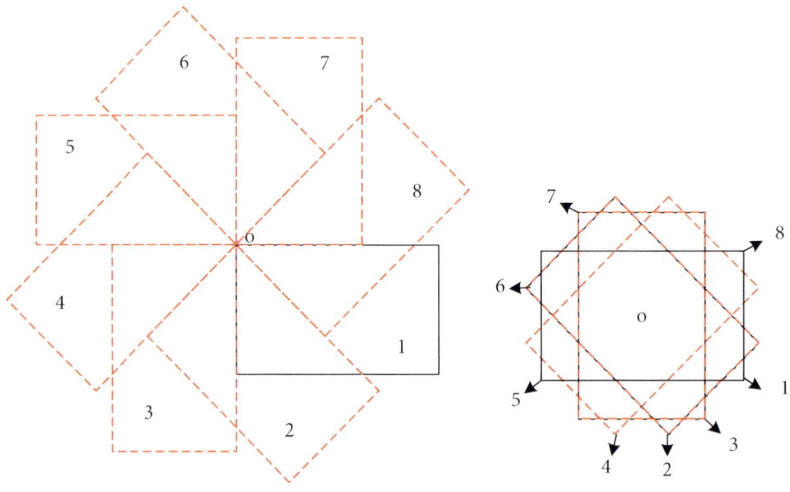

FIGURE 4: (a) The image rotates around the upper-left corner, and (b) the image rotates around the symmetry center.

In the above formula (7), A_i indicates the parameters $x_i, y_i, z_i, \text{roll}_i, \text{pitch}_i, \text{yaw}_i$ and i is the image number.

3.2. The Transformation between Different Rotation Systems. The remote sensing images collected based on the unmanned aerial vehicle rotate around their symmetry center theoretically, but the images georeferenced based on the world file rotate around their upper-left corner. Therefore, they have the different positions after the rotation. Figure 4(a) shows that the image rotates around the upper-left corner in different angles from position 1 to position 8, and Figure 4(b) indicates that the image rotates around the symmetry center according to the same angle form position 1 to position 8.

In Figure 4, when the image rotates around its upper-left corner and the image rotates around its symmetry center according to the same angle, the correspondent edges of images after rotation are parallel to each other. It indicates that the objects of the images have the same directions after the rotation. However, they have a translation gap between the images, in which we need to correct and improve the position accuracy. According to formula (2), after the image rotates around its symmetry center, the coordinates of the image are calculated as

$$x' = \left(j - \frac{w}{2}\right) \times \cos\theta + \left(i - \frac{h}{2}\right) \times \sin\theta + x_0,$$
$$y' = -\left(j - \frac{w}{2}\right) \times \sin\theta + \left(i - \frac{h}{2}\right) \times \cos\theta + y_0. \quad (8)$$

In formula (8), j is the pixel coordinate of the x direction, and i is the pixel coordinate of the y direction. θ is the rotation angle of the image rotating around the symmetry center. w is the width of the raw image, and h is the height of the image. When the image rotates around its symmetry center, (i, j) is calculated from $(-(w/2), h/2)$. When the image rotates around its upper-left corner, (i, j) is calculated from $(0, 0)$.

FIGURE 5: The SURF feature extraction and matching.

FIGURE 6: Fixed wing UAV.

TABLE 1: The specification of the fixed wing UAV.

UAV type	Captain/span	Flight weight	Control distance	Control manner	Cruising speed	Payload model
Fixed wing UAV	1.6 m/1.1 m	3 kg	10 km	Manual control/autopilot	60 km/h	SONY DSC-RX1R II camera

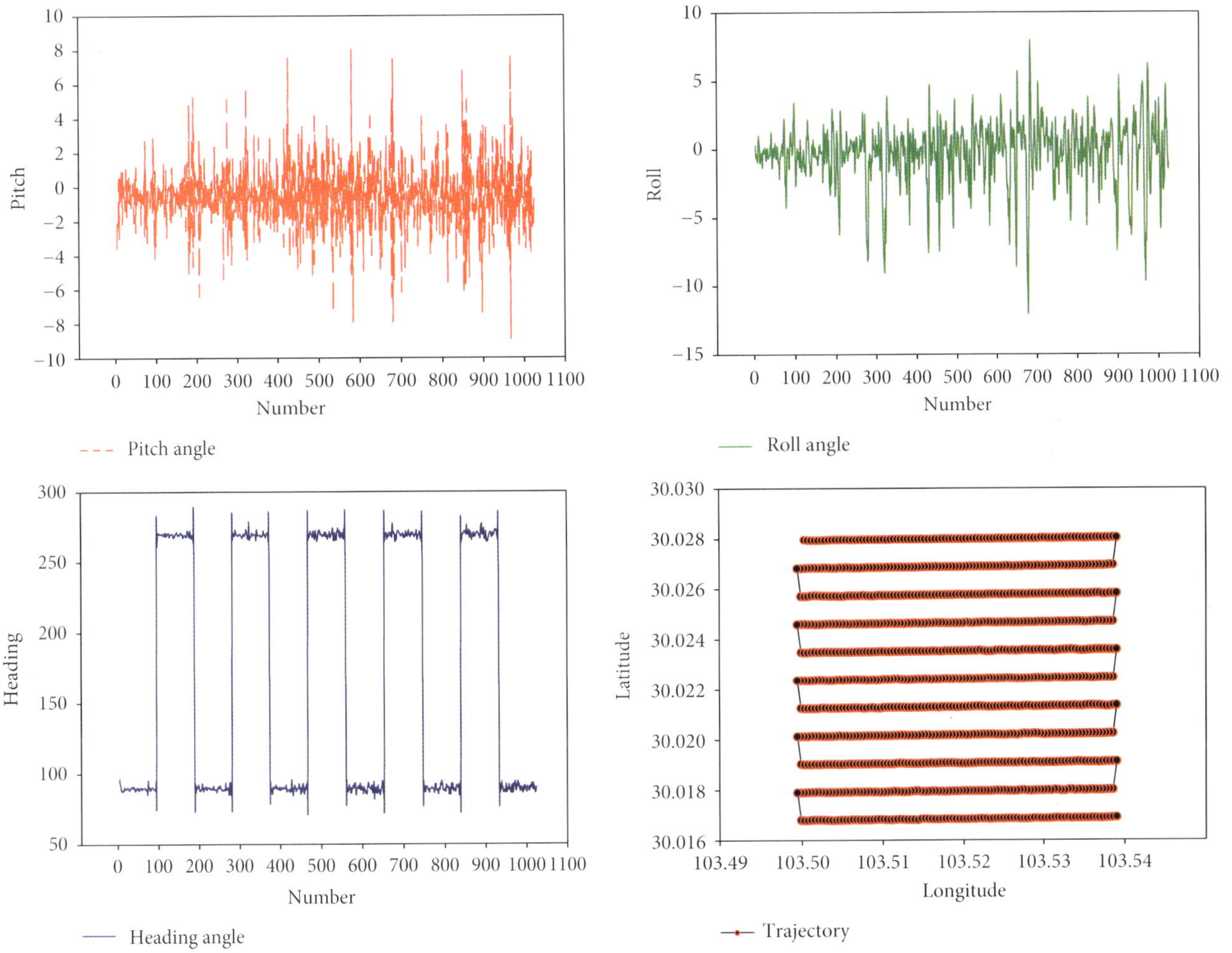

FIGURE 7: The attitude and trajectory of fixed wing UAV.

The translation amount caused by image rotation could be deduced from formula (8); the detailed form is as follows:

$$\Delta x = -\frac{w}{2} \times \cos\theta - \frac{h}{2} \times \sin\theta + \frac{w}{2},$$
$$\Delta y = \frac{w}{2} \times \sin\theta - \frac{h}{2} \times \cos\theta + \frac{h}{2}. \quad (9)$$

In formula (9), Δx and Δy are, respectively, the translation amounts on the x and y directions which indicate the position deviation that the image rotates around its symmetry center relative to the image rotating around the upper-left corner, θ is the image rotation angle that the image rotates around its symmetry center. w is the raw image width, and h is the raw image height. The translation amount is closely related to the rotation angle, image width, and image height.

4. The Rapid Georeference Accuracy Improvement Based on the Reference Image

The UAV image rapid georeference accuracy is dependent on the position data accuracy which is the output from the autopilot system. Generally, different types of UAV have different GPS modules, and the accuracies of GPS modules are different, especially the GPS module without differential positioning function. In order to improve the georeference accuracy and time efficiency further, the whole georeference accuracy improvement [illegible]

Figure 8: The rapid georeference without considering the angle information.

Figure 9: The rapid georeference application.

SURF (speeded up robust features) extraction and matching is proposed in this paper. Select one frame of the whole sequence to extract the SURF features and match with the reference image; then calculate the coordinates of the selected image upper-left corner through the feature matching. It reduces the symmetry error and improves the whole image georeference accuracy. SURF features have the robustness to image translation, rotation, scale, and the luminous changing. The feature extraction and matching could be up to subpixel accuracy. We do not discuss more details about the SURF features; the details could be referred in [23–26].

Since the reference image always covers a big area, the SURF features extracted from the UAV images to match the features extracted from the reference images will exhaust much time. To solve this problem, the features are extracted from the reference image in the given area which is initialized according to the boundary area of the georeferenced UAV images and match with the features extracted in the UAV image. When x equals $-(w/2)$ and y equals $h/2$, we put (x, y) into formula (4); the upper-left corner coordinates of the initially georeferenced UAV image will be acquired. When x equals $w/2$ and y equals to $-h/2$, we put (x, y) inputs into formula (4); the lower-right corner coordinates of the initially georeferenced UAV image will be acquired. Simultaneously, considering the initial georeference error δ, the searched feature area in the reference high-resolution image is as follows: $[(X_{\text{ul}} - \delta, Y_{\text{ul}} - \delta), (X_{\text{ul}} + \delta, Y_{\text{ul}} + \delta)]$.

Considering the time efficiency, the Gaussian probability model is employed in this paper to delete the wrong matched points. The distance from the feature in the UAV image to the feature in the reference image is $D_i (i = 1, 2, 3, \cdots, n)$; all the distances $D_i (i = 1, 2, 3, \cdots, n)$ between the matched features will be the same in the ideal conditions. All the distances D_i between matched features will satisfy the Gaussian normal distribution, the mean of distance is μ, and the variance is σ; the probabilities of samples which drop into the areas $[\mu - \sigma, \mu + \sigma]$, $[\mu - 2\sigma, \mu + 2\sigma]$, and $[\mu - 3\sigma, \mu + 3\sigma]$ are, respectively, 68.3%, 95.5%, and 99.7%, according to the Gaussian normal distribution model. If the distance D_i drops out of the confidence interval, we could recognize them as wrong matched features and delete them from the matched feature list. In Figure 5, we use the SURF feature extraction and matching to achieve the UAV image matching with a google image. It fulfils the purpose of the position error calculation between the initial georeferenced image and the reference image.

Figure 10: Quadrotor UAV.

The i_{th} feature in the UAV image is given, and its geographic coordinates are (X_i, Y_i). After the UAV image matches with the reference image, it gets the new geographic coordinates (X_i', Y_i'). The UAV image georeferenced error was computed as follows:

$$\Delta X = \sum_{i=1}^{n} \left(X_i' - X_i \right), \quad \Delta Y = \sum_{i=1}^{n} \left(Y_i' - Y_i \right). \tag{10}$$

5. Experiments and Analysis

Experiments adopt C# language combined with GDAL (Geospatial Data Abstraction Library) function to compile the rapid geographic reference software and achieve the UAV image rapid georeference. The rapid georeference software is installed on the IBM T430 computer as the experimental platform, with an Intel Core i5 2.60GHz CPU and a 4 GB size memory.

5.1. The Experiment of Rapid Georeference Based on the Fixed Wing UAV. In this experiment, the fixed wing UAV (http://

Table 2: The specification of the quadrotor UAV.

UAV type	Length/width/height	Flight weight	Control distance	Control manner	Cruising speed	Payload model
Quadrotor UAV	1.3 m/1.3 m/0.48 m	18 kg	10 km	Manual control/autopilot	0~50 km/h	SONY ILCE-6000 camera

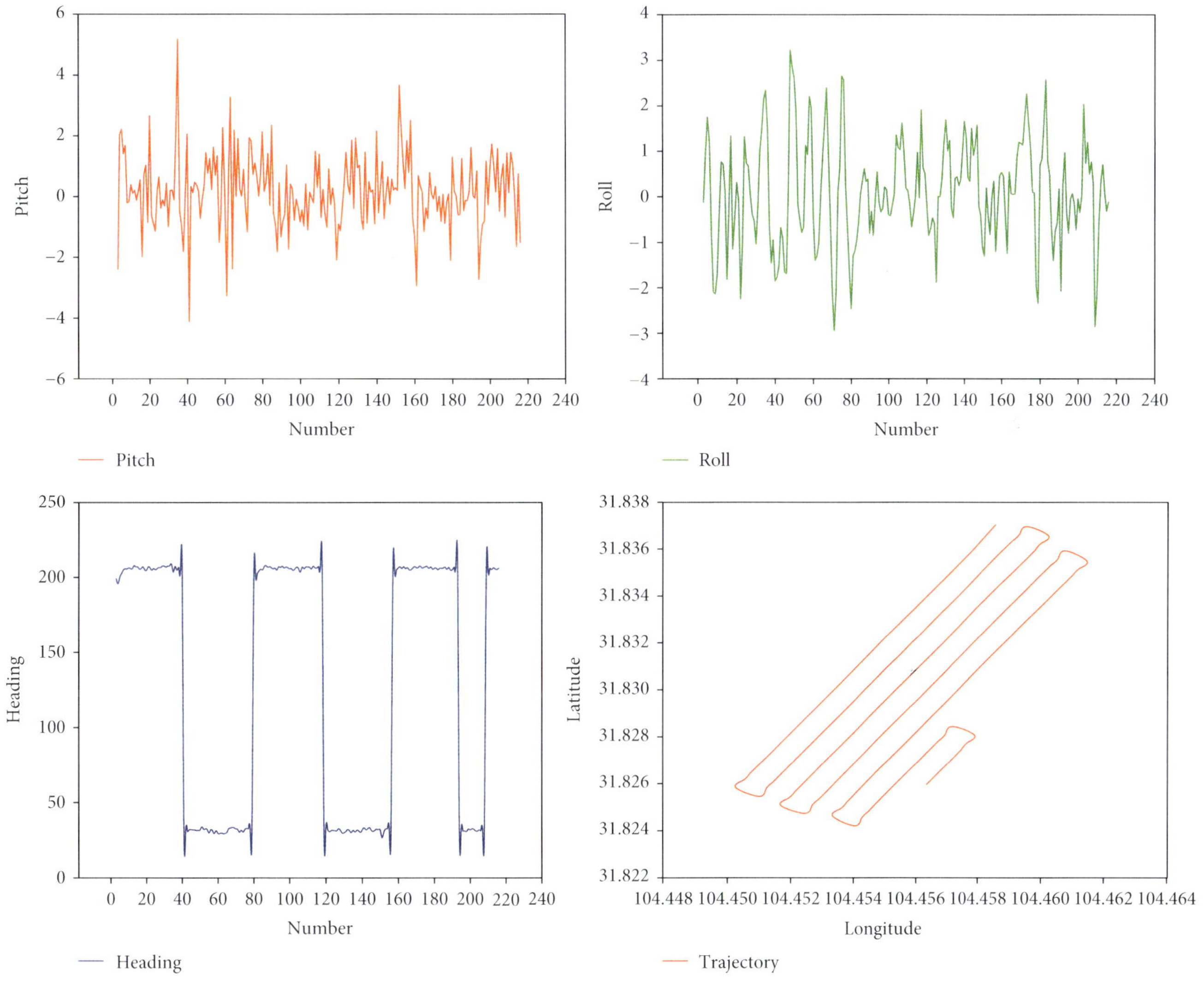

Figure 11: The attitude and trajectory of quadrotor UAV.

www.feimarobotics.com/) was used to collect the image data as shown in Figure 6. The specifications of the UAV are shown in Table 1. The UAV was equipped with a SONY DSC-RX1R II microsingle camera with a 20 mm focal length, of which CCD size is 23.5 × 15.6 mm. The resolution of the collected image is 7952 × 5304. The UAV autopilot system can output the longitude, latitude, elevation, pitch angle, roll angle, and heading angle. The GPS module of the autopilot system is without data differential function.

Taking the UAV remote sensing image data obtained from the Danling County in Sichuan Province, as an example, the average flying height is 250 m. Finally, 1025 images are obtained. The single image size is about 18 MB, and the total image size is about 18 GB. The attitude and position data are shown in Figure 7. All the images achieved rapid georeference within 3 minutes.

In Figure 8, the UAV images achieved the georeference without considering the image rotation and the error correction. The UAV image rapid georeference meets the time requirement, but we could not get the ground object distribution situation of the entire flight area. It is difficult to quickly select suitable images for disaster information submission and the interest object monitoring.

In Figure 9, it indicates the result of UAV image rapid georeference based on the algorithm proposed in this paper. All the images get the spatial reference. If they are loaded in

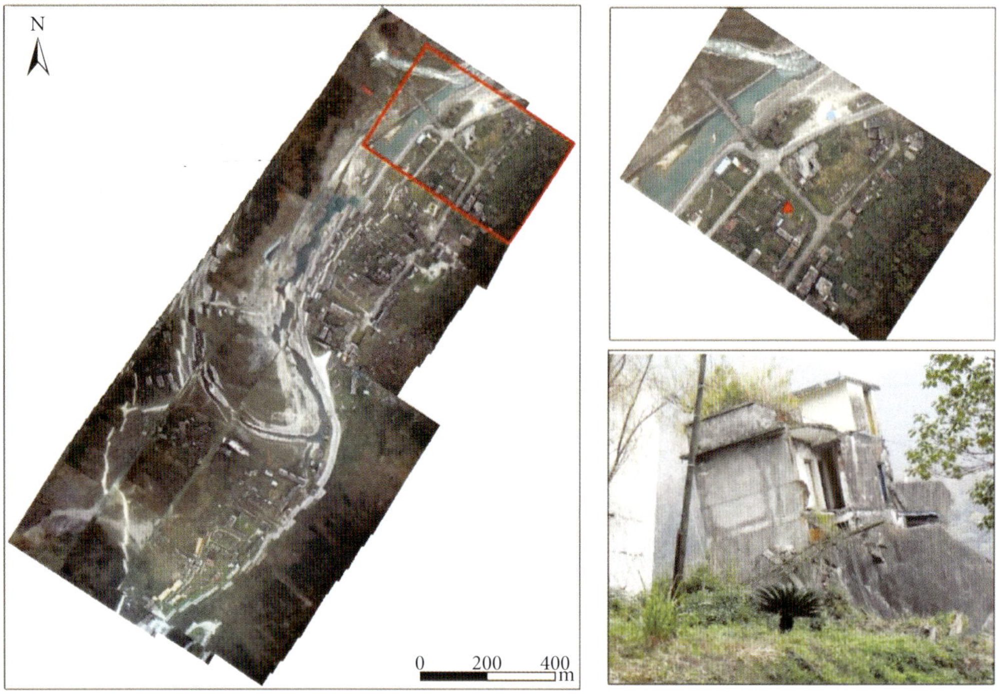

Figure 12: The image rapid georeference and selected damaged image submitted.

the same frame, it seems like a coarse panoramic image. Actually, they are separate layers. In the panoramic image, we can clearly know the object distribution on the ground surface in the flight area, such as the distribution of buildings, roads, and playground. And then, we could select the suitable images including disaster information to submit before the panoramic image stitching. It provides data support for disaster interpretation, disaster investigation, and disaster emergency response in the early response period.

If we adopt the software such as PhotoScan which will make the perfect result with a small stitching gap to generate the panoramic image, it will take more than three hours to generate the panoramic image. In the emergency response period, saving time is very important. If the emergency response commanders get the disaster information submitted through the rapid georeference, they will know the disaster location and make the correct rescue decision as soon as possible. In this situation, dozens of meter accuracy totally satisfy the emergency response requirements. And the panoramic image generated from PhotoScan could make the image map, which could be the next level image production and submitted to the emergency response commanders.

5.2. The Rapid Georeference Experiment Based on the Quadrotor UAV. In this experiment, we use the quadrotor UAV (http://www.flightwin.com/) to collect images as shown in Figure 10. The specifications of the quadrotor UAV are shown in Table 2. The UAV equipped with the Sony ILCE-6000 camera, with a 20 mm focal length, of which CCD size is 23.5 × 15.6 mm. The resolution of the collected image is 6000 × 4000.

Taking the UAV remote sensing image data obtained from the earthquake site in Beichuan County, Sichuan Province, as an example, the average flight height of the UAV is 450 m. Finally, 226 images are obtained. The single image size is about 10 MB, and the total image size is about 2.2 GB. The attitude and position data are shown in Figure 11. Achieve rapid geographic reference of all image data in less than 1 minute.

From Figure 12, we could know the whole situation of the area. All the images are separate layers. For example, the building is damaged by an earthquake. We select the suitable image and submit it to the response commanders before the panoramic image generated from the PhotoScan software. It is very important in the early period of emergency response. According to the rapid georeferenced result, the commanders will arrange workers to investigate the disaster information.

6. Conclusions

In this paper, the method and model of UAV image rapid georeference are discussed in detail and the algorithm is implemented through C# language combined with GDAL. The rapid georeference software can achieve one flying sortie UAV image referencing in a short time. It provides important technical support for grasping the overall disaster information and monitoring the key objects in the disaster area. At the same time, it can generate a vector outline of each georeferenced image, and then the vector outline is overlaid on the reference image to quickly select the image containing the disaster information to submit. It has an important application value in the postearthquake response period. In the future work, the image rapid georeference for the UAV with differential GPS module will be studied to improve the localization accuracy and reduce the relative error between adjacent images.

Conflicts of Interest

The authors declare that there is no conflict of interest regarding the publication of this paper.

Acknowledgments

This research was funded by the Fundamental Research Funds of the Institute of Earthquake Forecasting, China Earthquake Administration (2018IEF0204), and the National Key Research and Development Program of China (2017YFB0504104).

References

[1] A. Nedjati, B. Vizvari, and G. Izbirak, "Correction to: post-earthquake response by small uav helicopters," *Natural Hazards*, vol. 90, no. 1, pp. 511–511, 2018.

[2] D. Dominici, M. Alicandro, and V. Massimi, "UAV photogrammetry in the post-earthquake scenario: case studies in L'Aquila," *Geomatics, Natural Hazards and Risk*, vol. 8, no. 1, pp. 87–103, 2017.

[3] I. Aicardi, F. Chiabrando, A. M. Lingua, F. Noardo, and M. Piras, "Unmanned aerial systems for data acquisitions in disaster management applications," in *2014 Journal of universities and international development cooperation, in the III Congress of the University Network for Development Cooperation(CUCS) Congress*, pp. 164–171, Turin, Italy, 2013.

[4] Z. Xu, J. Yang, C. Peng et al., "Development of an UAS for post-earthquake disaster surveying and its application in Ms7. 0 Lushan earthquake, Sichuan, China," *Computers & Geosciences*, vol. 68, pp. 22–30, 2014.

[5] S. M. Adams, M. L. Levitan, and C. J. Friedland, "High resolution imagery collection for post-disaster studies utilizing unmanned aircraft systems (UAS)," *Photogrammetric Engineering & Remote Sensing*, vol. 80, no. 12, pp. 1161–1168, 2014.

[6] V. Baiocchi, D. Dominici, and M. Mormile, "UAV application in post-seismic environment," *The International Archives of the Photogrammetry, Remote Sensing and Spatial Information Sciences*, vol. 2, pp. 21–25, 2013.

[7] M. Brown and D. G. Lowe, "Automatic panoramic image stitching using invariant features," *International Journal of Computer Vision*, vol. 74, no. 1, pp. 59–73, 2007.

[8] A. Zomet, A. Levin, S. Peleg, and Y. Weiss, "Seamless image stitching by minimizing false edges," *IEEE Transactions on Image Processing*, vol. 15, no. 4, pp. 969–977, 2006.

[9] D. Turner, A. Lucieer, and C. Watson, "An automated technique for generating georectified mosaics from ultra-high resolution unmanned aerial vehicle (UAV) imagery, based on structure from motion (SfM) point clouds," *Remote Sensing*, vol. 4, no. 5, pp. 1392–1410, 2012.

[10] G. Zhou, "Near real-time orthorectification and mosaic of small UAV video flow for time-critical event response," *IEEE Transactions on Geoscience and Remote Sensing*, vol. 47, no. 3, pp. 739–747, 2009.

[11] I. Aicardi, F. Nex, M. Gerke, and A. Lingua, "An image-based approach for the co-registration of multi-temporal UAV image datasets," *Remote Sensing*, vol. 8, no. 9, p. 779, 2016.

[12] T. Rosnell and E. Honkavaara, "Point cloud generation from aerial image data acquired by a quadrocopter type micro unmanned aerial vehicle and a digital still camera," *Sensors*, vol. 12, no. 1, pp. 453–480, 2012.

[13] S. Harwin and A. Lucieer, "Assessing the accuracy of georeferenced point clouds produced via multi-view stereopsis from unmanned aerial vehicle (UAV) imagery," *Remote Sensing*, vol. 4, no. 6, pp. 1573–1599, 2012.

[14] F. Remondino and S. El-Hakim, "Image-based 3D modelling: a review," *The Photogrammetric Record*, vol. 21, no. 115, pp. 269–291, 2006.

[15] H. Fathi and I. Brilakis, "Automated sparse 3D point cloud generation of infrastructure using its distinctive visual features," *Advanced Engineering Informatics*, vol. 25, no. 4, pp. 760–770, 2011.

[16] J. Civera, A. J. Davison, J. A. Magallón, and J. M. M. Montiel, "Drift-free real-time sequential mosaicing," *International Journal of Computer Vision*, vol. 81, no. 2, pp. 128–137, 2009.

[17] S. Lovegrove and A. J. Davison, "Real-time spherical mosaicing using whole image alignment," in *European Conference on Computer Vision*, pp. 73–86, Springer, Berlin, Heidelberg, 2010.

[18] F. Remondino, L. Barazzetti, F. Nex, M. Scaioni, and D. Sarazzi, "UAV photogrammetry for mapping and 3d modeling–current status and future perspectives," *ISPRS - International Archives of the Photogrammetry, Remote Sensing and Spatial Information Sciences*, vol. XXXVIII-1/C22, no. 1, pp. 25–31, 2011.

[19] S. Bu, Y. Zhao, G. Wan, and Z. Liu, "Map2DFusion: real-time incremental UAV image mosaicing based on monocular slam," in *2016 IEEE/RSJ International Conference on Intelligent Robots and Systems (IROS)*, pp. 4564–4571, Daejeon, South Korea, October 2016.

[20] D. B. Barber, J. D. Redding, T. W. McLain, R. W. Beard, and C. N. Taylor, "Vision-based target geo-location using a fixed-wing miniature air vehicle," *Journal of Intelligent and Robotic Systems*, vol. 47, no. 4, pp. 361–382, 2006.

[21] L. Di, T. Fromm, and Y. Chen, "A data fusion system for attitude estimation of low-cost miniature UAVs," *Journal of Intelligent & Robotic Systems*, vol. 65, no. 1-4, pp. 621–635, 2012.

[22] H. Xiang and L. Tian, "Method for automatic georeferencing aerial remote sensing (RS) images from an unmanned aerial vehicle (UAV) platform," *Biosystems Engineering*, vol. 108, no. 2, pp. 104–113, 2011.

[23] H. Bay, T. Tuytelaars, and L. Van Gool, "Surf: speeded up robust features," in *Computer Vision - ECCV 2006*, pp. 404–417, Springer, Berlin, Heidelberg.

[24] H. Bay, A. Ess, T. Tuytelaars, and L. Van Gool, "Speeded-up robust features (SURF)," *Computer Vision and Image Understanding*, vol. 110, no. 3, pp. 346–359, 2008.

[25] E. Rublee, V. Rabaud, K. Konolige, and G. Bradski, "ORB: an efficient alternative to SIFT or SURF," in *2011 International Conference on Computer Vision*, pp. 2564–2571, Barcelona, Spain, November 2011.

[26] B. Sheta, M. Elhabiby, and N. El-Sheimy, "Assessments of different speeded up robust features (SURF) algorithm resolution for pose estimation of UAV," *International Journal of Computer Science & Engineering Survey*, vol. 3, no. 5, pp. 15–41, 2012.

4

A High-Precision Control Scheme based on Active Disturbance Rejection Control for a Three-Axis Inertially Stabilized Platform for Aerial Remote Sensing Applications

Xiangyang Zhou,[1,2] Chao Yang,[1] Beilei Zhao,[1] Libo Zhao,[2] and Zhuangsheng Zhu[1]

[1]*School of Instrumentation Science and Opto-electronics Engineering, Beihang University (BUAA), Beijing 100191, China*
[2]*State Key Laboratory for Manufacturing Systems Engineering, Xi'an Jiaotong University, Xi'an 710049, China*

Correspondence should be addressed to Xiangyang Zhou; xyzhou@buaa.edu.cn and Zhuangsheng Zhu; zszhu@buaa.edu.cn

Academic Editor: Biswajeet Pradhan

This paper presents a high-precision control scheme based on active disturbance rejection control (ADRC) to improve the stabilization accuracy of an inertially stabilized platform (ISP) for aerial remote sensing applications. The ADRC controller is designed to suppress the effects of the disturbance on the stabilization accuracy that consists of a tracking differentiator, a nonlinear state error feedback, and an extended state observer. By the ADRC controller, the effects of both the internal uncertain dynamics and the external multisource disturbances on the system output are compensated as a total disturbance in real time. The disturbance rejection ability of the ADRC is analyzed by simulations. To verify the method, the experiments are conducted. The results show that compared with the conventional PID controller, the ADRC has excellent capability in disturbance rejection, by which the effect of main friction disturbance on the control system can be weakened seriously and the stabilization accuracy of the ISP is improved significantly.

1. Introduction

It is of crucial importance for an aerial remote sensing system to obtain the high-resolution images under a nonideal multisource disturbance environment. The inertially stabilized platform (ISP) is a key component for an aerial remote sensing system, which is used to hold and control the line of sight (LOS) of the imaging sensors keeping steady relative to the tracking object in the inertial space [1, 2]. If the high-resolution images of the imaging sensor were obtained, the ISP should have the capability to effectively isolate the attitude changes of aircraft in three axes; besides, it is required to suppress the multisource disturbances whatever inside or outside of the aircraft body so that the high control precision can be gotten. Therefore, the most critical performance metric for the ISP is disturbance rejection.

Generally, the ISP is placed between the aircraft and imaging sensor served as a physical interface. It is a principal issue for the control system of the ISP to minimize the effects of disturbances introduced on the ISP [1]. In [2], a proportional integral control scheme is proposed to realize zero steady-state error of angular output when facing disturbances from carrier. In [3], a hybrid adaptive fuzzy proportional-integral-derivative (PID) controller is developed for a servo system with nonlinear property and uncertainties. In [4], a new current loop inside rate loop is particularly designed to suppress the influences of voltage fluctuation and motor back electromotive force on control precision. In [5], the model uncertainties and unmeasurable disturbances existing objectively are solved by introducing internal model control for a two-axis ISP system. In [6], an approach to compensate disturbing torque based on acceleration feedforward for the three-axis ISP is proposed. In [7], a parameter identification method for LuGre model based on multilevel coordinate search algorithm was presented. In [8], a compound control scheme based on both of the model reference adaptive control (MRAC) and PID control methods is proposed.

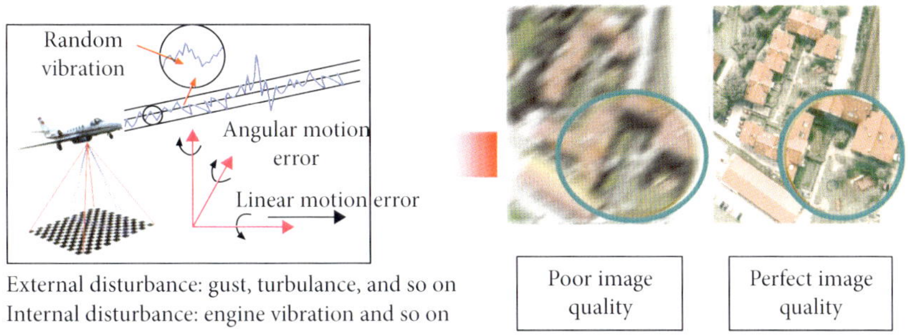

FIGURE 1: Schematic diagram: effect of the ISP on improving the images quality in an aerial remote sensing system.

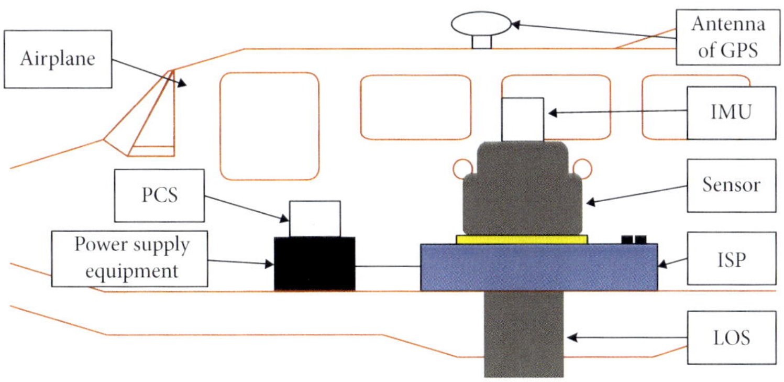

FIGURE 2: Schematic diagram of an aerial remote sensing system.

The active disturbance rejection control (ADRC) was proposed to deal with the nonlinear systems, where both uncertain dynamics and disturbances exist. The ADRC is composed of three components, that is, a tracking differentiator (TD), a nonlinear state error feedback (NLSEF), and an extended state observer (ESO). The key in the ADRC is to online estimate the total uncertainty, which lumps the internal uncertain dynamics and the external disturbances by the ESO. The ADRC can solve quickly and effectively the problems of nonlinear, strong coupling, and so on [9, 10]. The ADRC not only can improve the dynamic performance of the system but also can achieve good adaptability and robustness [11, 12]. A large number of simulation experiments show that the ADRC algorithm is simple to detect and compensate the disturbance of the control object automatically [13, 14]. This design methodology requires very little information about the plant dynamics, is very easy to tune, and has a very good disturbance rejection capability. In [15], a feedback linearization based the ESO control law is proposed for the trajectory tracking control of a flexible joint robotic system. In [16], the effectiveness of the ESO in the estimation of the states and uncertainties is demonstrated. In [17], an ADRC for the antenna pointing control of a large flexible satellite system is designed, by which the high pointing accuracy and rotation speed are achieved.

In this paper, to improve the ability of disturbance rejection of an aerial three-axis ISP, a disturbance rejection scheme based on the ADRC is proposed. To verify the method, the simulations and experiments are carried out, respectively. The rest of this paper is arranged as follows. Section 2 first introduces the background of this work. Section 3 explains the ADRC principle and algorithm that includes three parts: the TD, the NLSEF, and the ESO. Section 4 demonstrates the simulation analysis, including ADRC controller as well as the PID controller. The ADRC experiment is in Section 5, including the static base experiment and the moving base experiment. In the end, a brief conclusion is given in Section 6.

2. Background

2.1. Aerial Remote Sensing System. Figure 1 shows a schematic diagram to illustrate the important effect of the ISP on improving the image quality in an aerial remote sensing system. In general, an aerial remote sensing system consists of an aircraft vehicle, an imaging sensor, a three-axis ISP, and a position and orientation system (POS). During the work, the ISP is used as a physical interface to separate the imaging sensor and the aircraft. Depending on the ISP, the serious influences of various disturbances on the imaging sensors are effectively suppressed, by which the LOS can be kept stable all along and the high-resolution images are gotten eventually.

2.2. The Role of Three-Axis ISP System. Figure 2 shows the schematic diagram of an aerial remote sensing system. As shown in this figure, during application, the ISP is installed on the floor of the airplane. Meanwhile, the imaging sensor is installed on the inner azimuth gimbal of the ISP, and the POS is installed on the top of the imaging

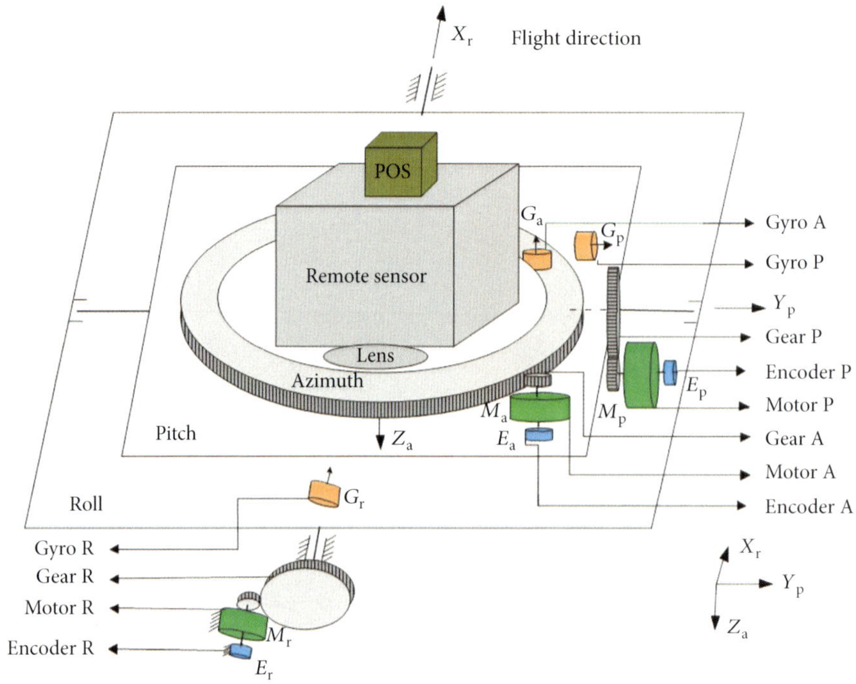

FIGURE 3: Schematic diagram of the three-axis ISP's principle.

sensor. When the aircraft rotates or jitters, the high-precision attitude reference information is measured by the POS in real time, which is then provided to the ISP control system to make the LOS of imaging sensor to achieve the high stabilization precision.

2.3. Working Principle of Three-Axis ISP System. Figure 3 shows the schematic diagram of the three-axis ISP's principle. As shown in the figure, the ISP consists of three gimbals, which are azimuth gimbal (A-gimbal), pitch gimbal (P-gimbal), and roll gimbal (R-gimbal) from the outside to the inside. Three rate gyros installed on the three different gimbals are used to measure inertial angular rate of P-gimbal, R-gimbal, and A-gimbal, which are used to improve the stabilization precision through compensating the difference between the rate command input and the angular rate of the gimbal. On the other aspects, three photoelectric encoders installed on different gimbals are used to measure the relative angular position changes between two adjacent gimbals. Three servo motors installed on different gimbals are used to drive R-gimbal, P-gimbal, and A-gimbal rotating inversely to compensate the attitude changes to three gimbals in real time.

3. Design of the ADRC Controller

3.1. Tracking Differentiator (TD). For a second-order system [18], we get

$$\dot{z}_1 = z_2, \quad \dot{z}_2 = f(z_1, z_2), \tag{1}$$

if the solution of (1) satisfies

$$\lim_{t\to\infty} z_1(t) = 0, \quad \lim_{t\to\infty} z_2(t) = 0. \tag{2}$$

So, for any boundary integrable variable, $v(t)$, and any constants, $T > 0$ and $r > 0$, the solution $x_1(t)$ of the system

$$\dot{x}_1 = x_2, \quad \dot{x}_2 = r^2 \cdot f\left(x_1 - v(t), \frac{x_2}{r}\right) \tag{3}$$

can meet the following relationship:

$$\lim_{r\to\infty} \int_0^T |x_1(r,t) - v(t)| dt = 0. \tag{4}$$

Therefore, assuming that a function $f(z_1, z_2)$ can be found to make (1) asymptotically stable at the origin, a nonlinear tracking differentiator of $v(t)$ in the form of (3) can be constructed, in which the same function $f(z_1, z_2)$ is used [19].

Kasim and Riyadh [20] have made some investigations on traditional structures and essential properties of the nonlinear TD. A kind of second-order nonlinear TD based on second-order bang-bang switch system has been proposed:

$$\dot{x}_1 = x_2, \quad \dot{x}_2 = -r \cdot \text{sign}\left(x_1 - v_0(t) + \frac{x_2|x_2|}{2r}\right), \tag{5}$$

where x_1 is the desired trajectory and x_2 is its derivative. Note that the parameter r is an application dependent and it is set accordingly to speed up or slow down the transient profile. Then, x_2 is denoted as the TD of $v_0(t)$.

In order to avoid chattering near the origin, changing the sign function to linear saturation function sat, then the modified TD is represented by [21–23]

$$\mathrm{sat}(x,\delta)=\begin{cases}\mathrm{sign}\ (x), & |x|\geq\delta,\\ \dfrac{x}{\delta}, & |x|<\delta.\end{cases} \quad (6)$$

The TD gets the differentiated signal by computing the integration of the relational viable. So it can efficiently weaken the effect of noise, while the classical differential component enlarges it and the differentiated signal obtained by this way is far from the real one. In some sense, the differential component of the TD is feasible physically, while differential components of the classical methods are not feasible physically [21].

3.2. Nonlinear State Error Feedback (NLSEF). For the first-order system [22],

$$\dot{x}=w(x,t)+u. \quad (7)$$

$w(x,t)$ is the sum of the perturbations of the system, and linear feedback is used for the control object:

$$u=-kx,\quad k>0. \quad (8)$$

Then, the closed-loop system can be expressed:

$$x=-kx+w(x,t). \quad (9)$$

The closed-loop equation of the finishing system is obtained [23]:

$$\frac{1}{2}\frac{dx^2}{dt}=-kx^2+xw=-k\left(x-\frac{w}{2k}\right)^2+\frac{w^2}{4k}. \quad (10)$$

If the state quantity x converges, then $dx^2/dt<0$ is needed, that is to say, it needs to be satisfied:

$$\left|x-\frac{w}{2k}\right|>\frac{|w|}{2k}. \quad (11)$$

It is easy to get

$$|x|>\frac{|w|}{k}. \quad (12)$$

As long as $|x|>|w|/k$, there must be $dx^2/dt<0$. The state parameter x converges, and the system disturbance $w(x,t)$ can also be well suppressed. But with the convergence of the state quantity x, it will gradually approach the range of $|x|\leq|w|/k$, which leads to the suppression of the disturbance of the system and the steady-state error of the system.

The ADRC reduces the steady-state error of the system by adopting appropriate nonlinear functions. According to the different error states, the nonlinear function can make the feedback gain appropriate increase and decrease, so as to reduce the steady-state error of the system, and effectively avoid the disadvantages caused by adopting the fixed feedback gain.

Nonlinear feedback is used to control objects:

$$u=-k|x|^{\alpha}\ \mathrm{sign}\ (x),\quad 0<\alpha<1. \quad (13)$$

Then, the controlled object is converted into a closed loop:

$$\dot{x}=w(x,t)-k|x|^{\alpha}\ \mathrm{sign}\ (x). \quad (14)$$

Then get

$$\frac{1}{1+\alpha}\frac{d|x|^{1+\alpha}}{dt}=-k\left(|x|^{\alpha}\ \mathrm{sign}\ (x)-\frac{w(x,t)}{2k}\right)^2+k\left(\frac{w(x,t)}{2k}\right)^2. \quad (15)$$

When $d|x|^{1+\alpha}/dt<0$, the state x is convergent, which requires

$$|x|^{\alpha}\ \mathrm{sign}\ (x)-\frac{w(x,t)}{2k}>\frac{w(x,t)}{2k}. \quad (16)$$

Then get

$$|x|>\left(\frac{|w(x,t)|}{k}\right)^{1/\alpha}. \quad (17)$$

According to the analysis of the upper form, when $|x|>(|w(x,t)|/k)^{1/\alpha}$, the state x converges, and with the further convergence, state x can converge to the range of $|x|\leq(|w(x,t)|/k)^{1/\alpha}$, so the steady-state error of the system will not be greater than $(|w(x,t)|/k)^{1/\alpha}$.

It is easier to know the linear feedback obtained the maximum steady-state error $|w(x,t)|/k$ compared with $(|w(x,t)|/k)^{1/\alpha}$that when $0<\alpha<1$ and $k>w(x,t)$, there are

$$\left(\frac{|w(x,t)|}{k}\right)^{1/\alpha}\ll\frac{|w(x,t)|}{k}. \quad (18)$$

For systems subject to uncertainties, the steady-state error obtained by using nonlinear feedback is much smaller than that by using linear feedback. In addition, in nonlinear feedback, if the α value is changed slightly, the $(|w(x,t)|/k)^{1/\alpha}$ can rapidly approach zero.

The TD generates the arranged transition process and its derivatives. The outputs of ESO estimate the states of the control system. By comparing the difference between the outputs of TD and those of ESO, the NLSEF control law $u_0(t)$ is used to drive the state trajectory to the desired reference signal. Its mathematic expression is given as [24]

$$\varepsilon_i=v_i-z_i\quad (i=1,2,\ldots,n). \quad (19)$$

Thus, the nonlinear combination of system state error feedback is formed:

$$u_0(t)=\sum_{i=1}^{n}\beta_i fal(\varepsilon_i,\alpha_i,\delta), \quad (20)$$

where α_i, β_i $(i=1,2,\ldots,n)$, and δ are adjustable parameters.

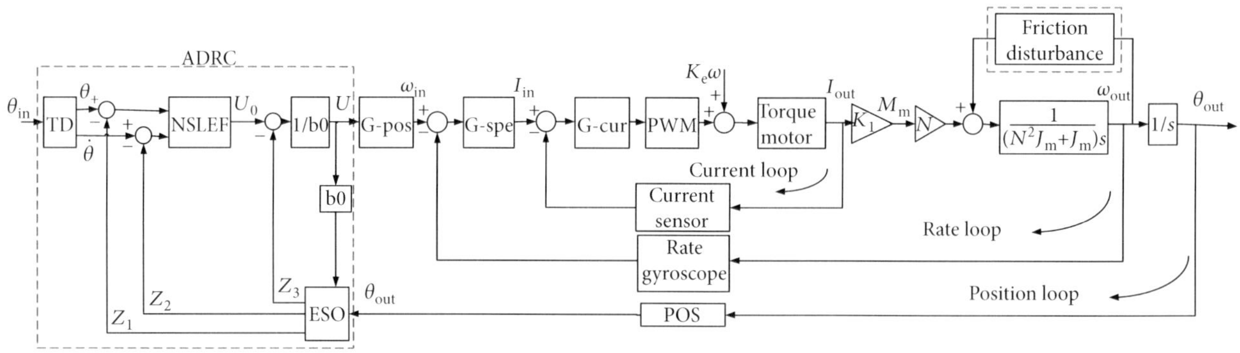

FIGURE 4: Block diagram of the ADRC-based three-loop compound controller.

$$\mathrm{fal}(e) = \begin{cases} |e|^{\alpha} \operatorname{sign}(e), & |e| > c, \\ \dfrac{e}{c^{1-\alpha}}, & |e| \le c, \end{cases} \tag{21}$$

where $fal(e)$ is defined as a modified exponential gain function [25].

3.3. Extended State Observer (ESO). The performance of the ESO is very important to the performance of the controller [18, 20]. The state observer is used to determine the internal state of the system by observing the external state of the system in different situations.

If the function $f(x_1, x_2)$ is not known and the above description is not valid, the way to expand the state observer is to expand $f(x_1, x_2)$ into a new state variable x_3:

$$\begin{aligned} x_3 &= f(x_1(t), x_2(t)), \\ \dot{x}_3 &= w(t). \end{aligned} \tag{22}$$

In this way, nonlinear objects can be regarded as new linear objects [25]:

$$\begin{aligned} \dot{x}_1 &= x_2, \\ \dot{x}_2 &= x_3 + bu, \\ \dot{x}_3 &= w(t), \\ y &= x_1. \end{aligned} \tag{23}$$

Then, the state observer is established:

$$\begin{aligned} e_1 &= z_1 - y, \\ \dot{z}_1 &= z_2 - \beta_{01} e_1, \\ \dot{z}_2 &= z_3 - \beta_{02} |e_1|^{1/2} \operatorname{sign}(e_1) + bu, \\ \dot{z}_3 &= -\beta_{03} |e_1|^{1/4} \operatorname{sign}(e_1). \end{aligned} \tag{24}$$

In this way, the provided parameters β_{01}, β_{02}, and β_{03} are selected properly, the state variables of the system can be estimated in real time by z_1, z_2, and z_3, and the real-time action amount of the acceleration $f(x_1, x_2)$ of the system is measured. Equation (25) can not only achieve good control effect but also avoid high-frequency tremor.

$$\mathrm{fal}(e, \alpha, \delta) = \begin{cases} \dfrac{e}{\delta^{\alpha-1}}, & |e| \le \delta, \\ |e|^{\alpha} \operatorname{sign}(e), & |e| > \delta, \end{cases} \tag{25}$$

where δ is the length of the linear section.

$$\begin{aligned} z_1(t) &\longrightarrow x_1(t), \\ z_2(t) &\longrightarrow x_2(t), \\ z_3(t) &\longrightarrow f(x_1(t), x_2(t)). \end{aligned} \tag{26}$$

If $f(x_1, x_2)$ contains time t and unknown disturbances $w(t)$, it is available [17]:

$$z_3 \longrightarrow x_3 = f[x_1(t), x_2(t), w(t)]. \tag{27}$$

The real-time action of acceleration acting on the system can be obtained:

$$a(t) = f[x_1(t), x_2(t), t, w(t)]. \tag{28}$$

Thus, (28) can become the extended state observer of the system. The ESO is a dynamic process that does not take advantage of system parameters and specific models but uses the information between inputs and outputs of the system to estimate the system state variables directly.

3.4. The ADRC-Based Three-Loop Compound Controller. Figure 4 shows the block diagram of the ADRC-based three-loop compound controller. As shown in this figure, the blocks of G-cur, G-spe, and G-pos indicate the controllers from the inner loop to the outer loop, which are the current loop, rate loop, and position loop, respectively; the PWM block indicates the pulse width modulation, which is used for the current amplification to drive the torque motor; the symbol L, R, K_t, and N are the inductance of a torque motor, resistance, motor torque coefficient, and transition ratio from the torque motor to the gimbal, respectively. Besides, J_m, and J_t indicate the moment of inertia of the motor and the moment of the gimbal inertia along the rotation axis, respectively.

As shown in this figure, in order to improve the control accuracy of the three-axis ISP system, the ADRC controller

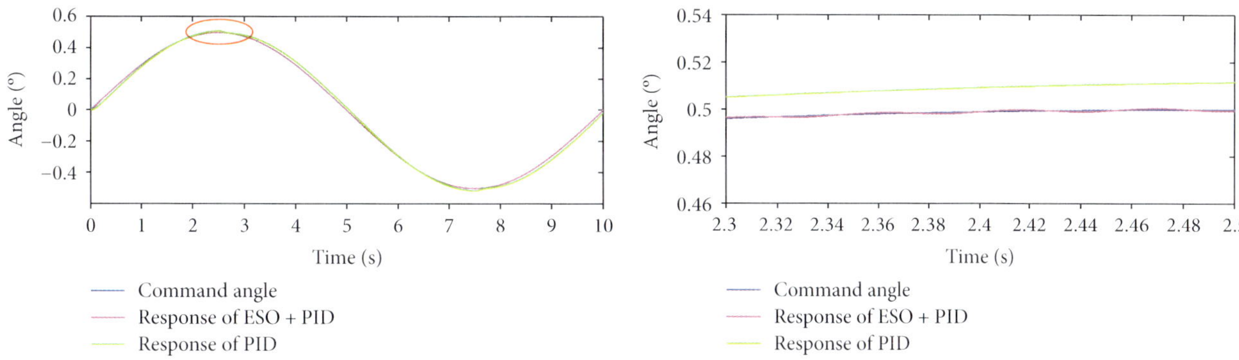

(a) Global curve (b) Local magnification curve

Figure 5: PID and ESO/PD sine angle position feedback contrast diagram.

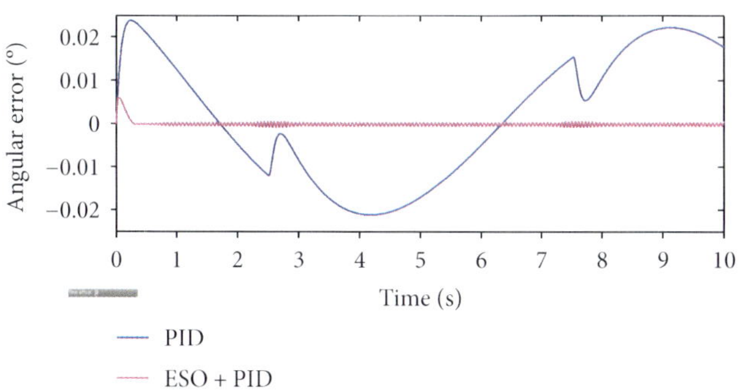

Figure 6: PID and ESO/PD angle position error contrast diagram.

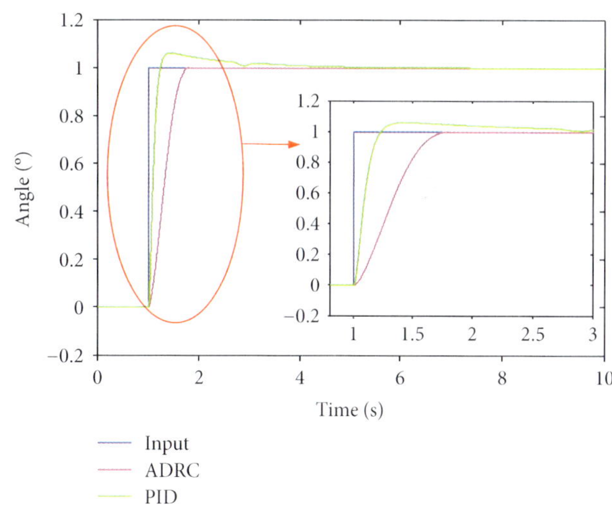

Figure 7: Comparison of position feedback between step angle of conventional PID and ADRC controller.

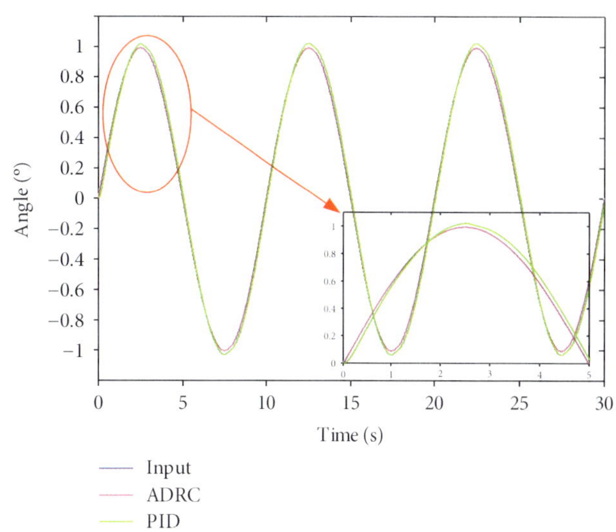

Figure 8: Comparison of position feedback to sinusoidal signal between conventional PID and ADRC controller.

is designed. One of the inputs of the ADRC is the tracking instruction of angular position of the ISP control system, θ_{in}, and another input of the ADRC, θ_{out}, is the real angular position of the ISP measured by the POS. The control amount u calculated by the ADRC is acted on the ISP position loop.

In Figure 4, the generalized inputs, θ and $\dot{\theta}$, are first obtained by the TD. The generalized outputs of the ISP control system, z_1 and z_2, are obtained by the ESO. z_1 is the angle value measured by POS in the position loop of the three-axis ISP system, z_2 is the derivative of the angle, and z_3 is the compensatory amount of the ESO to the disturbance in the ISP control system. The compared errors between the generalized inputs and the generalized outputs are inputted to the NLSEF to generate the control amount u_0, which are compared with z_3 to generate the control amount u. The control amount u after compensation is acted on the position loop of the ISP. Thus, the disturbance rejection ability and the control accuracy of the ISP control system are improved.

4. Simulation Analysis

4.1. Angular Position Response to Step Input. In order to verify the validity of ADRC algorithm, MATLAB/Simulink simulation is firstly carried out. The main parameters of the NLSEF control law are $\{\beta_1, \beta_2, \alpha_3, \alpha_4, \delta_2\}$. Normally, the value of α_3 is from 0.5 to 0.75, and the value of α_4 is from 1.25 to 1.5. The value of δ_2 is the same as the integral step

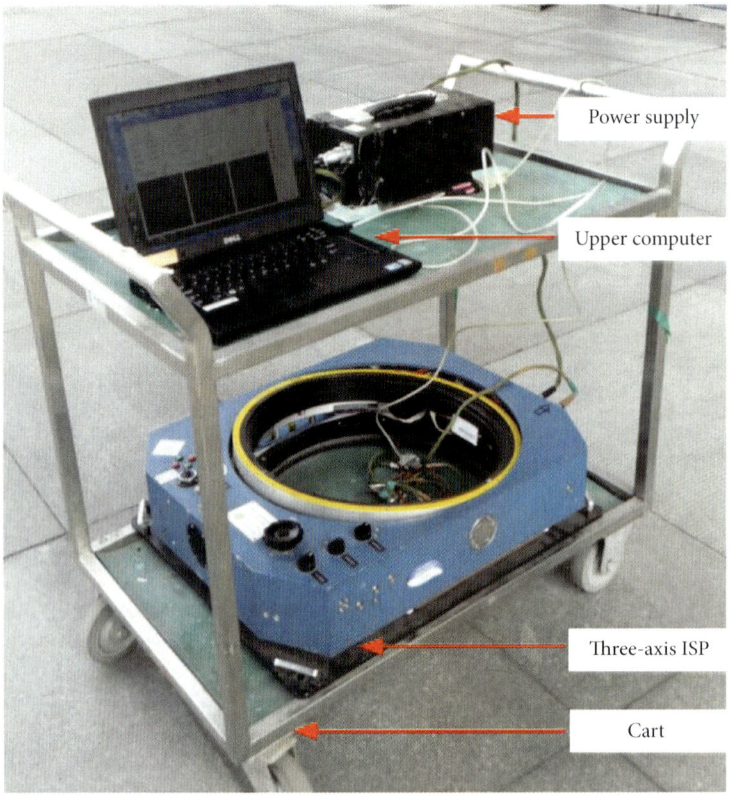

FIGURE 9: Picture of moving base experiment system.

length as 0.01 generally. The parameters tuned by the experiential method are as follows: $\beta_1 = 6.0$ and $\beta_2 = 1.5$.

In the simulation, the pitch gimbal system is chosen as the control object. When the given input angle instruction is $\theta_{\text{id}} = 0.5 \sin(\pi \cdot n \cdot T_s)$ deg, $T_d = 0$, the disturbance resistance performance of the PID and the ESO/PD controller is simulated by adding the disturbance model; the results are shown in Figure 5.

In Figure 5(a), the PID and the ESO/PD composite controllers can track the sinusoidal signals in different position precision. In the local magnification diagram of Figure 5(b), we can see that the PID has approximately 0.01° errors in tracking the sinusoidal instruction, but the ESO/PD compound controller tracking curve almost coincided with the sinusoidal instruction. Therefore, the ESO/PD controller has strong disturbance resistance ability, which can effectively suppress the disturbance effect, and its tracking error is almost zero.

Under the influence of disturbance, the comparison of angular position error between the PID and the ESO/PD controller is shown in Figure 6. From Figure 6, we can see that the angular error of PID fluctuates slowly from −0.02° to 0.025°. When the disturbance amplitude is large, the tracking error curve of the PID will increase correspondingly, while the angular error of the ESO/PD composite controller is stable at 0°. Therefore, the ESO/PD controller has better anti-interference ability than the PID controller.

Figure 7 shows the comparison of position feedback to step angle between conventional PID and ADRC controller. When the track reference is a step signal with magnitude 1° which lasts for 10 s, the ADRC controller has a good dynamic performance compared to the conventional PID controller. As shown in this figure, compared to the conventional PID controller, the ADRC controller has nearly no any overshoot and can more easily reach stabilization state with much shorter adjusting time.

4.2. Angular Position Response to Sinusoidal Input. Figure 8 shows the comparison of position feedback to sinusoidal signal between conventional PID and ADRC controller. When the track reference is sinusoidal signal with amplitude 1° and frequency 0.1 Hz, the ADRC controller can achieve better disturbance rejection performance compared to the conventional PID controller. As shown in this figure, compared to the conventional PID controller, the tracking curve of the ADRC controller coincides almost with the sine command.

5. Experimental Verification

In order to verify the validity and practicability of the ADRC, the experiments are carried out. Figure 9 shows the picture of moving base experiment system. In the experiments, the three-axis ISP is placed on a moving cart. The experimental system includes a three-axis ISP, a data acquisition and processing system (upper computer), and a 28 V DC power supply. The weight of the ISP is 40 kg and its maximum support load is 80 kg. The maximum leveling rotation

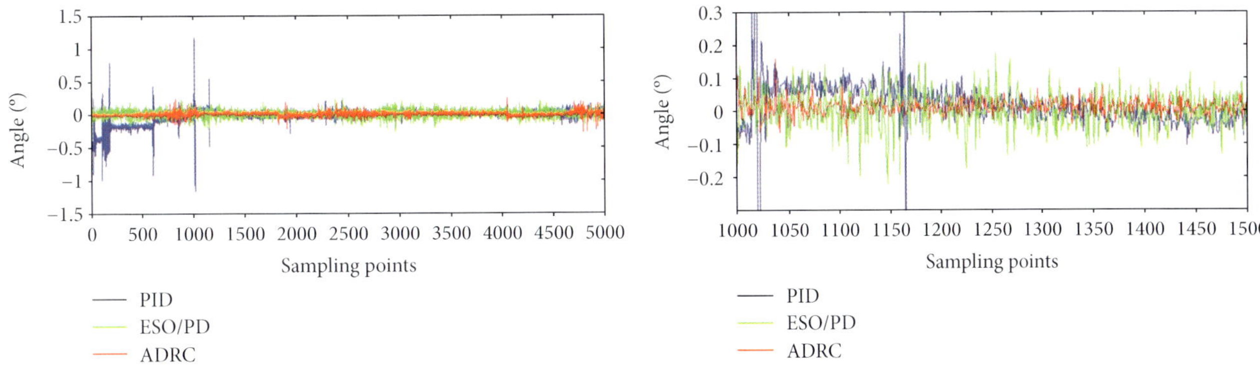

(a) Global curve

(b) Local enlarged drawing

FIGURE 10: Comparison of angle error of dynamic cart experiment.

angle range of the ISP is ±5°; the maximum heading rotation angle range is ±25°.

Set the tracking reference angle of the roll gimbal system as 0°; sampling points is 5000, and the sampling frequency is 50 Hz. As shown in Figure 10, in leveling experiments, the results obtained by three different methods are as follows: for the PID controller, the RMS (root mean square) error is 0.2673°; for the ESO/PD controller, the RMS error is 0.1315°, which is decreased up to 50.8% compared to the PID controller; for the ADRC controller, the RMS error is 0.0826°, which are decreased up to 69.1% and 37.2% compared to the PID controller and the ESO/PD controller, respectively. The results show that the ADRC control method has better control precision and interference suppression capability compared to the PID and the ESO/PD.

6. Conclusion

This paper investigates the high-precision control method for a three-axis inertially stabilized platform (ISP). Based on the analysis, a disturbance rejection control scheme based on the active disturbance rejection control (ADRC) is proposed to realize high-precision position servo control, which is robust to the multisource disturbances. Then, the numerical simulations and the experimental validations are conducted to analyze and verify the scheme, respectively. The results show that the ADRC control method can get a better control accuracy and have strong disturbance rejection ability than the conventional PID methods. Under the moving base experimental condition, the stabilization precision of the ISP obtained by the ADRC controller is improved up to 69.1% compared to the conventional PID control method.

Conflicts of Interest

The authors declare that there is no conflict of interest regarding the publication of this paper.

Acknowledgments

This project is supported in part by the National Natural Science Foundation of China (Grant nos. 51775017, 51375036), by the Beijing Natural Science Foundation (Grant no. 3182021), and by the Open Research Fund of the State Key Laboratory for Manufacturing Systems Engineering (sklms2018005).

References

[1] M. K. Masten, "Inertially stabilized platforms for optical imaging systems," *IEEE Control Systems Magazine*, vol. 28, no. 1, pp. 47–64, 2008.

[2] Y. Han, Y. Lu, and H. Qiu, "An improved control scheme of gyro stabilization electro-optical platform," in *2007 IEEE International Conference on Control and Automation*, pp. 346–351, Guangzhou, China, 2007.

[3] W. Ji, Q. Li, and B. Xu, "Design study of adaptive fuzzy PID controller for LOS stabilized system," in *Sixth International Conference on Intelligent Systems Design and Applications*, pp. 336–341, Jinan, China, 2006.

[4] X. Zhou, Y. Jia, Q. Zhao, and R. Yu, "Experimental validation of a compound control scheme for a two-axis inertially stabilized platform with multi-sensors in an unmanned helicopter-based airborne power line inspection system," *Sensors*, vol. 16, no. 3, p. 366, 2016.

[5] X. Zhou, H. Zhang, and R. Yu, "Decoupling control for two-axis inertially stabilized platform based on an inverse system and internal model control," *Mechatronics*, vol. 24, no. 8, pp. 1203–1213, 2014.

[6] Q. Mu, G. Liu, M. Zhong, and Z. Chu, "Imbalance torque compensation for three-axis inertially stabilized platform using acceleration feedforward," in *2012 8th IEEE International Symposium on Instrumentation and Control Technology (ISICT) Proceedings*, pp. 157–160, London, UK, 2012.

[7] Z. Li, D. Fan, and S. Fan, "LuGre-Model-based friction compensation in direct-drive inertially stabilized platforms," *IFAC Proceedings Volumes*, vol. 46, no. 5, pp. 636–642, 2013.

[8] X. Zhou, C. Yang, and T. Cai, "A model reference adaptive control/PID compound scheme on disturbance rejection for an aerial inertially stabilized platform," *Journal of Sensors*, vol. 2016, Article ID 7964727, 11 pages, 2016.

[9] W. Xue and Y. Huang, "Comparison of the DOB based control, a special kind of PID control and ADRC," in *Proceedings of the 2011 American Control Conference*, pp. 4373–4379, San Francisco, CA, USA, 2011.

[10] W. Xue, W. Bai, S. Yang, K. Song, Y. Huang, and H. Xie, "ADRC with adaptive extended state observer and its application to air–fuel ratio control in gasoline engines," *IEEE*

Transactions on Industrial Electronics, vol. 62, no. 9, pp. 5847–5857, 2015.

[11] Y. Huang and J. Han, "Analysis and design for the second order nonlinear continuous extended states observer," *Chinese Science Bulletin*, vol. 45, no. 21, pp. 1938–1944, 2000.

[12] Y. X. Su, C. H. Zheng, and B. Y. Duan, "Automatic disturbances rejection controller for precise motion control of permanent-magnet synchronous motors," *IEEE Transactions on Industrial Electronics*, vol. 52, no. 3, pp. 814–823, 2005.

[13] W. Xue and Y. Huang, "On performance analysis of ADRC for a class of MIMO lower-triangular nonlinear uncertain systems," *ISA Transactions*, vol. 53, no. 4, pp. 955–962, 2014.

[14] Q. Zheng, L. Dong, D. H. Lee, and Z. Gao, "Active disturbance rejection control for MEMS gyroscopes," *IEEE Transactions on Control Systems Technology*, vol. 17, no. 6, pp. 1432–1438, 2009.

[15] S. E. Talole, J. P. Kolhe, and S. B. Phadke, "Extended-state-observer-based control of flexible-joint system with experimental validation," *IEEE Transactions on Industrial Electronics*, vol. 57, no. 4, pp. 1411–1419, 2010.

[16] C. Zhao and Y. Huang, "ADRC based input disturbance rejection for minimum-phase plants with unknown orders and/or uncertain relative degrees," *Journal of Systems Science and Complexity*, vol. 25, no. 4, pp. 625–640, 2012.

[17] S. Li, X. Yang, and D. Yang, "Active disturbance rejection control for high pointing accuracy and rotation speed," *Automatica*, vol. 45, no. 8, pp. 1854–1860, 2009.

[18] Y. Tang, Y. Wu, M. Wu, X. Hu, and L. Shen, "Nonlinear tracking-differentiator for velocity determination using carrier phase measurements," *IEEE Journal of Selected Topics in Signal Processing*, vol. 3, no. 4, pp. 716–725, 2009.

[19] P. Wang and K. Zhang, "Research on line-of-sight rate extraction of strapdown seeker," in *Proceedings of the 33rd Chinese Control Conference*, pp. 859–863, Nanjing, China, 2014.

[20] I. Kasim and W. Riyadh, "On the improved nonlinear tracking differentiator based nonlinear PID controller design," *International Journal of Advanced Computer Science and Applications*, vol. 7, no. 10, 2016.

[21] H. Huang, L. Wu, J. Han, G. Feng, and Y. Lin, "A new synthesis method for unit coordinated control system in thermal power plant - ADRC control scheme," in *2004 International Conference on Power System Technology, 2004. PowerCon 2004*, pp. 133–138, Singapore, 2004.

[22] J. Han, "Nonlinear state error feedback control law—NLSEF," *Control and Decision*, vol. 3, 1995.

[23] J. Han, "From PID technique to active disturbances rejection control technique," *Control Engineering of China*, vol. 9, pp. 13–18, 2002.

[24] W. Jun, Y. J. Wang, and H. Jian, "Artificial bee colony algorithm based auto-disturbance rejection control for rehabilitation robotic arm driven by PM-TS actuator," in *2012 Proceedings of International Conference on Modelling, Identification and Control*, pp. 802–807, Wuhan, Hubei, China, 2012.

[25] Y. Bing, Z. H. Tian, and S. J. Shi, "Novel fault diagnosis scheme for HVDC system via ESO," *High Voltage Engineering*, vol. 11, 2007.

5

Detection of *Firmiana danxiaensis* Canopies by a Customized Imaging System Mounted on an UAV Platform

Chun Liu,[1] Mengchi Ai,[1] Zhuo Chen,[1] Yuan Zhou,[2] and Hangbin Wu[1]

[1]*College of Surveying and Geo-Informatics, Tongji University, Shanghai 200092, China*
[2]*Tongfang Surveying Engineering and Technology Co. Ltd., Shanghai 200092, China*

Correspondence should be addressed to Mengchi Ai; 1610954@tongji.edu.cn

Academic Editor: Zongyao Sha

The objective of this study was to test the effectiveness of mapping the canopies of *Firmiana danxiaensis* (FD), a rare and endangered plant species in China, from remotely sensed images acquired by a customized imaging system mounted on an unmanned aerial vehicle (UAV). The work was conducted in an experiment site (approximately 10 km^2) at the foot of Danxia Mountain in Guangdong Province, China. The study was conducted as an experimental task for a to-be-launched large-scale FD surveying on Danxia Mountain (about 200 km^2 in area) by remote sensing on UAV platforms. First, field-based spectra were collected through hand-held hyperspectral spectroradiometer and then analyzed to help design a classification schema which was capable of differentiating the targeted plant species in the study site. Second, remote-sensed images for the experiment site were acquired and calibrated through a variety of preprocessing steps. Orthoimages and a digital surface model (DSM) were generated as input data from the calibrated UAV images. The spectra and geometry features were used to segment the preprocessed UAV imagery into homogeneous patches. Lastly, a hierarchical classification, combined with a support vector machine (SVM), was proposed to identify FD canopies from the segmented patches. The effectiveness of the classification was evaluated by on-site GPS recordings. The result illustrated that the proposed hierarchical classification schema with a SVM classifier on the remote sensing imagery collected by the imaging system on UAV provided a promising method for mapping of the spatial distribution of the FD canopies, which serves as a replacement for field surveys in the attempt to realize a wide-scale plant survey by the local governments.

1. Introduction

Forests play a vital role on global carbon cycle. Understanding constitution and spatial distribution patterns of plant species in forestry is a major concern of botanists and environmentalists [1]. Great effort has been input to distinguish the plant species of forests and analyze their distribution. As an effective tool to acquire land coverage over vast regions, remote sensing technology has been applied on detection of forests. Several studies have conducted mapping of forest distribution and extraction of the species composition by applying remote sensing mounted on satellites and unmanned aerial vehicles (UAVs) [2]. The obvious advantage of satellite remote sensing is its capability in monitoring vegetation condition and forest resources over a large-scale area in a fast and cost-effective manner [3]. For instance, satellite remote sensing data has been approved effective in mapping the outline of a forest area in order to monitor the deforestation [4]. Remote sensing products supported by satellite platforms range from high spatial resolution to low resolution. Typical examples of high-resolution imagery from satellite platforms include Spot and WorldView-2 images which were used to identify urban tree species to assess the ecological services [5]. More coarse remote sensing imageries, such as the medium-resolution products from Landsat Thematic Mapper (TM) and high-resolution imagery Enhanced Thematic Mapper Plus (ETM+), were probably only suitable for mapping distribution of grouped plant species with an advantage of covering a larger area [6]. Recently, applications of remote sensing imagery acquired by an UAV have increased dramatically in various fields. Compared to satellite remote sensing, UAV remote sensing enhanced the ability to acquire high-spatial resolution images [7]. The application of UAV remote sensing included vegetation health monitoring, detailed composition analysis of plant species, and biomass estimation [8]. Further, remotely sensed images by UAVs provided important data resources for building high-resolution airborne orthophoto maps and point clouds which have been extensively applied in ecological studies [9].

In the area of forest monitoring, dynamic mapping of plant composition, properties of plant canopies, and forest health is an important task and remote sensing technology has been widely used as a data acquisition tool for such a purpose. For example, a metric camera mounted on an UAV was used to classify riparian forest species and to analyze the composition of forests [10]. Hill et al. examined whether such an UAV would be capable of creating accurate maps of the extent of patches of a single invasive plant, called IRIS (*Iris pseudacorus* L.) and they found that manual interpretation of the UAV-acquired imagery produced the most accurate maps, suggesting that more studies on automatic extraction of plant species are still needed to improve the classification performance from UAV images [11]. Other studies used the high-spatial resolution images acquired on UAV platforms to investigate detailed plant communities [12]. Recent advances in UAV-based remote sensing technology provide a way for obtaining highly densified point clouds from remotely sensed targets, which are useful for building a 3-D canopy structure for trees in forestry [13]. When 3-D data derived from images on UAV platforms and a digital elevation model were combined, it is possible to measure the height of plants and calculate the magnitude of changes in shape for the mapped plant species with great flexibility [14]. Zhang et al. designed a remote sensing system mounted on a light-weight drone for long-term forest monitoring and, based on the acquired data, built the canopy height model (CHM), which was useful for analyzing the species richness and abundance distributions [15]. In another application, an imaging system with near-infrared and RGB channels mounted on an UAV platform was used to investigate diseases and insect pests of vegetation [16]. Numerous studies have tested the effectiveness of mapping plant species of forests from remote sensing images. To better extract plant species and evaluate their health, Näsi et al. used RGB orthoimages, along with hyperspectral imagery collected on UAV platforms, to identify plants and map bark beetle damage on an individual tree level [17]. However, in some cases where vegetation species presented low dissimilarity in spectra and texture in remote sensing images, it was difficult to distinguish among them. For instance, *Firmiana danxiaensis* (FD), a Chinese unique and endangered Indus species, is one example that is difficult to be detected due to its growth condition and characteristics [18].

While recent advancement in remote sensing platforms provided possible data sources from which plant species could be potentially detected, an alternative approach for improving the result in vegetation mapping looks for advanced image classifiers. Many classification methods such as random forest (RF), artificial neural network (ANN), decision trees (DT), and support vector machine (SVM) have been proposed and applied for extracting vegetation cover from remotely sensing images. For example, Michez et al. used RF to detected riparian invasive plant species with unmanned aerial systems imagery [19]. Kwon et al. used ANNs to classify the forest vertical structure from digital images and Lidar data [20]. Frick et al. used a DT model to classify and analyze the vegetation composition in rewetted peat land from satellite imagery [21]. Han et al. detected the cropping area in the middle of Heihe River Basin based on SVM using multitemporal vegetation index data [22]. Compared to other classifiers, SVM-based classifiers are built on statistic learning theory which classify objects through the maximal margin hyperplane established from very limited input features called support vector [23]. A few studies have shown that in most cases, SVM was less sensitive to the training samples and the performance was better than other classifiers with the same number of training samples [24]. Therefore, SVM is often preferred for classification purposes when only very limited truth samples are available. The objective of this study is to verify the feasibility of mapping FD canopies from remotely sensed images acquired by a customized UAV imaging system and to test the effectiveness of a proposed image classifier, a hierarchical classification schema powered by SVM, to extract FD from the remotely sensed imagery.

The objective of this study is to verify the feasibility of mapping FD canopies from remotely sensed images acquired by a customized UAV imaging system and to test the effectiveness of a proposed image classifier, a hierarchical classification schema powered by SVM, to extract FD from the remotely sensed imagery.

2. Study Site

As the main habitat of FD, Danxia Mountain (located at 113°36′E, 24°51′N), Shaoguan City, Guangdong Province, China, was being studied (Figure 1). The total area of Danxia Mountain was around 200 km^2. The climate was characterized as subtropical monsoon, with an annual mean temperature of 19.7°C and annual precipitation of 1715 mm. The land surface was mainly covered by bare stones, conglomerates, and red soil. Dominant families of the flora included *Osteomeles subrotunda*, *Trichophorum subcapitatum*, *Engelhardtia chrysolepis*, *Lithocarpa uvariifolia*, *Acer oblongum*, *Helieia kwangtungensis*, resurrection plant, and *Firmiana danxiaensis* (FD). FD is a rare and endangered plant in China, and thus, protecting the plant species has become a significant task for both botanists and local governments.

FD mainly grows in shallow soil in Danxia Mountain (Figure 2). In fact, FD is the only tree species dominant in Mt. Danxia and was listed as a threatened species in China Species Red List [25]. Some characteristics about the species are listed in Table 1. The morphological form of FD canopies varies in shape, which makes them difficult to be extracted from other plant species purely through remote sensing images. FD reaches mature growing status during summertime which is also the season that an FD plant develops into its typical shape. Local governments as well as botanists are eager to map the spatial distribution of FD canopies and

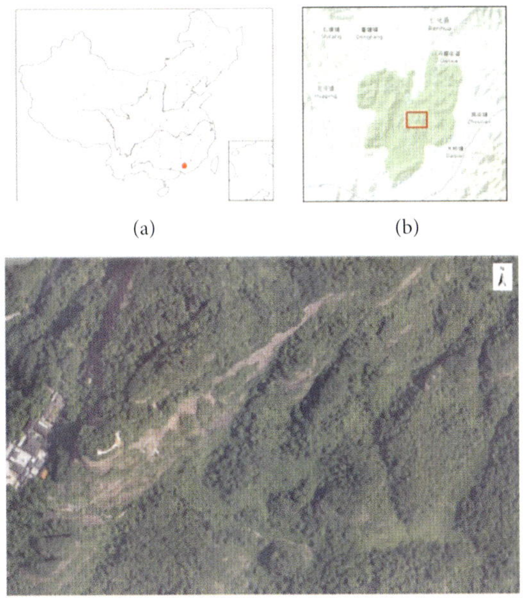

(a) (b) (c)

FIGURE 1: Study region. (a) Location of Danxia Mountain in the country, (b) outline of the study site, and (c) the experiment site.

FIGURE 2: Examples of FD distribution. The black squares denote a few examples of FD locations, and the typical canopy structure of the plant is illustrated in the inset figure.

TABLE 1: Physical descriptions of FD.

Characteristics	Description
Category	*Arbor*
Height	3 m–8 m
Canopy area	12.5–30 m^2
Blade shape (outline)	Mostly ellipse but with variations
Blade color	Turquoise

may design policies to protect the plant. But accurate positioning of the plant species is so far hard to obtain. The ground survey serves as a traditional method to map FD distribution but was limited due to the steep terrain, which makes some data impossible to be collected [26]. Therefore, alternative methods having more efficiency were desired. Satellite remote sensing could be too coarse to detect FD due to its small crowns, and plant texture information in imagery could not be captured, resulting in unsatisfactory extraction results of FD detection. Instead, data acquisition with UAVs could solve the issue because of the high-spatial resolution of UAV-collected images. However, considering the large area of the mountain (200 km^2), a typical experiment site with a small area (3 km × 4 km) was decided as a testing project before conducting wide-scale FD mapping by an UAV platform in this study.

3. Materials and Methods

3.1. Customized Multispectral Imaging System

3.1.1. Ground Spectra Collection and Selection of Spectral Bands. As a preliminary step, we distinguished the spectra among the plant species by acquiring on-site spectral signals of each plant species and the bare ground (denoted as Ground, hereafter) where there is no vegetation/plant covered. The on-site spectral signals were collected on July 11, 2016, using a field spectrometer (Avafield-2). This spectrometer recorded the spectral signals at wavelengths ranging from 350 nm to 1000 nm. The field of view was 25°. At each measuring unit, the signal of the reference panel was collected first for calibration. The reference panel was a sulfate plate with approximately 98% reflectance with the diameter size of 20 cm. The probe was put vertically down with the distance to the samples about 1 m. In order to avoid accidental error, each observation was repeated for five times. The averaged reflectance curves for each ground features were shown in Figure 3. Preliminary analysis indicated that Ground could be easily excluded by the signals in the near-infrared region. Four plant species, which was termed as high-spectral dissimilarity classes (HSD-classes hereafter), including *Osteomeles surotunda*, *Engelhardtia chrysolepis*, *Trichophorum subcapitatum*, and *Helieia kwangtungensis*, seem to have significant dissimilarity

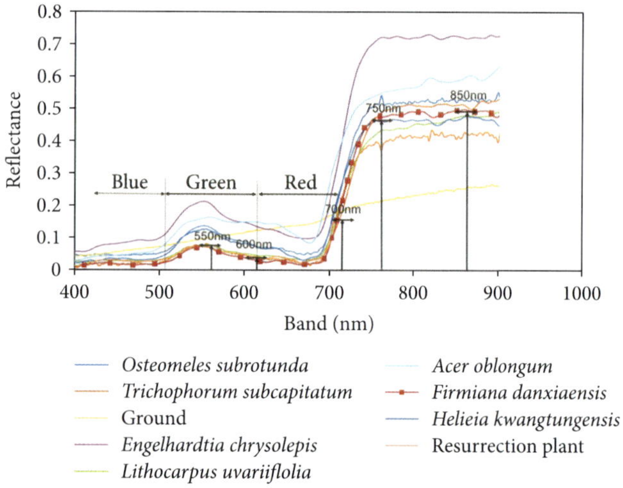

Figure 3: Mean spectral reflectance from field-collected spectra for eight vegetation species and the bare ground.

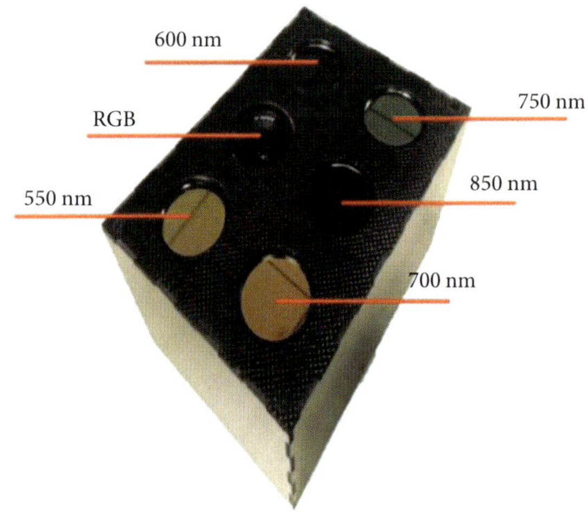

Figure 4: The customized imaging system with multispectral and RGB channels.

Table 2: Characteristics of the designed imaging system.

Parameters	Value
Bandwidth FWHM	10 nm
Exposure interval	4 s
Filter transmittance	95%
Frame image size	4000 × 6000
GND resolution (1000 m alt.)	12 cm

in spectra from the rest of the other four species. Therefore, it is possible to separate the HSD-classes and Ground from the other species using purely spectra information. The left four species, which was termed as low spectral dissimilarity class (LSD-classes hereafter), are difficult to be separated based on the spectra only. Thus, other geometric or texture information must be considered in the design of the classification schema.

3.1.2. Customized Multispectral Imaging System. Based on the analysis of the spectral signatures collected on-site as well as general knowledge of the spectral responses from green vegetation, a customized multispectral imaging system was designed for our study objective. The designed imaging system captured eight spectral bands, including three wide channels (by RGB camera) and five multispectral channels, as shown in Figure 4. Standard deviation (std) of reflectance for the selected wavelengths were calculated, resulting in 0.0623, 0.0512, 0.0838, 0.1881, and 0.1965 for 550 nm, 600 nm, 700 nm, 750 nm, and 850 nm, respectively. The imaging system was simplified by looking for suitable commercial cameras on the market. In the end, the camera Sony A 6000 was used directly for capturing the RGB channels. For the five multispectral bands, Sony A 6000 worked great after they were combined with specific band-pass filters, which realized the capturing signal channel at particular wavelengths. The main characteristics of the designed imaging system were shown in Table 2. The transmittance of the filter was 95%. Bandwidth FWHM (full width at half maximum) was 10 nm. The imaging system acquired six bands simultaneously with one exposure, and the time interval was 4 s. The spatial resolution was 12 cm when the flight height was set at 1000 m.

3.2. Remote Sensing Data Acquisition and Preprocessing

3.2.1. Imaging Data Acquisition. Image acquisition was carried out on July 14, 2016, a sunny day, using the customized imaging system mounted on an UAV platform. A ground control station software was used (Mission Planner, http://ardupilot.org/planner) to plan and control the flight. A route planning was designed based on the land terrain in order to balance the efficiency of data acquisition and the requirement of the overlapped area for building a fully qualified image dataset. In the end, images were acquired in a continuous mode for the whole experiment site (an area of about 10 km^2), resulting in 50% side overlap and 70% forward overlap at the altitude of 1000 m. The total number of exposures was 721 times. Each exposure produced eight images (including three RGB channels and 5 multispectral channels) which were saved in tiff format.

3.2.2. Image Data Preprocessing. Geometric distortion influenced mapping accuracy and must be diminished. Calibration on the digital numbers to radiance was necessary to extract the spectral reflectance from the plants and the ground. Further, calibrated images were used to generate a mosaicked orthoimage and a DSM map. Overall, the image preprocessing included three steps, optical calibration, radiometric calibration and matching, and image mosaicking.

Optical calibration processing included geometric calibration and removal of vignetting. Geometric calibration determined the lens distortion parameters and principal points of the imaging system. The coefficients of lens distortion included radial and tangential distortions. Radial distortion was the curving effect due to the subtle radial shift. The distortion brought the pincushion effect in value position. Tangential distortion affected a planar shift in the perspective of the image due to the nonalignment of the lens [27]. The calibration toolbox in Matlab was used for geometric calibration [28]. The calibration of vignetting used a look-up table which recorded the vignetting parameters obtained in the laboratory on a per-pixel basis.

Radiometric processing was used to transform DNs (digital numbers) to radiance (Rad) for each image band. Laboratory calibration and empirical line method were adopted to determine the quantum efficiency function and to calculate the necessary parameters [29]. Quantum efficiency defined the fraction of photon flux that contributes to the photocurrent in a photodetector or a pixel. The relationship between DNs and Rad can be expressed in

$$\mathrm{DN} = \mathrm{Rad} \times \mathrm{QE}_{\mathrm{band}}, \tag{1}$$

where DN is the gray value at a pixel in an image and Rad is the radiance value of the pixel. $\mathrm{QE}_{\mathrm{band}}$ is the quantum efficiency of an imaging system. In our study, laboratory calibration was used to derive $\mathrm{QE}_{\mathrm{band}}$. The experiment determined the parameters on a per-pixel basis by establishing a function between the luminance of standard light source with known radiance and DN of the imaging system. $\mathrm{QE}_{\mathrm{band}}$ was then used to transform DNs to radiance values.

RGB and multispectral images and point clouds were generated from the calibrated images using Pix4D software [30]. Those calibrated images had enough overlapped regions. Orthoimages were computed based on the RGB and multispectral channels separately and clipped to make a stacked image which covered the study area. The vegetation indices, as listed in Table 3, were derived from the stacked image. The DSM was generated from the point clouds by inverse distance weighing [31] and smoothed by a 5-by-5-pixel averaging filter.

3.3. Canopy Segmentation for UAV Imagery. Image segmentation is a fundamental task of image processing that partitions the image by grouping pixels into homogeneous regions or patches. Object-based image segmentation proved to be useful for complicated images [32]. To do so, it is necessary to design a combined homogeneity index (referred to as H) based on which the variance of the index presented in each patch can be evaluated. In this study, the geometry, texture, and spectral features of different plant species appearing in the acquired imagery were fully considered. Given a patch that consists of a set of image pixels, H is an index that comprehensively reflects the homogeneity of the patch. A bottom up region-merging algorithm was adopted to realize image segmentation. Region merging starts from a single pixel and looks for its surrounding pixels that must satisfy H index if they are merged. The criterion for object merging is to minimize H index. In each step, adjacent objects are selected based on the smallest growth of the heterogeneity. A predefined parameter, HA, is given based on a trial-and-error test. If the heterogeneity is higher than HA, the merging process will stop. H is defined as [33]

$$H = w_{\mathrm{sp}} \times \Delta h_{\mathrm{sp}} + w_{\mathrm{sh}} \times \Delta h_{\mathrm{sh}}, \tag{2}$$

where w_{sp} and w_{sh} are the weights of the spectral index and shape index and the sum of w_{sp} and w_{sh} is 1, Δh_{sp} is a spectral index which captures the spectra characteristics, and Δh_{sh} is a shape index reflecting the shape characteristics of a merged object. Spectral feature was defined by weighing each channel number of candidate objects n_{merge} and standard deviation σ. Shape feature was determined by compactness feature, smooth feature, and the ratio of them. The given object denoted as obj1 and another object denoted as obj2 to be merged to obj1, Δh_{sp}, and Δh_{sh} were calculated by (3) and (4):

$$\Delta h_{\mathrm{sp}} = \sum_{i=1}^{N} w_i \left(n_{\mathrm{merge}} \sigma_i^{\mathrm{merge}} - \left(n_{\mathrm{obj1}} \sigma_i^{\mathrm{obj1}} - n_{\mathrm{obj2}} \sigma_i^{\mathrm{obj2}} \right) \right), \tag{3}$$

where N is the number of layers (bands) that participated in segmentation and w_i is the weight assigned to band i to represent its importance ($\sum_{i=1}^{N} w_i = 1$), n_{merge} is the total number of pixels in the merged object, σ_i^x is standard deviation of spectra for pixels contained in an object (x is obj1, obj2, or the merged object), and n_{obj1} and n_{obj2} are the number of pixels in obj1 and obj2, respectively, before they are merged. Δh_{sp} indicates the dissimilarity of obj1 and obj2 from the spectral perspective.

$$\Delta h_{\mathrm{sh}} = w_{\mathrm{comp}} \bullet \Delta h_{\mathrm{comp}} + w_{\mathrm{sm}} \bullet \Delta h_{\mathrm{sm}}. \tag{4}$$

In (4), Δh_{comp} and Δh_{sm} were the index of the shape compactness and smoothness defined by (5) and (6), respectively.

$$\Delta h_{\mathrm{comp}} = n_{\mathrm{merge}} \bullet \frac{l_{\mathrm{merge}}}{\sqrt{n_{\mathrm{merge}}}} - \left(n_{\mathrm{obj1}} \bullet \frac{l_{\mathrm{merge}}}{\sqrt{n_{\mathrm{merge}}}} \right), \tag{5}$$

$$\Delta h_{\mathrm{sm}} = n_{\mathrm{merge}} \bullet \frac{l_{\mathrm{merge}}}{b_{\mathrm{merge}}} - \left(n_{\mathrm{obj1}} \bullet \frac{l_{\mathrm{obj1}}}{b_{\mathrm{obj1}}} + n_{\mathrm{obj2}} \bullet \frac{l_{\mathrm{obj2}}}{b_{\mathrm{obj2}}} \right). \tag{6}$$

In (5) and (6), l_x and b_x represent the perimeter and the bounding box of object x (x = obj1, obj2, or the merged object) and the other parameters are the same as in (3).

In this study, we used eCognition Developer (Trimble, 2015) to segment the processed image for the study area. The eight bands (RGB bands and multispectral bands) were involved and equally weighted ($w_i = 1/8$, $i = 1, 2, \ldots, 8$). A weight of 0.6 was given to the compactness weight (w_{comp}) and 0.4 was given to the smoothness (w_{sm}), considering that compactness before merging should be given more priority. An equal weight was given to the shape parameter (w_{sh}) and the spectral parameter (w_{sp}). One difficulty then was to decide the threshold (H_A) based on which the merging process could be stopped if the calculated H index was greater than the threshold. High H_A would result in larger size of patches which could include more than one plant species in the patches. Conversely, too small H_A might result in a single FD crown separated into several patches. Therefore, the determination of the threshold was realized by a trial-and-error strategy by setting a range of H_A and the one that segmented the area into patches making most FD located in a single patch, judged from field trothing data, was taken as the result. In the end, H_A was given to 35, which seemed

TABLE 3: Three vegetation indices used in the study.

Vegetation index	Equation
Normalized difference vegetation index (NDVI)	$(R_{850nm} - R_R)/(R_{850nm} + R_R)$
Structurally insensitive pigment index (SIPI)	$(R_{850nm} - R_B)/(R_{850nm} + R_B)$
Red edge normalized difference vegetation index (RENDVI)	$(R_{750nm} - R_R)/(R_{750nm} + R_R)$
Photochemical reflectance index (PRI)	$(R_G - R_{600nm})/(R_G + R_{600nm})$

appropriate in terms of FD grouping which appeared in the segmented patches.

3.4. Design of Image Classification Schema. In order to classify the canopy of FD, a hierarchical classification method, for discriminating nine classes (including Ground), was decided and established based on the analysis of the spectral characteristics of the targeted plant species as well as the typical morphological form of FD canopies. The hierarchical classification method simplified the classification process into three levels. Each level allowed using specific algorithm and rules to classify the desired classes. The flowchart of classification strategy was shown in Figure 5. At the first level, segmented objects (patches) were classified into Ground and non-Ground (vegetation) objects, which was implemented after computing vegetation indices (VIs), indicators of the green density, from remote sensing images. VIs have also been used to separate vegetation species in many other studies [34]. Out of the VIs, normalized difference vegetation index (NDVI) is probably the most popular one which is calculated by comparing the magnitude of spectral reflectance at wavelengths red and near infrared, given by [35]

$$\frac{R_{850\,\mathrm{nm}} - R_R}{R_{850\,\mathrm{nm}} + R_R}, \tag{7}$$

where $R_{850\,\mathrm{nm}}$ is the reflectance at wavelength 850 nm and R_R is the reflectance at wavelength 705 nm or of the red channel in a RGB image.

In addition to NDVI, there are other vegetation indices (VIs) which were proposed for the similar purpose (Table 3). Those VIs were formulated by taking the characteristics of spectral reflectance of green vegetation from different aspects. For example, the spectral reflectance of vegetation usually presents a green peak located at wavelength 550 nm in the spectrum and strong spectral absorption in the blue and red regions of the spectrum. Furthermore, the red edge effect from vegetation is typically observed for vegetation reflectance in that a rapid increase of reflectance occurs starting at the red region (around 700 nm in the spectrum) to the near-infrared region (usually at 750 nm or 850 nm). Those information provided the basis for designing the spectral channels of our customized multispectral imaging system on UAVs.

A simple filter based on the processed VIs was applied to exclude ground patches. At level 2, plant species were classified as four HSD-classes and an LSD-class. The HSD-classes included four plant species, namely, *Osteomeles surotunda*, *Engelhardtia chrysolepis*, *Trichophorum subcapitatum*, *and Helieiakwangtungensis*. The left plant species, including FD, were assigned as a combined LSD-class which could not be separated by purely spectral information due to low dissimilarity in spectra. At this level, SVM was used to classify the 5 classes (four HSD-classes and a combined LSD-class) using the spectral features (the eight bands and vegetation indices) only. The mean spectra value and vegetation indices (NDVI, SIPI, RENDVI, and PRI, in Table 3) of all the pixels located within each patch were calculated and used in the SVM classifier. At level 3, the combined LSD-class was further divided into four plant species, including FD, based on the geometric feature, terrain feature, and texture features by another SVM classifier. Here, SVM, a nonparameter-supervised classifier, was chosen as the classifier for both levels 2 and 3 because it has advantage achieving good classification results without sensitivity to sample size [36].

Terrain properties, geometric properties, and texture characteristics were used at level 3 to identify patches covered by FD canopies. Terrain properties, including slope, aspect, and DSM, were input to the classifier because the FD was observed to appear mainly at medium terrain height and southern slope. Because we did not have DEM data with spatial resolution comparable to the processed images, a smoothed DSM was used to process slope and aspect data.

Texture and geometric information denoted the shape of the targeted patch. In this study, Gray-level Co-occurrence Matrix (GLCM) was used to represent texture [37]. Four GLCM indices were included in classification (Table 4). For geometric properties, three indices were formulated and used in the classifier (Table 4).

Standard radial Gaussian was used as the kernel function for the SVM classifier. The main parameters to be decided in the classifier included C (the penalty parameter which stands for an error term) and γ (controlling training efficiency). As there was no prior knowledge which C and γ were acceptable, they were determined based on a recursive loop with different combinations. During the process, a good pair of C and γ could be identified [38].

Field trip was arranged to collect ground truth data for model training and result validation. The date of ground data collection was conducted on July 21, 2016, one week after the UAV flight. A GPS device (Trimble R2 GNSS) was used to record the coordinate of FD trunk location, and the general shape for each recorded FD canopy was drafted. RTK (Real-Time Kinematic) was set as the operating mode, producing an absolute positioning accuracy within 1-2 cm. Two thirds of the recordings were used for model training purposes, and the rest were used for result validation. eCognition Developer was used to implement the

TABLE 4: The equation and description of index.

Type	Index	Equation and description of index
Texture#	GLCM_homogeneity	$\sum_{i=0}^{I}\sum_{j=0}^{J} P\frac{(i,j)}{(1+(i-j)^2)}$
	GLCM_contrast	$\sum_{i=0}^{I}\sum_{j=0}^{J} P(i,j)\times(i-j)^2$
	GLCM_dissimilarity	$\sum_{i=0}^{I}\sum_{j=0}^{J} P(i,j)\times\lvert i-j\rvert$
	GLCM_Entropy	$-\sum_{i=0}^{I}\sum_{j=0}^{J} P(i,j)\times\ln P(i,j)$
Geometry*	Shape index	$s=\frac{e}{4\bullet\sqrt{A}}$
	Density	$d=\frac{\sqrt{n}}{1+\sqrt{\mathrm{Var}(X)+\mathrm{Var}(Y)}}$

$P_{i,j}$ is the probability of a gray level pair, i and j, in the GLCM defined by a position operator, and I and J are the maximum values given in each dimension of GLCM, respectively. *e and A were the perimeter and area, respectively, of the targeted patch, n was the number of pixels, and X and Y were the width and length of the bounding box of the patch.

hierarchical classification strategy. The segmented patches, a DSM layer, slope layer, and aspect layer participated in the classification process. Spectra, geometric, and texture indices were calculated during the classification process by customized functions.

4. Result

4.1. UAV-Collected Imagery. The orthoimage processed from the spectra collected by UAV sensors showed clear differentiation, in terms of spectral, texture, and geometry, in the distribution of different plant species and the bare ground (Figure 6). The DSM, ranging from 214 m to 302 m, showed the spatial variations of surface height. Such height reflects the changes over the space in both DEM and the plant species. When combined with other topography information such as the slope and aspect, better understanding of the preferred locations of each plant species could be analyzed. The acquired images from UAV allowed producing various vegetation indices which could be helpful data input for classifying the plant species and the ground. For example, red edge normalized difference vegetation index (RENDVI) processed from the multispectral band at 775 nm and red band of RGB showed clearly the differentiated patterns in the distribution of the bare ground and plant species. One advantage of applying UAV platform is that the high-spatial resolution of those images provided enough detail information for building advanced image classifiers, which otherwise is impossible for other satellite platforms.

4.2. Classification Result of FD Canopy. SVM, using radial Gaussian used as the kernel function, was used to classify HSD-classes and LSD-classes. At level 2, C and γ were determined as (5, 10) for the SVM classifier. At level 3, C and γ were determined as (2, 10) for the classifier. The different parameter settings allowed optimized classification result at different classification steps. The final classification result, denoted as FD species, non-FD species, and Ground, for the study site was shown in Figure 7(a).

The result indicated that the FD species only covered very small part of the area, which was in agreement with the reality. Though FD could spread over the whole study area, the spatial distribution of FD tended to cluster and was dominantly located close to the bare ground and the cliff (Figure 7(b)). This finding may serve as important clue for local botanists to evaluate the FD distribution.

4.3. Accuracy Test. The GPS recordings from the field trip were used to validate the classification result. Those points that were recorded as FD were overlaid on top of the classified segments (Figure 8). The locations of FD species generally matched well with the segments assigned to FD which, after merged with adjacent ones, had an area ranging from 10 m^2 to 32 m^2 (Figure 8(b)), comparable to the statistics of the canopy size from field samples (Table 1). Classification results by the proposed hierarchical classifier showed that the overall classification accuracy at levels 2 and 3 reached 87.4% and 75.3%, respectively (Table 5). For the SVM classifier at level 2, there were 17.6% cases from HSD-classes that were mistakenly assigned to LSD classes, though most LSD-classes were labeled correctly (accuracy = 92.4%). There were 331 cases classified as LSD-class, including 53 cases which were actually HSD-class. Out of the 52 FD samples in the testing set, 6 cases were wrongly mapped as HSD-class (Table 5(a)) and the remaining 46, along with randomly selected 51 non-FD samples, were entered into the next SVM classifier at level 3 for testing. The result indicated that the SVM classifier at level 3 performed slightly worse than that at level 2, mainly due to the fact that almost one-third of non-FD samples were wrongly classified as FD. The hierarchical classification powered by SVM classifier for FD species reached an overall accuracy of 76.9%, a reasonable result considering the complex terrain and multiple plant species.

5. Discussion

Remote sensing image acquisition by sensors mounted on UAV platforms has been widely used in the classification of forests [39]. There are several advantages by applying UAV-based remote sensing. Firstly, UAV platforms can fly at low-altitude and thus images having high-spatial resolution can be acquired. Low-altitude remote sensing decrease the distance between the target objects and the sensors and thus the quantity of acquired images is improved. Because of the high-spatial resolution, extracting FD from UAV images also has less impact from mixed pixel issue. Furthermore, the collected images are less affected by the atmosphere and luminous impact. Secondly, the flight route and flight height can be customized based on the actual need and thus UAV platforms provide more flexibility for data acquisition. Sensors mounted on UAVs can be used to map most places on the land surface, and setting an appropriate flight height can balance

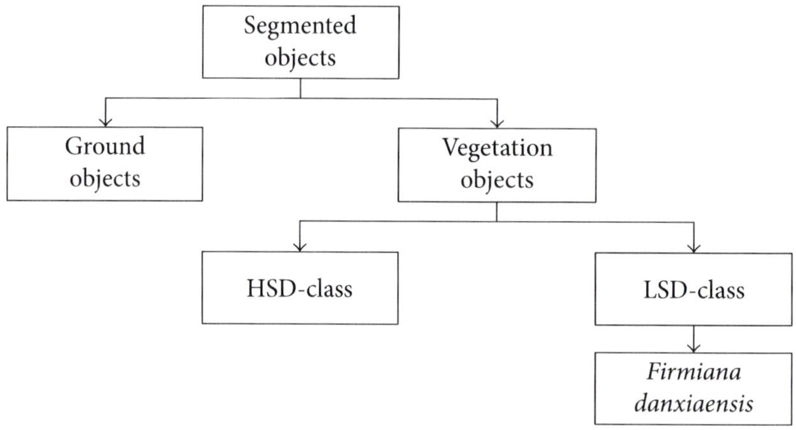

Figure 5: The flow chart of the classification strategy. The hierarchical classification strategy included three levels. At the last level (level 3), FD canopies were identified from the segmented objects.

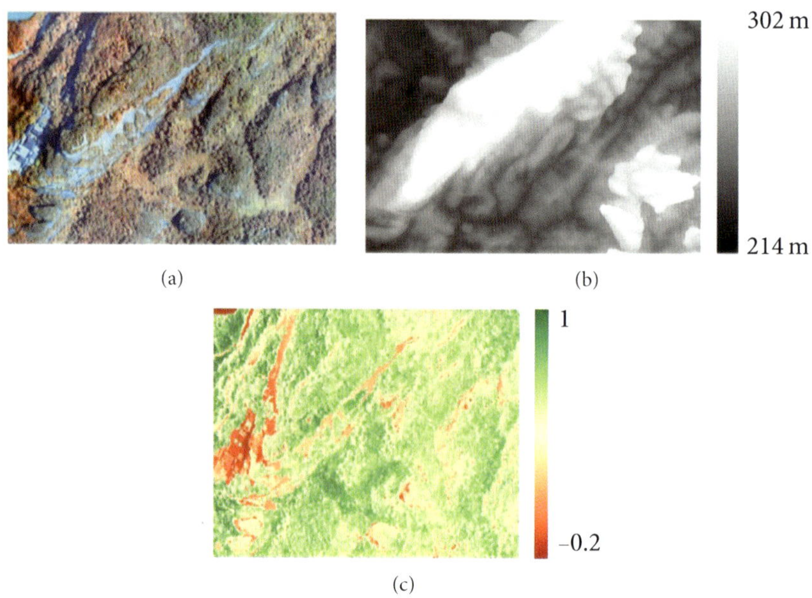

Figure 6: Example of data products computed from the spectra collected by the image system on UAVs. (a) Mosaicked orthoimage, (b) the smoothed DSM, and (c) NDVI index. One of the vegetation indices is used to distinguish vegetation area and soil area and to input to the SVM classifier at level 2 in the classification schema.

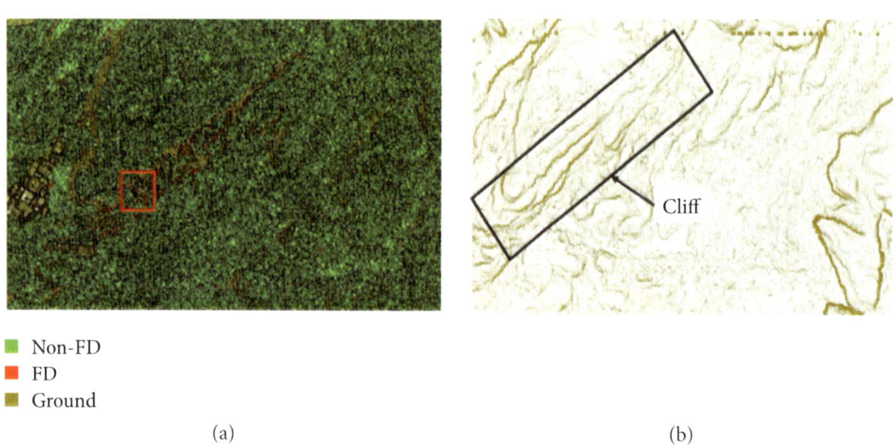

Figure 7: Classification result for the segmented patches of the image collected on an UAV platform. (a) Overall classification map for the study site. (b) The corresponding DSM contour line showing the cliff area (steep terrain) in the black rectangle where FD densely populated.

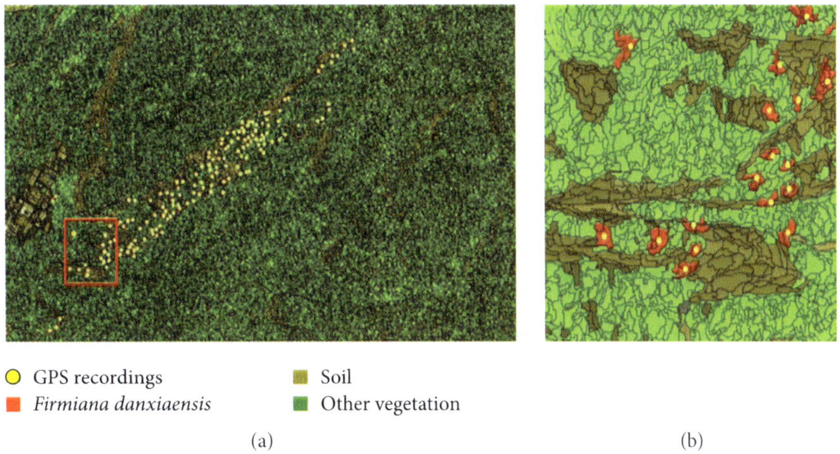

Figure 8: Validation of the classification result based on field trip. Yellow points are GPS recordings showing the FD locations. (a) Overall classification map overlaid by FD locations from GPS recordings and (b) detail of FD labeled segments/patches (in red) and the ground truth/GPS points (yellow dots).

Table 5: Accuracy assessment of the FD classification for the segmented patches.

(a) Confusion matrix at level 2

Reference class	Map class		Producer's accuracy
	HSD-class	LSD-class	
HSD-classes	248	53	82.4%
LSD-classes	23 (6*)	278	92.4%
User's accuracy	91.5%	84.0%	87.4%

*Number of FD samples located in the cell. Kappa = 0.748.

(b) Confusion matrix at level 3

Reference class	Map class		Producer's accuracy
	FD	Non-FD	
FD	40	6	87.0%
Non-FD	18	33	64.7%
Accuracy	67.0%	84.7%	75.3%

Total number of FD: 52; number of FD correctly labeled: 40; accuracy for FD extraction: 76.9; Kappa = 0.535.

the coverage and spatial resolution of the acquired data; thus, the efficiency of data collection can be improved because the flight range can be determined as required. Thirdly, UAV platforms can support any cameras designed to collect images for particular spectral wavelengths. As the study conducted here, RGB channels and 5 multispectral channels were designed by using particular band-pass filters on the commonly commercial cameras. By collecting and analyzing the diversity of spectra, imaging systems can be customized to collect spectra based on the mapping purposes. However, unlike satellite remote sensing platforms which can cover a large area in one glance, UAVs are usually suitable for mapping small area only due to the limited viewing angles and relatively low flight height. Further, the demand for georegistering and mosaicing multiple images collected at adjacent flight routes also introduces extra input for remote sensing by UAVs. Possible solutions may integrate UAV images with other imagery products from satellite remote sensing.

In our study, we proposed a hierarchical classification approach after the analysis of the spectral features of the plant species. Such strategy allowed us to separate different classes in a progressive manner and proved to be suitable for extracting FD from the acquired imagery. For example, the bare ground could be easily excluded based on the computed VIs simply by using threshold filters, namely, segmented patches with VIs lower than predefined thresholds were labeled as Ground. SVM has been proved effective in classifying remote sensing imagery when a limited number of samples are available [40]. Several studies have been using the hierarchical classification method and SVM algorithm to classify specific plants [41]. For example, SVM was used as a binary classifier to detect seminatural habitats [42]. The combination of a hierarchical classification schema and SVM algorithm has three main advantages. Firstly, the process for building such classification structure is readily available in several existing software packages such as eCognition and Matlab [43]. After necessary datasets, including the orthoimages, DSM, and segmented patches in this study, are ready, the classifier could be quickly setup with those input datasets. The main work of building the classifier involved establishment of the rules to label the input patches based on related data (e.g., DSM and the orthoimages in this work). The second advantage is that the classification schema can simplify the classification process. The hierarchical classification approach separates the candidate classes into several levels, and thus, the classification can be divided into a few subobjectives. When combined by appropriate algorithms at each level, improving the overall classification accuracy can be realized. Lastly, the hierarchical classification combined with SVM is efficient for extraction of a single class (e.g., FD in this study) from multiple class labels. The basic SVM classifier only supports binary classification issues, namely, 0-1 cases. SVM also supports classifying multiple classes called K-class classifier which split the process into k steps [44]. In the current study, SVM was applied to classify

LSD-class and non-LSD-class at level 2 and to classify FD and non-FD at level 3, both of which were binary classification, though there were multiple plant species involved. In fact, there was no need to classify the species within HSD-classes or to classify each species within the LSD-classes, considering the very specific objective of the study.

There are two steps which may introduce uncertainties in the ultimate classification result based on the proposed hierarchical classification schema; they are image segmentation and classification of segmented patches. First, detection of FD canopy is highly dependent on the result of image segmentation which was performed based on a bottom up region-merging algorithm in this study. A homogeneity index, H, was used to control the process of object merging. It is clear that the result of image classification depends on the criteria adopted for the H index. Factors involving the spectra and geometry of produced patches were considered in computing the index. The weight assignment for each of the factors is critically important. In addition, the threshold value (H_A) in the study determined the process of segmentation: either continuing the merging process or stopping it. The factors and their weighing schema as well as the threshold all determine the size of the segmented patches. Though we used a trial-and-error strategy to determine H_A, we observed that there was still a small part of the patches (which were FD canopy tested by field GPS recordings) which were either too small or too big, resulting in mislabeling in the next hierarchical classification step by the SVM classifier. The next step, hierarchical classification for the segmented patches, also introduced certain uncertainties. There were three levels in the proposed hierarchical classification schema. We observed that there was very few mislabeled cases at the first level. However, the SVM classifiers at levels 2 and 3 did produce mislabeled patches, including cases that either non-FD patch was labeled as FD or FD patch was labeled otherwise. While SVM is regarded as an advanced classifier that has been successfully applied in many image classification studies [45], the data quality used to train the model and the data quality in the testing samples can surely have much impact on the result. Spectra as well as spectra-derived vegetation indices were taken as input in the SVM classifier at the second level. It was pointed out that mixed spectra in remote sensing imagery would inevitably lead to side effect on the classification result [46]. Still, though the study took terrain properties, geometric properties, and texture characteristics in the SVM classifier at the third level, certain confusion between FD and other plant species might also exist, making the wrong assignment of labels to some segmented patches. Terrain properties were probably the most important factors that should be taken to differentiate LSD-classes, as FD are distributed mainly close to cliff area where the slope index played a vital role in the classifier. Other factors related to geometry and texture could be examined and selected in more detail for the SVM classifier. Nevertheless, our work demonstrated the effectiveness of applying the designed object-based image segmentation and hierarchical classification schema in the extraction of FD canopy from remotely sensed images on the UAV platform.

6. Conclusion

Recent applications of UAVs make the equipment rapidly evolve into easy-to-use platforms for sensors deployable to acquire fine-scale vegetation data over large areas in a timely and cost-effective way. Customized imaging systems mounted on UAV platforms possess the capability and flexibility of collecting low-altitude remote sensing data for classifying land cover with plant species. In this work, we designed a customized imaging system mounted on UAV and acquired RGB and multispectral images covering part of Danxia Mountain where most of *Firmiana danxiaensis* (FD), a rare and endangered plant in China, are distributed. The identification of FD distribution has been desired by local practitioners as well as researchers in academia for a long time. However, large-scale FD mapping, based on traditional field survey, was called a halt due to the high cost and even safety issues associated with the work. With advances in remote sensing technology, especially UAV applications, this study tried to test the possibility of mapping FD through the customized imaging system mounted on an UAV platform. To this end, a limited number of field spectra collection were conducted first to help analyze the characteristic of FD and other plant species, which enabled the design of a reasonable hierarchical classification schema for the acquired UAV images. Various image products, including DSM and vegetation indices, were derived from the imaging system and involved in the classification process. Support vector machine was integrated to achieve good extraction of FD canopies from the UAV imagery. Spectra, terrain, geometric, and texture properties were fed into the proposed classifier so that the plant species could be separated. The accuracy of detection was evaluated by the ground truth GPS recordings. The study proved that UAV could be a promising platform to conduct FD surveying and mapping, which provides fundamental data to protect the rare plant species. Future work may focus on fusing UAV-based images with other products from satellite remote sensing so that the efficiency for mapping FD could be further improved.

Conflicts of Interest

The authors declare that there is no conflict of interest regarding the publication of this paper.

Acknowledgments

This research is funded by the National Natural Science Foundation of China (no. 41771481, no. 41671451, and no. 41701524), the Fundamental Research Funds for the Central Universities and China Postdoctoral Science Foundation (Grant no. 2017M621530), and Shanghai Municipal Commission of Economy and Informatization (no. JMRH-2018-1016). The authors thank Tongfan Surveying Engineering and Technology Co. Ltd. and Aerospace ShuWei High Tech. Co. Ltd. for their help in data measurements.

References

[1] X. Chen and B. L. Li, "Spatial distribution of forest biome energetics in China," *Forestry*, vol. 78, no. 5, pp. 461–469, 2005.

[2] C. Toth and G. Jóźków, "Remote sensing platforms and sensors: a survey," *ISPRS Journal of Photogrammetry and Remote Sensing*, vol. 115, pp. 22–36, 2016.

[3] M. Fagan, R. DeFries, S. Sesnie et al., "Mapping species composition of forests and tree plantations in northeastern Costa Rica with an integration of hyperspectral and multitemporal Landsat imagery," *Remote Sensing*, vol. 7, no. 5, pp. 5660–5696, 2015.

[4] S. Zaman and M. Katoh, "Assessment of forest covers change in tropical moist deciduous forest in Thakurgaon, Bangladesh using ALOS data (multipurpose forest management)," *Journal of Forest Planning*, vol. 16, pp. 285–292, 2017.

[5] Y. T. Mustafa and H. N. Habeeb, "Object based technique for delineating and mapping 15 tree species using VHR WorldView-2 imagery," in *Proceedings Volume 9239, Remote Sensing for Agriculture, Ecosystems, and Hydrology XVI, 92390G*, Amsterdam, Netherlands, October 2014.

[6] M. Ozdogan, "A practical and automated approach to large area forest disturbance mapping with remote sensing," *PLoS One*, vol. 9, no. 4, article e78438, 2014.

[7] I. Colomina and P. Molina, "Unmanned aerial systems for photogrammetry and remote sensing: a review," *ISPRS Journal of Photogrammetry and Remote Sensing*, vol. 92, pp. 79–97, 2014.

[8] B. Lu, Y. He, and H. Liu, "Investigating species composition in a temperate grassland using unmanned aerial vehicle-acquired imagery," in *2016 4th International Workshop on Earth Observation and Remote Sensing Applications (EORSA)*, pp. 107–111, Guangzhou, China, July 2016.

[9] O. Nevalainen, E. Honkavaara, S. Tuominen et al., "Individual tree detection and classification with UAV-based photogrammetric point clouds and hyperspectral imaging," *Remote Sensing*, vol. 9, no. 3, p. 185, 2017.

[10] R. Dunford, K. Michel, M. Gagnage, H. Piégay, and M. L. Trémelo, "Potential and constraints of unmanned aerial vehicle technology for the characterization of Mediterranean riparian forest," *International Journal of Remote Sensing*, vol. 30, no. 19, pp. 4915–4935, 2009.

[11] D. J. Hill, C. Tarasoff, G. E. Whitworth, J. Baron, J. L. Bradshaw, and J. S. Church, "Utility of unmanned aerial vehicles for mapping invasive plant species: a case study on yellow flag iris (*Iris pseudacorus* L.)," *International Journal of Remote Sensing*, vol. 38, no. 8–10, pp. 1–23, 2017.

[12] J. W. Karl, J. K. Gillan, N. N. Barger, J. E. Herrick, and M. C. Duniway, "Interpretation of high-resolution imagery for detecting vegetation cover composition change after fuels reduction treatments in woodlands," *Ecological Indicators*, vol. 45, pp. 570–578, 2014.

[13] K. Lim, P. Treitz, M. Wulder, B. St-Onge, and M. Flood, "LiDAR remote sensing of forest structure," *Progress in Physical Geography: Earth and Environment*, vol. 27, no. 1, pp. 88–106, 2003.

[14] M. Dalponte, L. Bruzzone, and D. Gianelle, "Fusion of hyperspectral and LIDAR remote sensing data for classification of complex forest areas," *IEEE Transactions on Geoscience and Remote Sensing*, vol. 46, no. 5, pp. 1416–1427, 2008.

[15] J. Zhang, J. Hu, J. Lian, Z. Fan, X. Ouyang, and W. Ye, "Seeing the forest from drones: testing the potential of lightweight drones as a tool for long-term forest monitoring," *Biological Conservation*, vol. 198, pp. 60–69, 2016.

[16] J. Albetis, S. Duthoit, F. Guttler et al., "Detection of *Flavescence dorée* grapevine disease using unmanned aerial vehicle (UAV) multispectral imagery," *Remote Sensing*, vol. 9, no. 4, p. 308, 2017.

[17] R. Näsi, E. Honkavaara, P. Lyytikäinen-Saarenmaa et al., "Using UAV-based photogrammetry and hyperspectral imaging for mapping bark beetle damage at tree-level," *Remote Sensing*, vol. 7, no. 11, pp. 15467–15493, 2015.

[18] S. F. Chen, M. W. Li, H. J. Jing et al., "De novo transcriptome assembly in *Firmiana danxiaensis*, a tree species endemic to the Danxia landform," *PLoS One*, vol. 10, no. 10, article e0139373, 2015.

[19] A. Michez, H. Piégay, L. Jonathan, H. Claessens, and P. Lejeune, "Mapping of riparian invasive species with supervised classification of unmanned aerial system (UAS) imagery," *International Journal of Applied Earth Observation & Geoinformation*, vol. 44, pp. 88–94, 2016.

[20] S.-K. Kwon, H.-S. Jung, W.-K. Baek, and D. Kim, "Classification of forest vertical structure in South Korea from aerial orthophoto and Lidar data using an artificial neural network," *Applied Sciences*, vol. 7, no. 10, p. 1046, 2017.

[21] A. Frick, P. Steffenhagen, S. Zerbe, T. Timmermann, and K. Schulz, "Monitoring of the vegetation composition in rewetted peatland with iterative decision tree classification of satellite imagery," *Photogrammetrie - Fernerkundung - Geoinformation*, vol. 2011, no. 3, pp. 109–122, 2011.

[22] H. Han, M. Ma, and X. Wang, "Classifying cropping area of middle Heihe River Basin in China using multitemporal normalized difference vegetation index data," *Journal of Applied Remote Sensing*, vol. 8, no. 1, article 083654, 2014.

[23] G. M. Foody and A. Mathur, "Toward intelligent training of supervised image classifications: directing training data acquisition for SVM classification," *Remote Sensing of Environment*, vol. 93, no. 1-2, pp. 107–117, 2004.

[24] G. Mountrakis, J. Im, and C. Ogole, "Support vector machines in remote sensing: a review," *ISPRS Journal of Photogrammetry & Remote Sensing*, vol. 66, no. 3, pp. 247–259, 2011.

[25] S. Wang and Y. Xie, *China Species Red List*, vol. 1, Higher Education Press, Beijing, China, 2004.

[26] J. Ouyang, H. Peng, X. Luo, Z. Chen, A. Zhang, and Y. Ma, "Environmental features of the micro-landforms of the spatial distribution of the national rare species of *Firmiana danxiaensis* on the Danxiashan Mountain," *Scientia Geographica Sinica*, vol. 37, no. 10, pp. 1585–1592, 2017.

[27] J. Kelcey and A. Lucieer, "Sensor correction of a 6-band multispectral imaging sensor for UAV remote sensing," *Remote Sensing*, vol. 4, no. 5, pp. 1462–1493, 2012.

[28] J. Y. Bouguet, "Camera Calibration Toolbox for Matlab," *High Speed Vision System & Object Tracking*, 2013, http://www.vision.caltech.edu/bouguetj/calib_doc/index.html.

[29] D. Olsen, C. Dou, X. Zhang, L. Hu, H. Kim, and E. Hildum, "Radiometric Calibration for AgCam," *Remote Sensing*, vol. 2, no. 2, pp. 464–477, 2010.

[30] F. J. Mesas-Carrascosa, J. Torres-Sánchez, I. Clavero-Rumbao et al., "Assessing optimal flight parameters for generating accurate multispectral orthomosaicks by UAV to support site-specific crop management," *Remote Sensing*, vol. 7, no. 10, pp. 12793–12814, 2015.

[31] R. Buffat, T. Sarjakoski, M. Santos, and L. Sarjakoski, "Feature-aware surface interpolation of rooftops using low-density Lidar data for photovoltaic applications," in *Geospatial Data in a Changing World*, Lecture Notes in Geoinformation and Cartography, Springer, Cham, 2016.

[32] V. Dey, Y. Zhang, and M. Zhong, "A review on image segmentation techniques with remote sensing perspective," *Pattern Recognition*, vol. 38, no. 9, pp. 1277–1294, 2010.

[33] M. Baatz and A. Schäpe, "Multiresolution segmentation: an optimization approach for high quality multi-scale image segmentation," in *Angewandte Geographische Informations-Verarbeitung XII*, J. Strobl, T. Blaschke, and G. Griesbner, Eds., pp. 12–23, Wichmann Verlag, Karlsruhe, Germany, 2000.

[34] R. M. Igamberdiev, R. Bill, H. Schubert, and B. Lennartz, "Analysis of cross-seasonal spectral response from kettle holes: application of remote sensing techniques for chlorophyll estimation," *Remote Sensing*, vol. 4, no. 11, pp. 3481–3500, 2012.

[35] R. S. Defries and J. R. G. Townshend, "NDVI-derived land cover classifications at a global scale," *International Journal of Remote Sensing*, vol. 15, no. 17, pp. 3567–3586, 1994.

[36] Z. Sha and Y. Bai, "Mapping grassland vegetation cover based on support vector machine and association rules," in *2013 Ninth International Conference on Natural Computation (ICNC)*, pp. 44–49, Shenyang, China, July 2014.

[37] Q. Yu, P. Gong, N. Clinton, G. Biging, M. Kelly, and D. Schirokauer, "Object-based detailed vegetation classification with airborne high spatial resolution remote sensing imagery," *Photogrammetric Engineering & Remote Sensing*, vol. 72, no. 7, pp. 799–811, 2006.

[38] D. Giveki, M. A. Soltanshahi, and G. A. Montazer, "A new image feature descriptor for content based image retrieval using scale invariant feature transform and local derivative pattern," *Optik - International Journal for Light and Electron Optics*, vol. 131, pp. 242–254, 2017.

[39] Y. Yuan and X. Hu, "Random forest and objected-based classification for forest pest extraction from Uav aerial imagery," in *ISPRS - International Archives of the Photogrammetry, Remote Sensing and Spatial Information Sciences, XLI-B1*, pp. 1093–1098, Prague, Czech Republic, July 2016.

[40] L. Bruzzone, M. Chi, and M. Marconcini, "A novel transductive SVM for semisupervised classification of remote-sensing images," *IEEE Transactions on Geoscience and Remote Sensing*, vol. 44, no. 11, pp. 3363–3373, 2006.

[41] B. Lu and Y. He, "Species classification using unmanned aerial vehicle (UAV)-acquired high spatial resolution imagery in a heterogeneous grassland," *ISPRS Journal of Photogrammetry and Remote Sensing*, vol. 128, pp. 73–85, 2017.

[42] C. Schuster, T. Schmidt, C. Conrad, B. Kleinschmit, and M. Förster, "Grassland habitat mapping by intra-annual time series analysis – comparison of RapidEye and TerraSAR-X satellite data," *International Journal of Applied Earth Observation and Geoinformation*, vol. 34, no. 34, pp. 25–34, 2015.

[43] M. J. Huang, *Development of Knowledge-based Classification System: Urban-feature Extraction Using Aerial Imagery with Airborne LiDAR Data*, VDM Verlag Dr. Müller, 2009, http://www.vdm-verlag.de.

[44] Z. Li, S. Tang, and S. Yan, "Multi-class SVM classifier based on pairwise coupling," in *Pattern Recognition with Support Vector Machines, vol 2388*pp. 321–333, Springer, Berlin, Heidelberg.

[45] P. Du, J. Xia, W. Zhang, K. Tan, Y. Liu, and S. Liu, "Multiple classifier system for remote sensing image classification: a review," *Sensors*, vol. 12, no. 4, pp. 4764–4792, 2012.

[46] C. Hecker, M. van der Meijde, H. van der Werff, and F. D. van der Meer, "Assessing the influence of reference spectra on synthetic SAM classification results," *IEEE Transactions on Geoscience and Remote Sensing*, vol. 46, no. 12, pp. 4162–4172, 2008.

An Unsupervised Algorithm for Change Detection in Hyperspectral Remote Sensing Data using Synthetically Fused Images and Derivative Spectral Profiles

Youkyung Han,[1] Anjin Chang,[2] Seokkeun Choi,[3] Honglyun Park,[3] and Jaewan Choi[3]

[1] *School of Convergence & Fusion System Engineering, Kyungpook National University, Sangju 37224, Republic of Korea*
[2] *School of Engineering and Computing Sciences, Texas A&M University-Corpus Christi, 6300 Ocean Dr., Corpus Christi, TX 78412, USA*
[3] *School of Civil Engineering, Chungbuk National University, 1, Chungdae-ro, Seowon-gu, Cheongju, Chungbuk 28644, Republic of Korea*

Correspondence should be addressed to Jaewan Choi; jaewanchoi@chungbuk.ac.kr

Academic Editor: Hyung-Sup Jung

Multitemporal hyperspectral remote sensing data have the potential to detect altered areas on the earth's surface. However, dissimilar radiometric and geometric properties between the multitemporal data due to the acquisition time or position of the sensors should be resolved to enable hyperspectral imagery for detecting changes in natural and human-impacted areas. In addition, data noise in the hyperspectral imagery spectrum decreases the change-detection accuracy when general change-detection algorithms are applied to hyperspectral images. To address these problems, we present an unsupervised change-detection algorithm based on statistical analyses of spectral profiles; the profiles are generated from a synthetic image fusion method for multitemporal hyperspectral images. This method aims to minimize the noise between the spectra corresponding to the locations of identical positions by increasing the change-detection rate and decreasing the false-alarm rate without reducing the dimensionality of the original hyperspectral data. Using a quantitative comparison of an actual dataset acquired by airborne hyperspectral sensors, we demonstrate that the proposed method provides superb change-detection results relative to the state-of-the-art unsupervised change-detection algorithms.

1. Introduction

Hyperspectral imaging based on spaceborne or airborne imagery has strong potential for applications in remote sensing because each pixel of the hyperspectral data, which are composed of a continuous spectral profile, includes a detailed description of the spectral features of the image. These descriptions allow for an analysis of the specific differences in the characteristics of the earth's surface. Several satellite-based hyperspectral sensors, such as the Hyperion spectrometer of the National Aeronautics and Space Administration (NASA), and various airborne sensors, including the Airborne Imaging Spectrometer for Applications (AISA), Compact Airborne Spectrographic Imager (CASI), and HyMap, are currently available. In addition, several hyperspectral sensors are planned for future launches, including the Hyperspectral Imager SUIte (HISUI), Hyperspectral Precursor of the Application Mission (PRISMA), HYPererspectral-X IMagery (HYPXIM), and the Environmental Mapping and Analysis Program (ENMAP) [1]. Thus, the development of core technologies for various applications of future hyperspectral sensor systems is necessary.

Change detection is one of the most commonly used applications in remote sensing. In general, change detection is defined as the use of remotely sensed data for the same area at different times to identify altered areas on the earth's surface. When studying human-induced or natural disasters such as floods, earthquakes, landslides, oil spills, and industrial accidents, change-detection technologies based on remotely sensed data can be effectively used to detect and estimate the extent of the damaged area. Several researchers have analyzed the existing change-detection methods for applications in the remote sensing data, including image differencing, image rationing, vegetation index, regression, data transformation, change vector, postclassification comparison, and GIS-based methods [2–5]. According to the literature, change-detection techniques comprise two types: supervised change detection and unsupervised change detection [5]. In particular, unsupervised change-detection techniques are preferred for remote sensing applications because the training data do not need to be manually collected, even though supervised methods can provide change aspects of specific land-cover classes and "from-to" information of change detection [3].

Traditional unsupervised change-detection algorithms are focused on proposing change-detection indices and applying automatic thresholding methods for generating a binary map of the changed area. For example, change-detection indices such as change vector analysis (CVA), image correlation, and *Relative Dimensionless Global Error* (ERGAS) have been proposed [6, 7]. In addition, various thresholding techniques have been developed and applied to multitemporal remote sensing data [8]. Molina et al. [9] integrated different change-detection indices to improve the statistical properties of indices for single change detection. Bruzzone and Prieto proposed modified image differencing and thresholding of adaptive parcels based on a homogeneous region [10]. Markov random fields (MRFs), expectation-maximization (EM) algorithms, and neural networks have been used to determine optical thresholds for automatic change detection [11, 12]. In addition, Bruzzone and Bovolo [13] analyzed change-detection techniques for very-high-resolution (VHR) satellite imagery and developed a novel framework using the top-down approach considering the radiometric characteristics of multitemporal images.

With the launches and high availability of hyperspectral sensors, new change-detection methodologies have been developed, and existing methods have been modified to make them appropriate for hyperspectral imagery. Eismann et al. [14] proposed an algorithm based on linear predictors to detect subtle targets against a complex background, and Kim [15] modified matched filtering using target signal exclusion. Song et al. [16] proposed unsupervised change-detection algorithms using spectral unmixing and iterative error analysis (IEA). Hao et al. [17] applied hyperspectral data to detect changes in urban forest resources in natural disaster zones. Image transformation techniques, such as multivariate alteration detection (MAD) and principal component analysis (PCA), which can be applied to general multispectral images have been extended to include hyperspectral data [18, 19]. In particular, iterative regularized multivariate alteration detection (IR-MAD) is considered a state-of-the-art change-detection algorithm due to its stability and outstanding change-detection results [20–22]. However, dimensionality reduction through PCA or the minimum noise fraction (MNF) is required prior to the application of IR-MAD-based algorithms. Meola et al. [23] proposed a model-based change-detection approach that reduces false alarms caused by shadow differences using calibrated hyperspectral data, and Liu et al. [24] proposed a hierarchical change-detection algorithm using endmember detection and cluster merging in multitemporal hyperspectral images. Júnior et al. [25] used CVA based on distance (Euclidean distance and Mahalanobis distance) and similarity (spectral angle mapper, SAM, and spectral correlation mapper, SCM) measurements and concluded that CVA using the Euclidean distance and SAM is superior to other measures. Wu et al. [26] developed a subspace-based algorithm using two types of information from Hyperion and from the Chinese HJ-1A satellite, proving that measurements based on the SAM are identical to the orthogonal subspace projection- (OSP-) based measurements. Based on this concept, the authors have proposed local and adaptive measures based on OSP to minimize registration errors. Other advanced algorithms based on target detection, anomaly detection, and change detection have also been developed recently [27–29].

These related approaches have generally focused on the development of change-detection algorithms that increase accuracy by effectively using hyperspectral bands to maximize the advantages of hyperspectral sensors. In other words, these approaches have focused on using abundant spectral information. However, the dissimilarity of radiometric and geometric properties between multitemporal data based on the acquisition time or position of the sensors should be resolved to enable hyperspectral imagery for detecting changes in natural and human-impact areas. Moreover, few considerations related to data noise have been included in the spectrum of hyperspectral imagery, potentially decreasing the change-detection accuracy when general change-detection algorithms are applied to hyperspectral images without dimensionality reduction.

Most studies have also considered hyperspectral data with low or medium spatial resolutions due to the technical limitations of generating hyperspectral sensors with high spatial resolutions. However, considering the future development of hyperspectral sensors, hyperspectral data with high spatial resolutions are necessary. Thus, approaches for change detection should consider high spectral and spatial resolutions simultaneously. High-resolution images acquired at various times demonstrate local geometric location differences and spectral variation by phenomenological or temporal differences between dates of imagery, even when the data are georeferenced. These geometric differences can cause poor change-detection results based on false alarms of the changed area caused by the differences [30]. Information exclusively obtained from the data acquired at two times, that is, before and after the change occurs, is insufficient for accurate change detection.

The main objective of this study is to develop a change-detection index suitable for hyperspectral imagery. We applied unsupervised change detection using an image fusion technique and an ordered combination of similarity measures to identify the changed regions. To minimize the radiometric and geometric dissimilarities between multitemporal images with high spatial and spectral resolutions, a temporal-data-based synthetic image fusion method is applied which generates combined images at two data points at different times. A combination of the synthetically fused images is then used

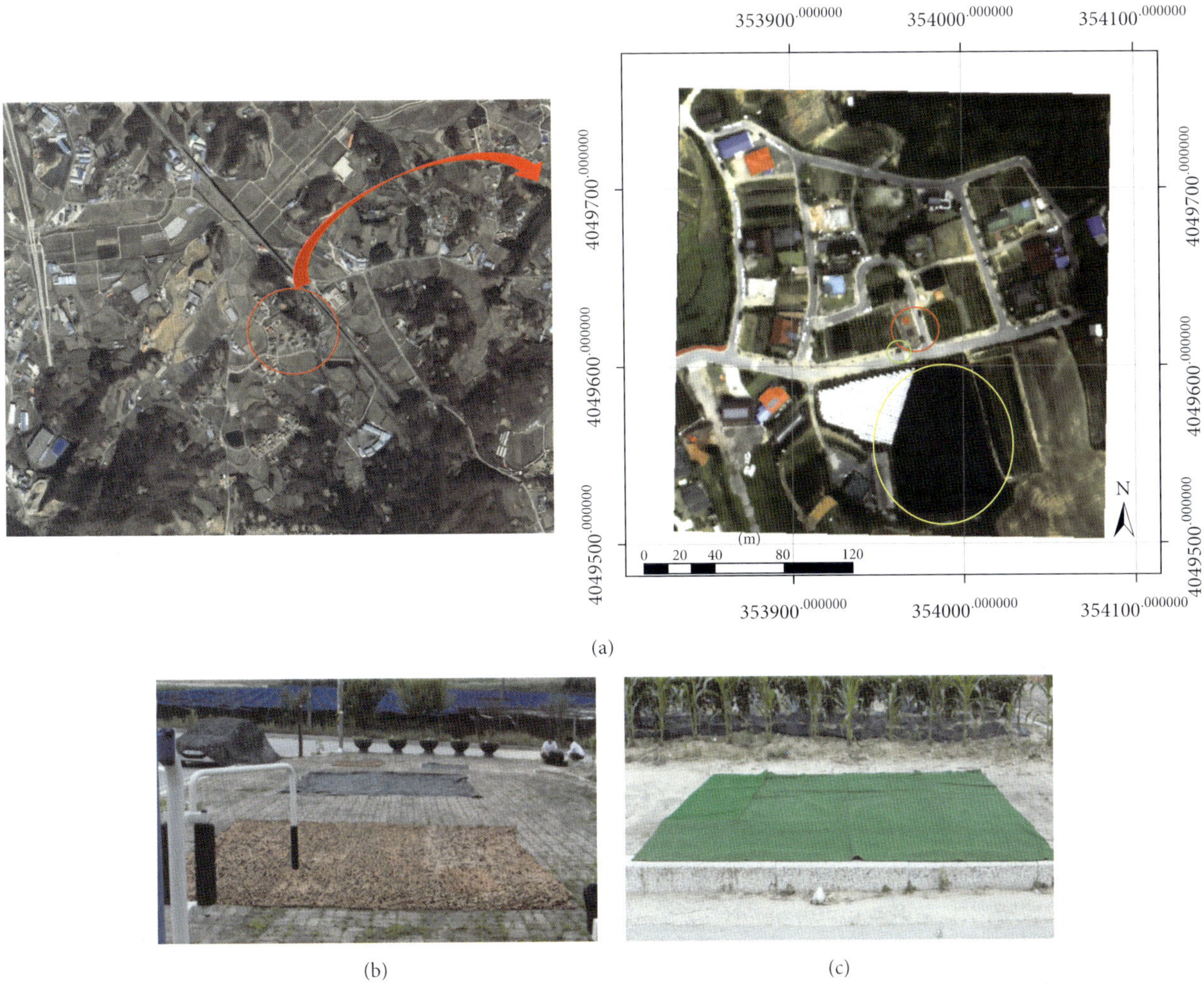

Figure 1: Study site with targets for change detection: (a) CASI image of study site generated with an RGB composite (red: 657 nm, green: 557 nm, and blue: 471 nm), (b) example of the area changed by camouflage nets (red circle of Figure 1(a)), and (c) example of the area changed by artificial turf (green circle of Figure 1(a)).

to optimize the change-detection results. We propose a new measure for change detection that is appropriate for hyperspectral data without the loss of dimensionality. To evaluate the proposed method, we install a test-bed acquired from an airborne hyperspectral sensor taken at different times. Finally, the obtained experimental results are compared with the performance of the state-of-the-art algorithms.

Table 1: CASI data specifications.

Sensor	CASI
Spatial resolution	0.5 m
Bands	48
Wavelength	370–1044 nm
Spectral resolution (nm @ FWHM)	3.6
Flight height	750 m

2. Study Site and Dataset Description

The proposed change-detection algorithm was applied to airborne hyperspectral images collected by CASI. The CASI images for this experiment were collected over Gangnae-myeon, Cheongwon-gun, and Chungcheongbuk-do, Korea, on 22 June 2013 (127° 22′ E, 36° 34′ N). The data specifications are described in Table 1.

A total of 45 spectral bands with spectral wavelength of 413.4–1044.9 nm, excluding noise bands, were used, and a subset of 500 × 500 image sizes was extracted for the experiments. The constructed study site is shown in Figure 1(a). To artificially construct the changed areas within the study site, we installed two types of targets: camouflages (4.5 m × 4.5 m, 1.5 m × 1.5 m, and 0.5 m × 0.5 m) and an artificial turf (4.5 m × 4.5 m) that have colors similar to the vegetation and ground in the study site (Figures 1(b) and 1(c)). In addition, we covered a car using a camouflage net, as shown in the left side of Figure 1(b). The ginseng field (yellow circle of Figure 1(a)) includes an inclined canvas, giving rise to different reflectance values depending on the position of the airborne hyperspectral sensor.

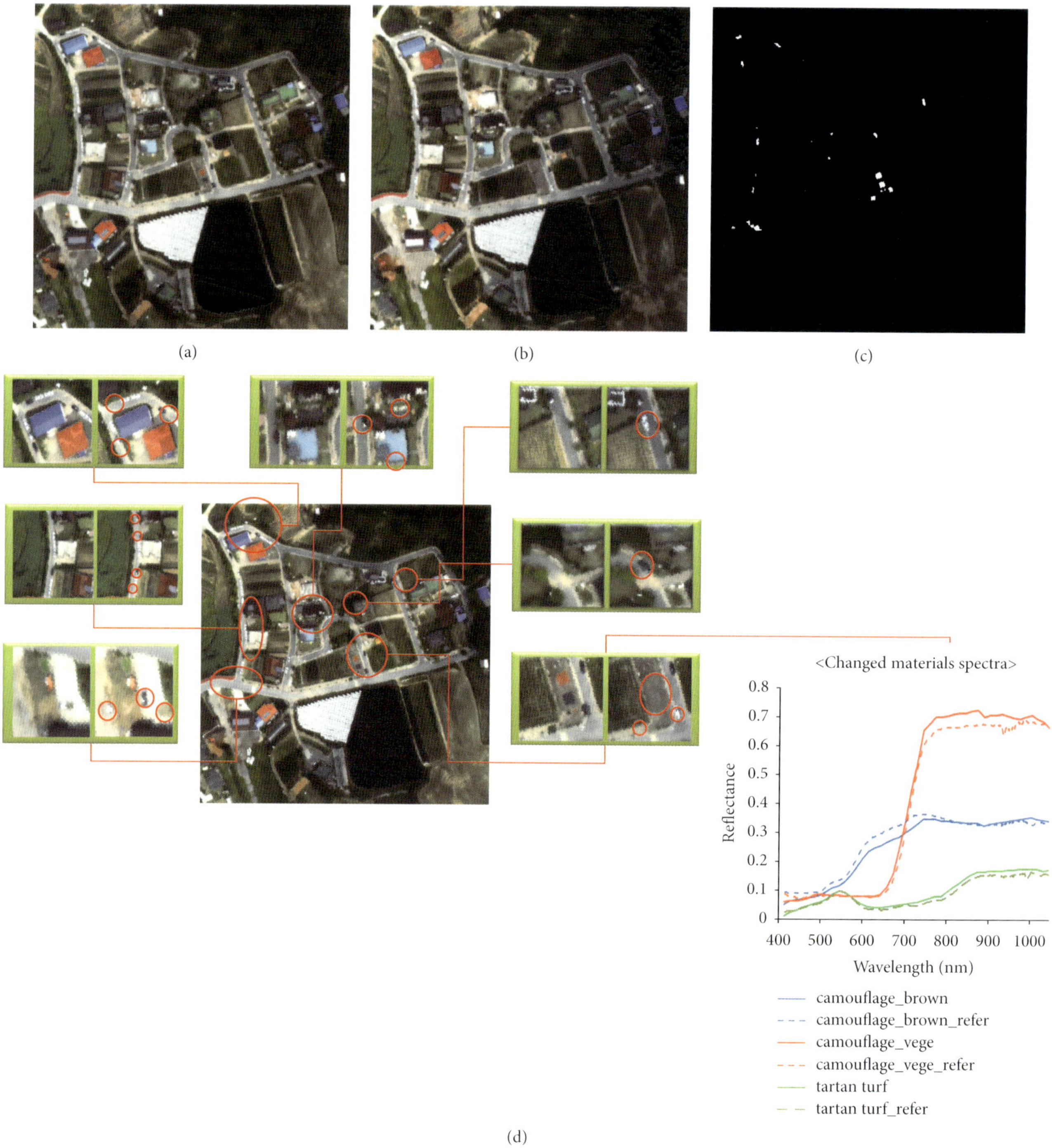

FIGURE 2: Study area: (a) reference data, (b) target data, (c) ground-truth data, and (d) specification of changed area in the study area.

Figures 2(a) and 2(b) show the reference and target data for the experiment, and Figures 2(c) and 2(d) show the ground-truth data of the altered regions for the evaluation of the accuracy using image interpretation [31].

3. Methodology

The workflow of our method is illustrated in Figure 3. Following geometric and radiometric preprocessing of hyperspectral data acquired at different times with deliberate differences, synthetic image fusion is applied. Then, a spectral similarity is calculated using the fused images. Finally, the changed area is detected by the proposed unsupervised similarity measure. A detailed explanation of each step is provided below.

3.1. Preprocessing. Prior to change detection, geometric and radiometric corrections must be applied to the multitemporal

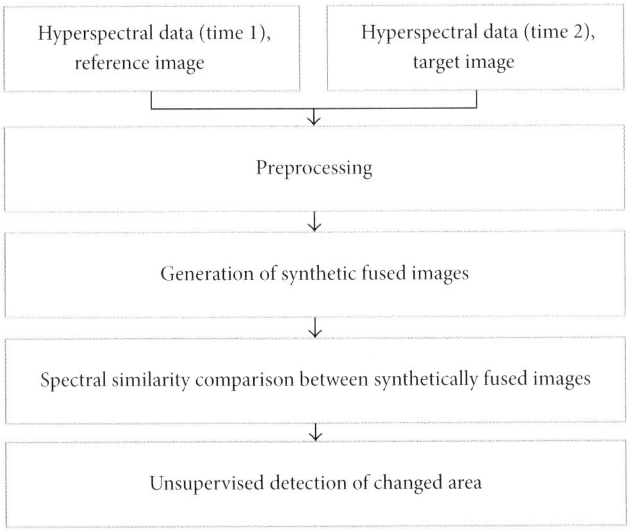

FIGURE 3: Workflow of the proposed method.

images. In our proposed method, image-to-image registration using a manually extracted ground control point (GCP) is used to obtain georeferenced images. Registered hyperspectral images are atmospherically corrected using the atmospheric and topographic correction (ATCOR) module and then radiometrically adjusted using an empirical line-calibration algorithm based on linear predictions [32]. Four tarps with different reflectance (3.5%, 23%, 35%, and 53%) are used as reference targets for the radiometric calibration. During the preprocessing step, spectral information for the same location may include identical characteristics even though partial noise data are presented.

3.2. Generation of Synthetically Fused Images. Synthetically fused images are generated based on the cross-sharpening methodology that was first developed for VHR satellite-based multitemporal data to minimize spatial dissimilarities while preserving spectral distortion [33]. A general pansharpening algorithm for VHR images is applied to obtain a high-spatial-resolution multispectral image using high-spatial-resolution panchromatic and low-spatial-resolution multispectral images acquired at the same time and at the same sensor position. The cross-sharpening algorithm is a modified version of the algorithm of Chen et al. [33, 34]. Cross-sharpened images are effective for minimizing change-detection errors caused by geometric displacement and spectral variation, including the noise in multitemporal images [33]. The cross-sharpening algorithm uses panchromatic and multispectral datasets acquired at different times and at different sensor positions. However, general airborne hyperspectral datasets do not include high-spatial-resolution panchromatic data. Therefore, to apply the cross-sharpening algorithm to the hyperspectral dataset, we generate a synthetic dataset with high-spatial-resolution multispectral images and low-spatial-resolution hyperspectral images and then employ a block-based fusion algorithm for sharpening the hyperspectral images using the multispectral images. Initially, the hyperspectral imagery obtained from CASI sensor is grouped in three blocks: 413.4–585.9 nm (corresponding to bands 1–13 of the hyperspectral images from CASI), 600.3–686.4 nm (corresponding to bands 14–20 of the hyperspectral images from CASI), and 700.7–1044.9 nm (corresponding to bands 21–45 of the hyperspectral images from CASI). Then, we generate a synthetic dataset with high-spatial-resolution multispectral images and low-spatial-resolution hyperspectral images corresponding to each block [35, 36]. Here, $HS^h_{t_1}$ and $HS^h_{t_2}$ are the original hyperspectral images with high spatial resolutions before the change (time t_1) and after the change (time t_2), respectively. The workflow for the synthetic dataset generation is shown in Figure 4.

The multispectral data with high spatial resolution ($MS^h_{t_1}$) at time t_1 are obtained by spectrally degrading $HS^h_{t_1}$ by averaging all bands in each block of the hyperspectral image [37]. Then, hyperspectral data with low spatial resolution ($HS^l_{t_1}$) are created by downscaling $HS^h_{t_1}$ via a Gaussian point spread function (PSF) [38]. In the downscaling process, the original hyperspectral data with 0.5 m spatial resolution are downscaled to 1 m spatial resolution. By applying this process to $HS^h_{t_2}$, we obtain two synthetic multitemporal datasets, $MS^h_{t_2}$ and $HS^l_{t_2}$. After the synthetic multitemporal images are generated, a cross-sharpening algorithm is applied. The cross-sharpening method is defined as special image fusion that is accomplished using hyperspectral and multispectral images obtained at the same or at different times using [35]

$$f\left(MS^h_{t_1}, HS^l_{t_1}\right) = F^h_{t_1 t_1}, \quad (1)$$

$$f\left(MS^h_{t_2}, HS^l_{t_2}\right) = F^h_{t_2 t_2}, \quad (2)$$

$$f\left(MS^h_{t_1}, HS^l_{t_2}\right) = F^h_{t_1 t_2}, \quad (3)$$

$$f\left(MS^h_{t_2}, HS^l_{t_1}\right) = F^h_{t_2 t_1}, \quad (4)$$

where f is the specific fusion algorithm and F^h_{mn} is the synthetically fused image with high spatial resolution generated using the multispectral image at time m and the hyperspectral image at time n. To generate the cross-sharpened hyperspectral image, we employ a block-based fusion algorithm using a multispectral band corresponding to the range of wavelengths of hyperspectral bands [35, 36, 39]. In contrast to typical fusion processing with multispectral and panchromatic data, block-based fusion algorithm considers the multispectral bands as a panchromatic band and hyperspectral bands as multispectral bands. Each multispectral band is regarded as the panchromatic image, and the corresponding hyperspectral bands are considered the multispectral dataset. As mentioned above, we divide the wavelength range into three blocks. Accordingly, the multispectral image $MS^h_{t_1}$ is composed of three bands, $MS^{h(1)}_{t_1}$, $MS^{h(2)}_{t_1}$, and $MS^{h(3)}_{t_1}$, and each band of $MS^h_{t_1}$ is regarded as a panchromatic image for pansharpening. The hyperspectral image $HS^l_{t_1}$ is also partitioned as three hyperspectral images $HS^{l(1)}_{t_1}$, $HS^{l(2)}_{t_1}$, and $HS^{l(3)}_{t_1}$,

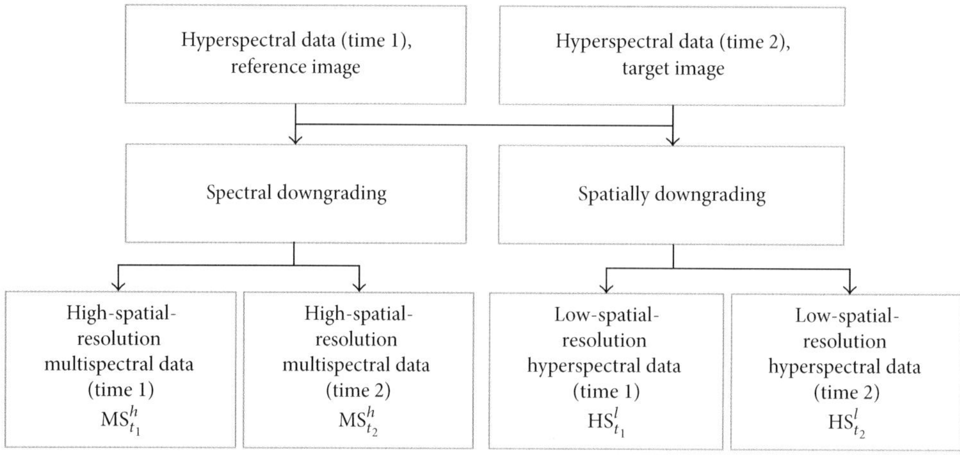

FIGURE 4: Workflow of the synthetic multitemporal dataset generation corresponding to each block of hyperspectral imagery.

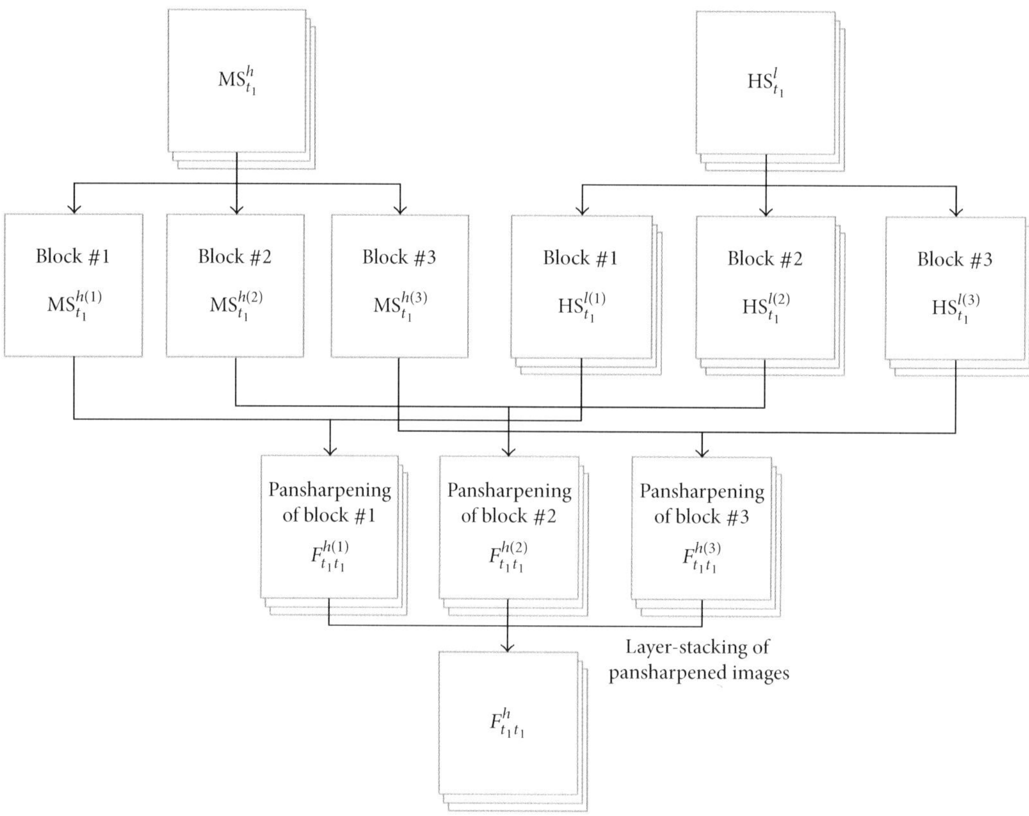

FIGURE 5: Workflow of the block-based pansharpening algorithm.

corresponding to the wavelength of each band of $MS^h_{t_1}$. The pansharpening algorithm is applied using generated panchromatic and partitioned hyperspectral images. After applying the block-based pansharpening approach, fused multispectral bands of each block are integrated as $F^h_{t_1t_1}$. Figure 5 represents the workflow of block-based fusion using multispectral bands $MS^h_{t_1}$ and hyperspectral bands $HS^l_{t_1}$.

In the case of the pansharpening algorithm, we use the Gram-Schmidt (GS) method, which is a representative and efficient pansharpening algorithm [40]. In (1)–(4), the spatial resolution, the number of bands, and spectral wavelength of the synthetically fused images are equivalent to those of the original hyperspectral image. In addition, fused images corresponding to the same multispectral images have similar spatial characteristics, indicating that some of the geometric errors in the change detection can be minimized using a pair of fused images with similar spatial information. Figure 6 shows an example in which the geometric error is minimized in the synthetically fused images. The spatial characteristics

(a) (b) (c)

FIGURE 6: Example of an image fusion result using synthetic hyperspectral images: (a) $F^h_{t_1t_1}$, (b) $F^h_{t_2t_2}$, and (c) $F^h_{t_1t_2}$.

of $F^h_{t_1t_1}$ in the tiled roof of the house are more similar to $F^h_{t_1t_2}$ than to $F^h_{t_2t_2}$.

3.3. Spectral Similarity Measure for Detecting Change. In this study, two representative normalized spectral measures are used. First, the spectral angle distance (SAD) measure is used to calculate the spectral similarity between the spectra of two pixels in the original multitemporal hyperspectral images [41]. We assume that $x = (x_1, x_2, \ldots, x_L)$ is a spectrum of the reference data and that $y = (y_1, y_2, \ldots, y_L)$ is a spectrum of the target data in a hyperspectral dataset with L bands. According to these spectral vectors, SAD is calculated as follows:

$$\mathrm{SAD}(x, y) = \cos^{-1}\left(\frac{\langle x, y\rangle}{\|x\|\,\|y\|}\right), \tag{5}$$

where $\langle x, y\rangle$ represents the inner product between the two spectrum vectors and $\|x\|$ and $\|y\|$ represent the magnitudes of spectra x and y, respectively. The SAD range is [0-1].

Having obtained the SAD measure, the Euclidean distance (ED) is also used to measure the spectral dissimilarity between the pixels. The ED, a representative similarity distance measure, is defined as the square root of the sum of the squared differences between the corresponding spectra. The ED between two spectra is expressed as

$$\mathrm{ED}(x, y) = \left[\frac{1}{L-1}\sum_{i=1}^{L}(x_i - y_i)^2\right]^{1/2}. \tag{6}$$

The ED is mathematically simple, conducive for rapid processing, and sensitive to the absolute reflectance values. In addition, the values are not normalized. Therefore, the values of ED between the two spectra are normalized by the magnitude of the spectral vector as follows:

$$\widehat{\mathrm{ED}} = \frac{\mathrm{ED} - \mathrm{ED}_{\min}}{\mathrm{ED}_{\max} - \mathrm{ED}_{\min}}, \tag{7}$$

where $\mathrm{ED}_{\max}$ and $\mathrm{ED}_{\min}$ are the maximum and minimum values of the calculated ED, respectively. According to (7), the ED value can be normalized as [0-1].

3.4. Unsupervised Change Detection Using a Derivative Spectral Profile. When change detection between $F^h_{t_1t_1}$ and $F^h_{t_2t_2}$ of (1)-(2) is applied based on SAD and ED measures, some unchanged areas may be identified as changed areas given the dissimilarity of the geometric and radiometric characteristics of the features. However, as noted in Section 3.2, the synthetically fused image $F^h_{t_1t_2}$ has a tendency to adhere to the spatial properties of the multispectral image at time 1 ($\mathrm{MS}^h_{t_1}$) and to adhere to the spectral properties of the hyperspectral image at time 2 ($\mathrm{HS}^l_{t_2}$). Therefore, if the fused image $F^h_{t_1t_2}$ is used with the fused image at time 1, that is, $F^h_{t_1t_1}$, for change detection, then the errors from the geometric difference between the two images are minimized while maintaining the spectral difference between the two sets of data. This characteristic is more significant in the derivative pixel spectra [42]. Figure 7 illustrates the spectral profile (Figure 7(b)) and second-derivative spectral profile (Figure 7(c)) of the pixel that is actually changed (Figure 7(a)), corresponding to a camouflage net that exists only at time 1. Comparing the second-derivative spectral profiles of fused images $F^h_{t_1t_1}$ and $F^h_{t_2t_2}$, we find that $F^h_{t_1t_2}$ exhibits a high profile variation, which is beneficial for the identification of the changed region.

Therefore, the SAD and ED measures are integrated to generate a new similarity distance (integrated similarity distance, ISD) based on the combination of the original and derivative spectral profiles of the synthetically fused images. The ISD is defined as

$$\begin{aligned}\mathrm{ISD} &= \mathrm{SAD}\left(F^h_{t_1t_1}, F^h_{t_2t_2}\right)\\ &\quad * \min\left(\widehat{\mathrm{ED}}\left(F''^h_{t_1t_1}, F''^h_{t_1t_2}\right), \widehat{\mathrm{ED}}\left(F''^h_{t_2t_2}, F''^h_{t_2t_1}\right)\right),\end{aligned} \tag{8}$$

where F'' is the second-derivative spectral profile, min is the minimum value, and $*$ is the convolution filter. If a pixel has a larger ISD value than a predefined threshold, it is determined to be a changed pixel. In the evaluation of the original spectral profile, SAD measurements between two hyperspectral datasets tend to exhibit different spectral and spatial characteristics. However, these results may include

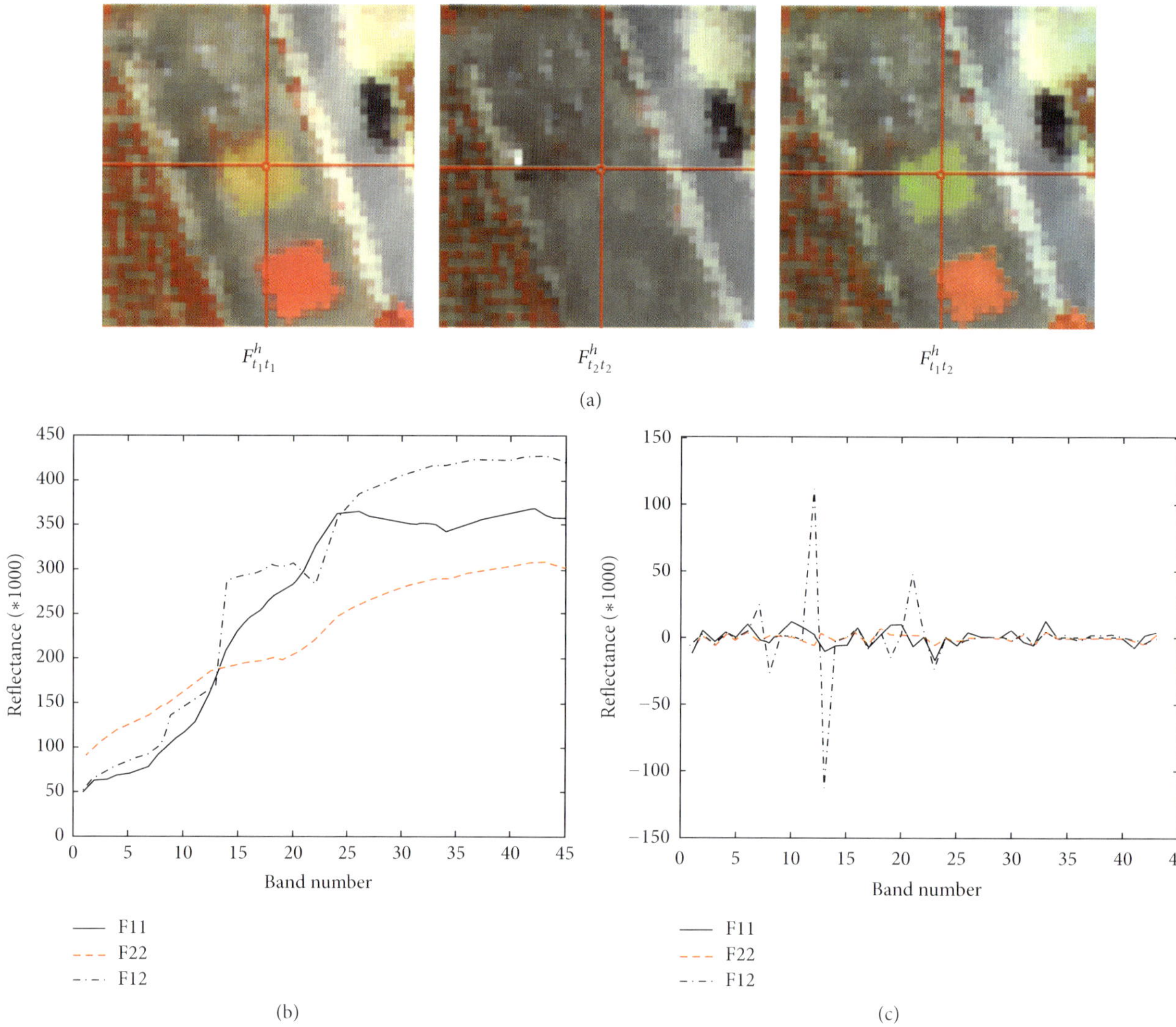

FIGURE 7: Spectral property comparison of fused images for detecting changed areas: (a) sample target of fused images $F^h_{t_1t_1}$, $F^h_{t_2t_2}$, and $F^h_{t_1t_2}$, (b) spectral profile of pixel values, and (c) spectral profile of the second-derivative pixel values (black line: $F^h_{t_1t_1}$, red dot line: $F^h_{t_2t_2}$, and blue dot line: $F^h_{t_1t_2}$).

false-positive regions of change detection. The ED of the derivative spectral profile between the synthetically fused image pair demonstrates similar spatial characteristics; thus, we calculated the ED through second-derivative spectral profiles of the hyperspectral data. In particular, the overall information acquired from the synthetic image fusion pairs $((F^h_{t_1t_1}, F^h_{t_1t_2})$ and $(F^h_{t_2t_2}, F^h_{t_2t_1}))$ is used for change detection to optimize the accuracy of the change-detection results.

4. Results and Discussion

4.1. Change-Detection Results. To estimate the performance of the proposed ISD measure, various unsupervised state-of-the-art change-detection algorithms are applied. The magnitude of the change vector (Euclidean distance of CVA or CVAED) and the spectral information divergence (SID) similarity measures are applied to detect the changed area [24, 41]. Subspace-based change-detection (SCD) algorithms, that is, original SCD, adaptive SCD (ASCD), and local SCD (LSCD), are also employed [26]. ASCD and LSCD are applied to estimate the effects of varying sensor positions or misregistration between the multitemporal hyperspectral data. Finally, IR-MAD is chosen as the change-detection algorithm for quantitative estimations because it is frequently cited as a representative unsupervised change-detection algorithm. In the case of ASCD and LSCD, a window size of 3 × 3 is selected while considering the size of the camouflage nets at the test site. To apply IR-MAD, PCA is applied as preprocessing for dimensionality reduction, and the top five principal components, including approximately 99% of the information, are used for calculating IR-MAD.

Each change-detection result map is calculated from two corresponding hyperspectral datasets and then normalized over a range of [0-1]. Figure 8 presents the calculated results

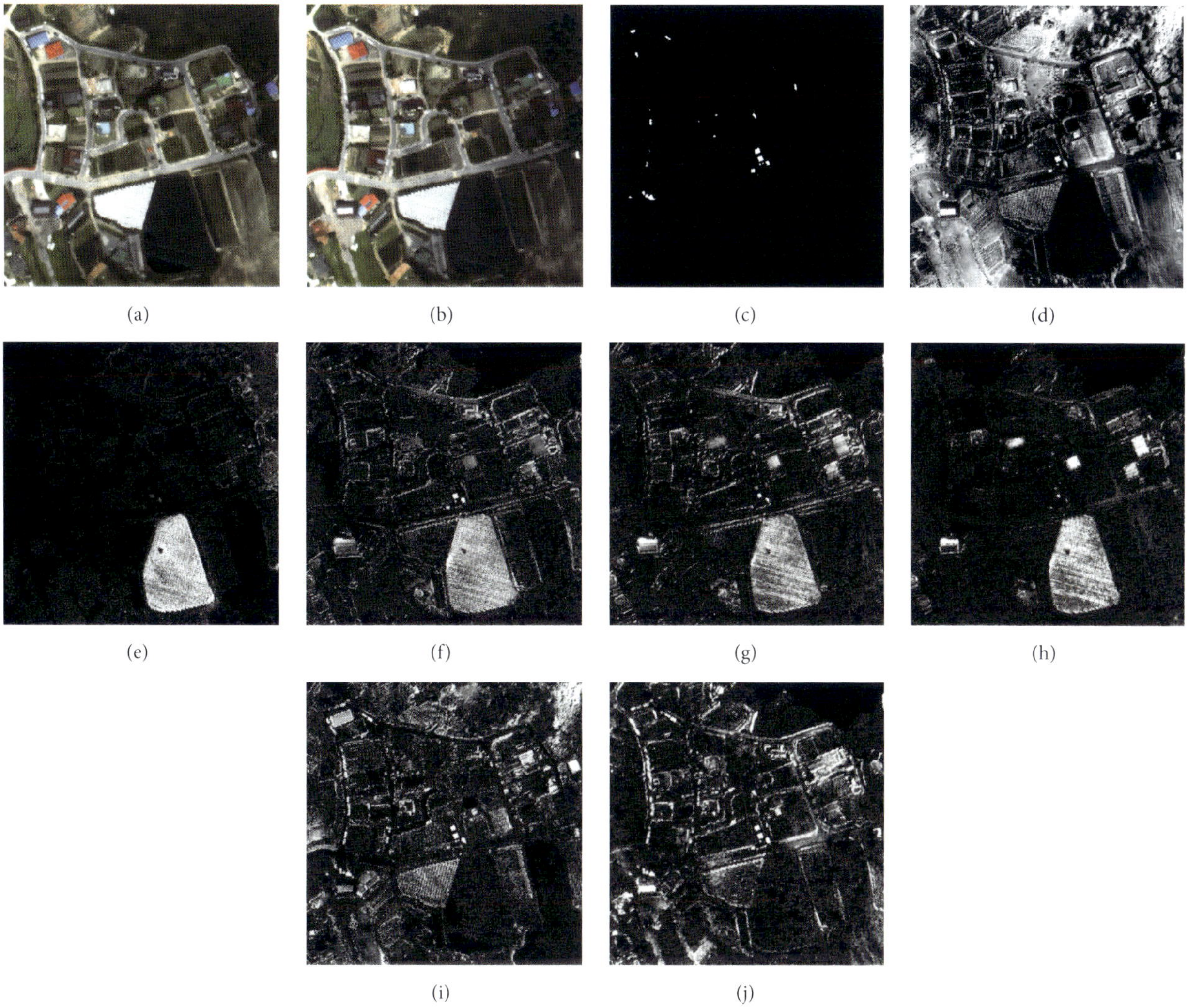

Figure 8: Comparison of the change-detection results: (a) reference hyperspectral image, (b) target hyperspectral image, (c) ground-truth image, (d) result using the CVAED, (e) result using the SID, (f) result using the SCD, (g) result using the ASCD, (h) result using the LSCD, (i) result using the IR-MAD, and (j) result using the proposed algorithm.

of the state-of-the-art and proposed methods; these are displayed using a gamma correction. The change-detection results obtained from the CVAED (Figure 8(d)) have higher values in the mountains than the other change-detection algorithms. The estimated results using SID (Figure 8(e)) and SCD-based algorithms (Figures 8(f)–8(g)) show highly dissimilar values in the ginseng field in the lower part of the site. In the case of ASCD and LSCD, some unchanged areas of edge or linear features were efficiently removed relative to the result obtained by SCD. However, small changed areas that have similar size with the sliding window represent lower values than other unchanged areas because these changed areas are blurred by the effect of local processing based on the moving window of ASCD and LSCD. In addition, the proposed algorithm (Figure 8(j)) shows more distinguishable and contrasting values over the camouflage nets and artificial turf relative to the results obtained using other change-detection methods, including the IR-MAD algorithm (Figure 8(i)), even though the result obtained by the IR-MAD shows better quality of change-detection output than that of the other existing algorithms.

The change-detection results can be quantitatively evaluated by estimating the receiver operating characteristic (ROC) curve and by evaluating the binary change map using a thresholding technique. The ROC curve provides a graphical plot for the estimating the performance and selecting an optimal model from the class distribution [26, 43, 44]. The ROC curve is composed of the cumulative distribution function of the detection rate versus that of the false-alarm rate. From the curve, an appropriate threshold for separating changed and unchanged areas can be selected. Using the ROC curve, the area under the curve (AUC) can be calculated. The AUC describes the probability that the change-detection algorithm will rank a randomly chosen positive data point higher than a randomly chosen negative data point [44]. Figure 9 shows the ROC curves corresponding to each algorithm, showing that the proposed algorithm can detect changed regions better than the other methods regardless

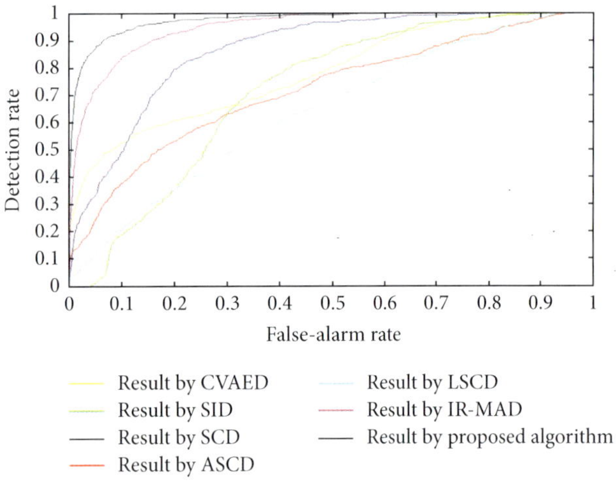

FIGURE 9: ROC curves corresponding to each similarity measure for change detection.

TABLE 2: AUC values corresponding to each similarity measure for change detection.

Method	AUC value
CVAED	0.7776
SID	0.7112
SCD	0.8628
ASCD	0.7153
LSCD	0.6385
IR-MAD	0.9562
Proposed algorithm	0.9776

of the threshold used for the unsupervised change-detection process. Poor results of ASCD and LSCD relative to the SCD result are due to the blurring effect of the process based on a sliding window, as shown in Figures 8(g) and 8(h). IR-MAD shows a higher value than the other existing algorithms but requires the use of a dimensionality reduction process. Moreover, the AUC by IR-MAD is lower than that obtained using the proposed algorithm. The efficiency of our algorithm can be proven based on its high AUC value without the use of dimensionality reduction techniques such as PCA and MNF, as shown in Table 2. The ROC analysis indicates that the proposed method can optimize the change-detection results while minimizing the false-alarm rate.

An automatic thresholding technique is applied based on the unsupervised change-detection results presented in Figure 8. The threshold determination is based on Rosin's threshold algorithm [45]. Rosin's threshold, which is a representative unimodal threshold, assumes that the histogram of the similarity distance image is a proportional unimodal function and that a nonchanged class of pixels (e.g., the background) is considerably larger than the changed class of the pixels in the image. Rosin's threshold fits a straight line from the peak of the histogram. Then, the point of the maximum deviation between the line and the histogram curve is selected as the threshold. Using the estimated threshold values, we extract the binary images corresponding to the changed area. After the thresholding, a 3-by-3 median filter is applied to consider the minimum size of the changed regions.

Figure 10 depicts the change-detection results using the existing and proposed algorithms with automatic thresholding for unsupervised change detection. Figure 11 presents the magnified results near the targets for a visual assessment. As shown in Figures 10(a) and 11(c), ground-truth data are generated, including a total of 863 changed pixels. Referring to the change-detection results using CVAED, the nearly changed regions are represented as nonchanged areas (Figure 10(b)). The change-detection results based on the SID- and SCD-based algorithms are presented in Figures 10(c)–10(f). In these results, the actual changed regions, for example, the camouflage net and the artificial turf in the middle of the site and the ginseng field in the lower part of the site, are detected as changed areas. This result is attributed to the inclined canvas on the ginseng field, which exhibits varying radiometric and spatial properties based on the off-nadir angles of the given scene. Thus, many false alarms arise in the change-detection results. In addition, the region corresponding to the brown camouflage net that has spectral properties similar to the neighboring pixels is not detected as a changed region. In the results of the IR-MAD algorithm, the artificial turf and certain building roofs (left side of Figure 10(a)) are not detected as changed areas. The result of the proposed algorithm integrated by convoluting the similarity measures between the synthetically fused images is presented in Figure 10(h). The ginseng field is considered to be an unchanged area, whereas the camouflage net and artificial turf regions are effectively detected as changed areas (Figure 11(j)) relative to the IR-MAD result (Figure 11(i)). However, in all cases, the small targets of the camouflage net are not detected due to the limited spatial resolution of the data. With the exception of the small targets, all of the changed regions are detected using the proposed algorithm, and some regions that did not actually change are also detected. These errors are attributed to the atmospheric correction and weather conditions. As shown in Figure 10(b), cloud shadows occurred on the right side of the datasets. Thus, some change-detection errors occurred based on these areas. To quantitatively evaluate the change-detection results, the detected changes, overall accuracy, completeness, correctness, and false-alarm rate are calculated based on the ground-truth data. Evaluation factors are estimated with the number of pixels, and the results are presented in Table 3. The change-detection result based on the proposed algorithm shows higher completeness than the results of the existing algorithms. Therefore, our algorithm detected the highest portion of the actually changed areas. The CVAED, IR-MAD, and proposed algorithms showed better overall accuracy, correctness, and false-alarm rates than the SID- and SCD-based algorithms. However, the high correctness and low false-alarm rates of the CVAED and IR-MAD results indicate that these algorithms could not adequately detect the changed areas, in contrast to the proposed algorithm. Therefore, we conclude that our algorithm detects changed areas efficiently and minimizes the overall error in the change detection without the use of dimensionality reduction.

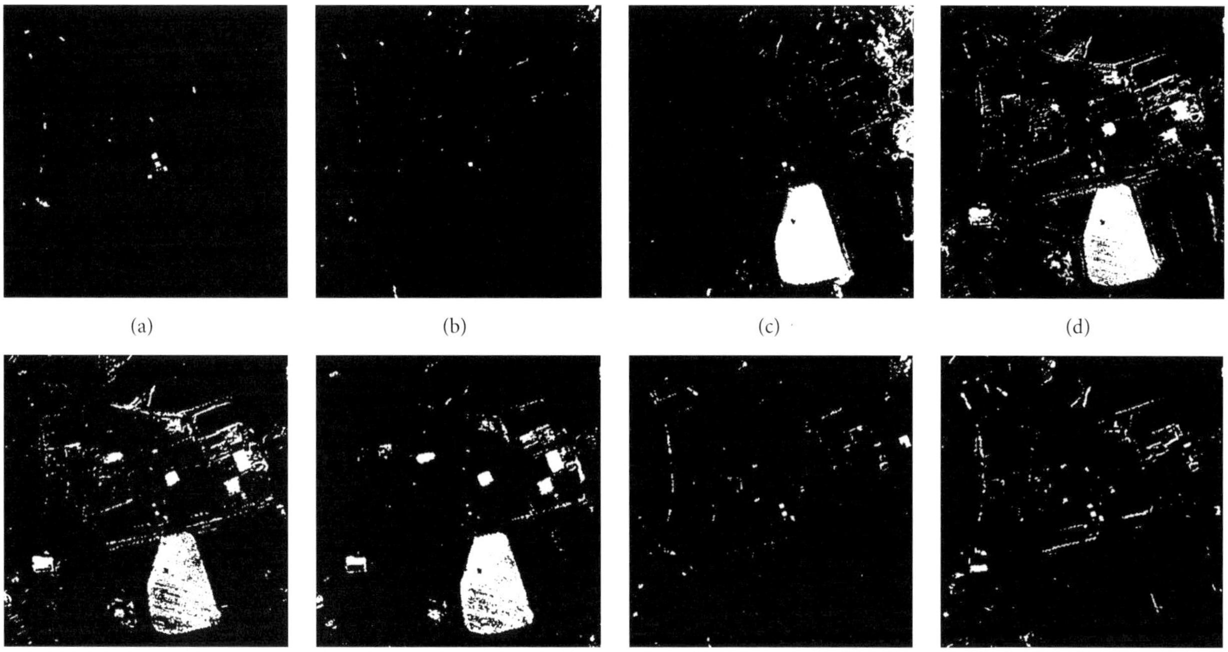

FIGURE 10: Comparison of the change-detection maps: (a) ground-truth image, (b) result using the CVAED, (c) result using the SID, (d) result using the SCD, (e) result using the ASCD, (f) result using the LSCD, (g) result using the IR-MAD, and (h) result using the proposed algorithm.

TABLE 3: Accuracy of the change-detection maps using the existing and proposed algorithms.

	Overall accuracy	Completeness	Correctness	False-alarm rate
CVAED	0.995	0.154	0.109	0.003
SID	0.905	0.184	0.005	0.093
SCD	0.899	0.538	0.012	0.100
ASCD	0.912	0.328	0.009	0.087
LSCD	0.917	0.152	0.004	0.081
IR-MAD	0.991	0.461	0.127	0.007
Proposed algorithm	0.983	0.768	0.097	0.017

5. Conclusions

In this study, we proposed a novel algorithm appropriate for detecting changes in hyperspectral data without the dimensionality reduction step that can lose specific information of the data. Synthetically fused images that preserve spectral differences while minimizing the spatial differences between the two images acquired at different times were used for the change-detection procedure. Based on the proposed similarity distance and the application of the second-derivative spectral profiles of pixels, most of the changed regions, except for certain small targets, were effectively detected. The change-detection results using our algorithm showed the highest AUC value relative to the results from the other state-of-the-art algorithms. The high AUC value indicates that the change-detection result can be optimized while minimizing the false-alarm rate. In the case of the evaluation from the binary change-detection map, the CVAED and IR-MAD methods showed higher overall accuracies of the change-detection results than that of the proposed approach. These results arose due to the higher correctness and false-alarm rates of these methods. The proposed method produced the highest completeness values, implying that the proposed algorithm detected the largest portion of truly changed areas. Therefore, we conclude that our methodology can be used to detect subtle changes in hyperspectral data acquired at different times using the properties of the derivative spectral profiles in pixels.

However, certain regions that were not changed between the two data were incorrectly detected due to the spectral dissimilarity caused by atmospheric effects and cloud shadows. Moreover, small targets were also not detected due to the limited spatial resolution of the acquired hyperspectral data. This problem will be addressed in our future work using various hyperspectral data together with modifications of the similarity distance measure.

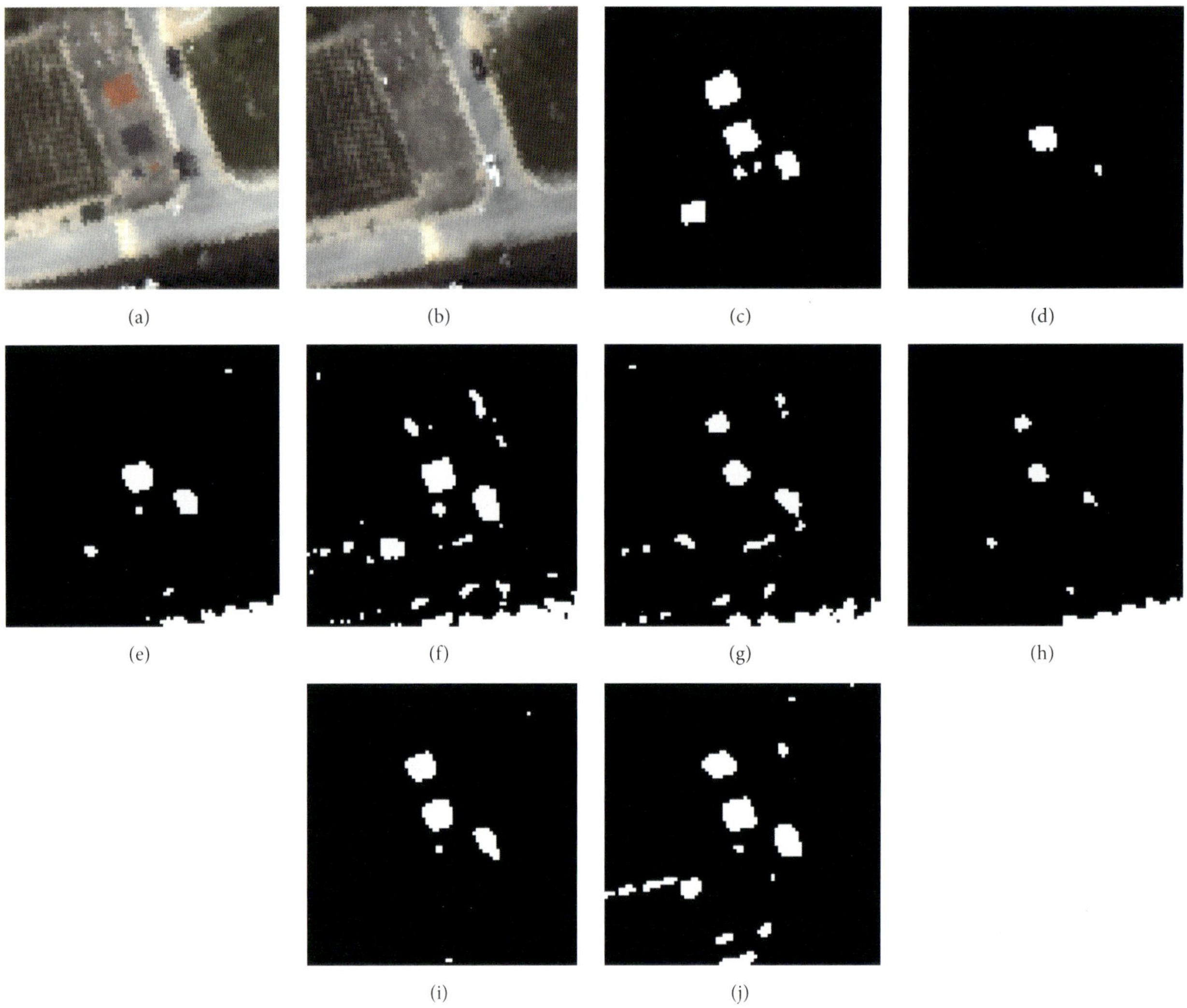

FIGURE 11: Comparison of the magnified change-detection maps: (a) reference hyperspectral image, (b) target hyperspectral image, (c) ground-truth image, (d) result using the CVAED, (e) result using the SID, (f) result using the SCD, (g) result using the ASCD, (h) result using the LSCD, (i) result using the IR-MAD, and (j) result using the proposed algorithm.

Conflicts of Interest

The authors declare that they have no conflicts of interest.

Acknowledgments

This work was supported by the Space Core Technology Development Program through the National Research Foundation of Korea (NRF) and was funded by the Ministry of Science, ICT & Future Planning (NRF-2012M1A3A3A02033469 and NRF-2014M1A3A3A03034798).

References

[1] H. Yamamoto, R. Nakamura, and S. Tsuchida, "Radiometric calibration plan for the hyperspectral imager suite (HISUI) instruments," in *Proceedings of the Multispectral, Hyperspectral, and Ultraspectral Remote Sensing Technology, Techniques and Applications IV Conference*, 85270V, Kyoto, Japan, October 2012.

[2] R. J. Radke, S. Andra, O. Al-Kofahi, and B. Roysam, "Image change detection algorithms: a systematic survey," *IEEE Transactions on Image Processing*, vol. 14, no. 3, pp. 294–307, 2005.

[3] D. Lu, P. Mausel, E. Brondízio, and E. Moran, "Change detection techniques," *International Journal of Remote Sensing*, vol. 25, no. 12, pp. 2365–2407, 2004.

[4] M. Hussain, D. Chen, A. Cheng, H. Wei, and D. Stanley, "Change detection from remotely sensed images: from pixel-based to object-based approaches," *ISPRS Journal of Photogrammetry and Remote Sensing*, vol. 80, pp. 91–106, 2013.

[5] A. Singh, "Digital change detection techniques using remotely-sensed data," *International Journal of Remote Sensing*, vol. 10, no. 6, pp. 989–1003, 1989.

[6] D. Renza, E. Martinez, and A. Arquero, "A new approach to change detection in multispectral images by means of ERGAS index," *IEEE Geoscience and Remote Sensing Letters*, vol. 10, no. 1, pp. 76–80, 2013.

[7] J. Im and J. R. Jensen, "A change detection model based on neighborhood correlation image analysis and decision tree

classification," *Remote Sensing of Environment*, vol. 99, no. 3, pp. 326–340, 2005.

[8] S. Patra, S. Ghosh, and A. Ghosh, "Histogram thresholding for unsupervised change detection of remote sensing images," *International Journal of Remote Sensing*, vol. 32, no. 21, pp. 6071–6089, 2011.

[9] I. Molina, E. Martinez, A. Arquero, G. Pajares, and J. Sanchez, "Evaluation of a change detection methodology by means of binary thresholding algorithms and informational fusion processes," *Sensors*, vol. 12, no. 3, pp. 3528–3561, 2012.

[10] L. Bruzzone and D. F. Prieto, "An adaptive parcel-based technique for unsupervised change detection," *International Journal of Remote Sensing*, vol. 21, no. 4, pp. 817–822, 2000.

[11] L. Bruzzone, "Automatic analysis of the difference image for unsupervised change detection," *IEEE Transactions on Geoscience and Remote Sensing*, vol. 38, no. 3, pp. 1171–1182, 2000.

[12] G. Pajares, "A hopfield neural network for image change detection," *IEEE Transactions on Neural Networks*, vol. 17, no. 5, pp. 1250–1264, 2006.

[13] L. Bruzzone and F. Bovolo, "A novel framework for the design of change-detection systems for very-high-resolution remote sensing images," *Proceedings of the IEEE*, vol. 101, no. 3, pp. 609–630, 2013.

[14] M. T. Eismann, J. Meola, and R. C. Hardie, "Hyperspectral Change Detection in the Presence of Diurnal and Seasonal Variations," *IEEE Transactions on Geoscience and Remote Sensing*, vol. 46, no. 1, pp. 237–249, 2008.

[15] K. Kim, "Study on Improving Hyperspectral Target Detection by Target Signal Exclusion in Matched Filtering," *Korean Journal of Remote Sensing*, vol. 31, no. 5, pp. 433–440, 2015.

[16] A. Song, J. Choi, A. Chang, and Y. Kim, "Change Detection Using Spectral Unmixing and IEA(Iterative Error Analysis) for Hyperspectral Images," *Korean Journal of Remote Sensing*, vol. 31, no. 5, pp. 361–370, 2015.

[17] Z. Hao, H.-J. Song, and B.-C. Yu, "Application of hyper spectral remote sensing for urban forestry monitoring in natural disaster zones," in *Proceedings of the 2011 International Conference on Computer and Management, CAMAN 2011*, May 2011.

[18] A. A. Nielsen, "The regularized iteratively reweighted MAD method for change detection in multi- and hyperspectral data," *IEEE Transactions on Image Processing*, vol. 16, no. 2, pp. 463–478, 2007.

[19] T. Celik, "Unsupervised change detection in satellite images using principal component analysis and κ-means clustering," *IEEE Geoscience and Remote Sensing Letters*, vol. 6, no. 4, pp. 772–776, 2009.

[20] M. J. Canty and A. A. Nielsen, "Automatic radiometric normalization of multitemporal satellite imagery with the iteratively re-weighted MAD transformation," *Remote Sensing of Environment*, vol. 112, no. 3, pp. 1025–1036, 2008.

[21] P. R. Marpu, P. Gamba, and M. J. Canty, "Improving change detection results of ir-mad by eliminating strong changes," *IEEE Geoscience and Remote Sensing Letters*, vol. 8, no. 4, pp. 799–803, 2011.

[22] M. J. Canty and A. A. Nielsen, "Linear and kernel methods for multivariate change detection," *Computers and Geosciences*, vol. 38, no. 1, pp. 107–114, 2012.

[23] J. Meola, M. T. Eismann, R. L. Moses, and J. N. Ash, "Detecting changes in hyperspectral imagery using a model-based approach," *IEEE Transactions on Geoscience and Remote Sensing*, vol. 49, no. 7, pp. 2647–2661, 2011.

[24] S. Liu, L. Bruzzone, F. Bovolo, and P. Du, "Hierarchical unsupervised change detection in multitemporal hyperspectral images," *IEEE Transactions on Geoscience and Remote Sensing*, vol. 53, no. 1, pp. 244–260, 2015.

[25] O. A. C. Júnior, R. F. Guimarães, A. R. Gillespie, N. C. Silva, and R. A. T. Gomes, "A new approach to change vector analysis using distance and similarity measures," *Remote Sensing*, vol. 3, no. 11, pp. 2473–2493, 2011.

[26] C. Wu, B. Du, and L. Zhang, "A subspace-based change detection method for hyperspectral images," *IEEE Journal of Selected Topics in Applied Earth Observations and Remote Sensing*, vol. 6, no. 2, pp. 815–830, 2013.

[27] A. Brook and E. Ben-Dor, "Advantages of the boresight effect in hyperspectral data analysis," *Remote Sensing*, vol. 3, no. 3, pp. 484–502, 2011.

[28] A. Averbuch and M. Zheludev, "Two linear unmixing algorithms to recognize targets using supervised classification and orthogonal rotation in airborne hyperspectral images," *Remote Sensing*, vol. 4, no. 2, pp. 532–560, 2012.

[29] Y. Yuan, Q. Wang, and G. Zhu, "Fast hyperspectral anomaly detection via high-order 2-d crossing filter," *IEEE Transactions on Geoscience and Remote Sensing*, vol. 53, no. 2, pp. 620–630, 2015.

[30] J. R. G. Townshend, C. Gurney, J. McManus, and C. O. Justice, "The impact of misregistration on change detection," *IEEE Transactions on Geoscience and Remote Sensing*, vol. 30, no. 5, pp. 1054–1060, 1992.

[31] "Construction and data analysis of test-bed by hyperspectral imagery," http://www.hyperspectral-testbed.com/.

[32] A. Brook and E. B. Dor, "Supervised vicarious calibration (SVC) of hyperspectral remote-sensing data," *Remote Sensing of Environment*, vol. 115, no. 6, pp. 1543–1555, 2011.

[33] B. Wang, S. Choi, Y. Byun, S. Lee, and J. Choi, "Object-based change detection of very high resolution satellite imagery using the cross-sharpening of multitemporal data," *IEEE Geoscience and Remote Sensing Letters*, vol. 12, no. 5, pp. 1151–1155, 2015.

[34] Z. Chen, J. Chen, P. Shi, and M. Tamura, "An IHS-based change detection approach for assessment of urban expansion impact on arable land loss in China," *International Journal of Remote Sensing*, vol. 24, no. 6, pp. 1353–1360, 2003.

[35] B. Wang, S.-K. Choi, Y.-K. Han, S.-K. Lee, and J.-W. Choi, "Application of IR-MAD using synthetically fused images for change detection in hyperspectral data," *Remote Sensing Letters*, vol. 6, no. 8, pp. 578–586, 2015.

[36] Y. Kim and J. Choi, "Evaluation of block-based sharpening algorithms for fusion of Hyperion and ALI imagery," *Journal of the Korean Society of Surveying, Geodesy, Photogrammetry and Cartography*, vol. 33, no. 1, pp. 63–70, 2015.

[37] M. T. Eismann and R. C. Hardie, "Hyperspectral resolution enhancement using high-resolution multispectral imagery with arbitrary response functions," *IEEE Transactions on Geoscience and Remote Sensing*, vol. 43, no. 3, pp. 455–465, 2005.

[38] N. Yokoya, T. Yairi, and A. Iwasaki, "Coupled nonnegative matrix factorization unmixing for hyperspectral and multispectral data fusion," *IEEE Transactions on Geoscience and Remote Sensing*, vol. 50, no. 2, pp. 528–537, 2012.

[39] D. Sylla, A. Minghelli-Roman, P. Blanc, A. Mangin, and O. Hembise Fanton D'Andon, "Fusion of multispectral images by extension of the pan-sharpening ARSIS method," *IEEE Journal of Selected Topics in Applied Earth Observations and Remote Sensing*, vol. 7, no. 5, pp. 1781–1791, 2014.

[40] B. Aiazzi, S. Baronti, F. Lotti, and M. Selva, "A comparison between global and context-adaptive pansharpening of multispectral images," *IEEE Geoscience and Remote Sensing Letters*, vol. 6, no. 2, pp. 302–306, 2009.

[41] F. A. Kruse, A. B. Lefkoff, J. W. Boardman et al., "The spectral image processing system (SIPS)-interactive visualization and analysis of imaging spectrometer data," *Remote Sensing of Environment*, vol. 44, no. 2-3, pp. 145–163, 1993.

[42] J. Zhang, B. Rivard, and A. Sanchez-Azofeifa, "Derivative spectral unmixing of hyperspectral data applied to mixtures of lichen and rock," *IEEE Transactions on Geoscience and Remote Sensing*, vol. 42, no. 9, pp. 1934–1940, 2004.

[43] L. C. Alatorre, R. Sánchez-Andrés, S. Cirujano, S. Beguería, and S. Sánchez-Carrillo, "Identification of mangrove areas by remote sensing: The ROC curve technique applied to the northwestern Mexico coastal zone using Landsat imagery," *Remote Sensing*, vol. 3, no. 8, pp. 1568–1583, 2011.

[44] T. Fawcett, "An introduction to ROC analysis," *Pattern Recognition Letters*, vol. 27, no. 8, pp. 861–874, 2006.

[45] P. L. Rosin and E. Ioannidis, "Evaluation of global image thresholding for change detection," *Pattern Recognition Letters*, vol. 24, no. 14, pp. 2345–2356, 2003.

Convolution Neural Network based on Two-Dimensional Spectrum for Hyperspectral Image Classification

Hongmin Gao,[1] **Shuo Lin**,[1] **Yao Yang**,[1] **Chenming Li**,[1] **and Mingxiang Yang**[2]

[1]*College of Computer and Information Engineering, Hohai University, Nanjing 211100, China*
[2]*State Key Laboratory of Simulation and Regulation of Water Cycle in River Basin, China Institute of Water Resources and Hydropower Research, Beijing 100038, China*

Correspondence should be addressed to Chenming Li; lichenming55@163.com

Academic Editor: Stelios M. Potirakis

Inherent spectral characteristics of hyperspectral image (HSI) data are determined and need to be deeply mined. A convolution neural network (CNN) model of two-dimensional spectrum (2D spectrum) is proposed based on the advantages of deep learning to extract feature and classify HSI. First of all, the traditional data processing methods which use small area pixel block or one-dimensional spectral vector as input unit bring many heterogeneous noises. The 2D-spectrum image method is proposed to solve the problem and make full use of spectral value and spatial information. Furthermore, a batch normalization algorithm (BN) is introduced to address internal covariate shifts caused by changes in the distribution of input data and expedite the training of the network. Finally, Softmax loss models are proposed to induce competition among the outputs and improve the performance of the CNN model. The HSI datasets of experiments include Indian Pines, Salinas, Kennedy Space Center (KSC), and Botswana. Experimental results show that the overall accuracies of the 2D-spectrum CNN model can reach 98.26%, 97.28%, 96.22%, and 93.64%. These results are higher than the accuracies of other traditional methods described in this paper. The proposed model can achieve high target classification accuracy and efficiency.

1. Introduction

Hyperspectral images (HSIs) are typically composed of hundreds of spectral data channels in the same scene. HSIs can provide continuous data in space and spectrum through combined imaging and spectrum technology. Hyperspectral data are important in monitoring information of the Earth's surface because the spectral information provided by the hyperspectral sensor increases the accuracy of the resolution of target materials and thus improves classification accuracy [1].

At first, scholars mainly use artificial extraction of image features for object identification classification of remote sensing images using a local binary pattern, histogram of oriented gradient [2], and Gabor filter [3]. However, this method is ineffective in processing hyperspectral data with the increase in dimension. Thus, feature extraction and classifier are combined, thereby yielding a satisfactory classification effect. Methods for feature extraction include principal component analysis (PCA) [4], independent component analysis (ICA) [5], and linear discriminant analysis (LDA) [6] and robust PCA. The classifier went through a process from fuzzy K-nearest neighbor algorithm [7], naive Bayes with deep feature weighting [8], and logistic regression [9] to support vector machine (SVM) [10]. SVM improves classification performance by extending classification kernel [11]. However, these combinatorial methods demonstrate the following significant limitations. (1) Feature extraction uses linear transformations to extract potentially useful features from the input data. Hyperspectral data are essentially nonlinear considering a complex light-scattering mechanism [12]. (2) Most traditional classification methods only consider single-layer processing, which reduces the capability of feature learning, and are unsuitable for high-dimensional data.

Neural networks (NNs) with multiple layers and hidden nodes are more suitable than shallow classifiers, such as

SVM, in building an HSI data model [13]. The NNs, including multilayer perceptron [14] and radial basis function [15], have been studied for classifying remote sensing data. Researchers have proposed a semisupervised NN framework for large-scale HSI classification [16]. Various deep NNs (DNN) have been developed according to system architecture and activation functions; these networks include deep belief network (DBN) [17], deep Boltzmann machine [18], and AutoEncoder (AE) [19]. In 2014, a stacked AE (SAE) was used for HSI classification [20]. An improved AE based on sparse constraints was then proposed [21]. The DBN is another DNN model that was proposed in 2015 [22]. The depth model can extract robust features and is superior to other methods in terms of classification accuracy.

Convolution NN (CNN) [23] uses local receptive fields in efficiently extracting spatial information and sharing weights to significantly reduce the number of parameters. CNNs are used to extract the spatial spectral features of hyperspectral images for classification [24], and their performance was better than that of traditional classifiers such as SVM. In addition, a method using a virtual sample enhanced to limited labeled samples was proposed in [25]. A previous study proposed the use of a greedy layer unsupervised pretraining to form a CNN model [26]. However, the application technology of CNN in hyperspectral classification remains imperfect, and several shortcomings, such as easy saturation of the training gradient, low classification accuracy, and poor model generalization, should be addressed.

The spectral values of the HSIs in the third dimension are approximately continuous, and the curves of each feature possess a unique spectral plot that is different from those of other classes. In the traditional classification methods, one-dimensional spectral vectors are used as the final form of input data [27, 28] or neighboring pixels are used to form small regional pixel blocks as input data [29, 30]. Although the former simplifies the complexity of deep learning network training, it omits spatial dimension information of spectral values at the same time. The latter combines multiple pixels into one sample, which introduces heterogeneous noises and aggravates the problem of missing hyperspectral data.

Compared with the traditional CNN methods, this study designs a 2D-spectrum CNN model as follows:

(i) Hyperspectral pixels have rich spectral information. The traditional data processing methods which use small area pixel block or one-dimensional spectral vector as input unit bring many heterogeneous noises. In this paper, we convert the spectral value vector to 2D-spectrum image, so that the optimization of all CNN model parameters (including the BN parameters) is based on the spectral values of the pixel points and spectral space information. The target of fully extracting spectral spatial information can be achieved while heterogeneous noises are also avoided. In addition, a multilevel BN algorithm is achieved for the first time, and the effect of network acceleration is obvious.

(ii) A BN algorithm is introduced to reduce the vanishing gradient problem and dynamically accelerate the training speed of the DNN by reducing the scaling and initialization of the dependent parameters. A small area pixel block was selected as the input unit. Liu et al. [30] used the BN algorithm to the CNN for the HIS. However, the introducing heterogeneous noises and wasting scarce samples will weaken the BN algorithm's role in network regularization and accelerated training.

(iii) Softmax loss models are used instead of combining Softmax regression and multinomial logistic loss models; thus, the output of the last layer competes with one another to improve the classification accuracy. The experimental results show that the proposed CNN-based HSI classification model exhibits high accuracy and efficiency in the HSI dataset.

2. CNN-Based Classification Model

The researchers found that the human visual system can effectively solve the problem of classification, detection, and identification, with the rapid development of modern nervous systems. This development motivates researchers on biological visual systems to establish advanced data processing methods [31]. Cells in the cortex of a human visual system are only susceptible to small areas, and accepting cells in the field can exploit the local spatial correlation in the image.

The CNN architecture uses two special methods, namely, local receptive field and shared weights. The activation value of each convolution neuron is calculated by multiplying the local input with weight W, which is shared in the entire input space (Figure 1). Neurons that belong to the same layer share the same weight. The use of specific architectures, such as local receptive field and shared weights, reduces the total number of training parameters and facilitates the development of an efficient training model.

The complete CNN architecture consists of convolution and pooling layers. The convolution layer alternates with the pooling layer, thereby mimicking the properties of complex and simple cells in the mammalian visual cortex [32]. In the CNN, the input data are a matrix or tensor with a 3D spatial structure, where (H, W), (H', W'), and (H'', W'') represent the size of the spatial dimension of input data, convolution kernel, and output data, respectively. The number of convolution kernel feature channels is represented by D, and D'' represents the 3D data.

$$\begin{aligned} x &\in \mathbb{R}^{H\times W\times D}, \\ f &\in \mathbb{R}^{H'\times W'\times D\times D''}, \\ y &\in \mathbb{R}^{H''\times W''\times D''}, \end{aligned} \tag{1}$$

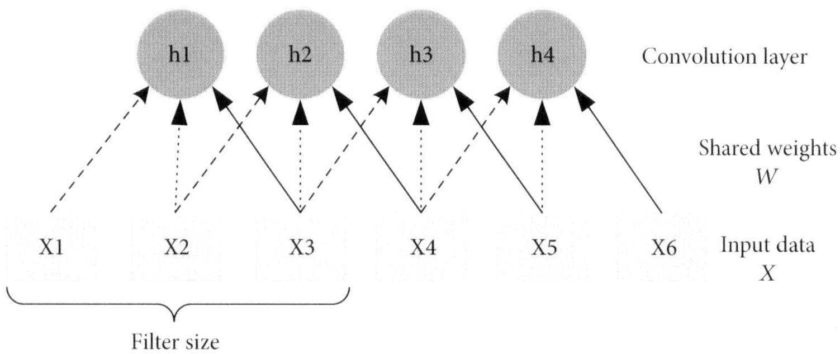

FIGURE 1: Local receptive field and shared weights.

where x is the input data, f is the convolution filter, and y is the output data. The 1D signal x is convoluted by filter f to calculate signal y as follows:

$$y_{i''j''d''} = b_{d''} + \sum_{i'=1}^{H'} \sum_{j'=1}^{W'} \sum_{d'=1}^{D} f_{i'j'd} \times x_{i''+i'-1,j''+j'-1,d',d''}, \tag{2}$$

where $b_{d''}$ is the neuron offset and $f_{i'j'd}$ is the convolution kernel matrix of the dth $i' \times j'$.

The pooling layer is typically obtained after the convolution layer. The most common pooling function is max pooling. This function calculates the maximum response of each feature channel in the $H' \times W'$ region. The feature map becomes robust to the distortion of the data and achieves a high invariance through the pooling. The pooling layer can also decrease the size of the feature map, thereby reducing computational burden.

$$y_{i''j''d''} = \max_{1 \le i' \le H', 1 \le j' \le W'} x_{i''+i'-1,j''+j'-1,d}. \tag{3}$$

The traditional CNN processing flow is designed in accordance with the data structure characteristics of the HSI classification task (Figure 2). The training data are obtained through forward propagation to determine the actual output and then compared with the real data tag competition. Stochastic gradient descent (SGD) is used to modify the synapses and parameters of the network structure, and several iterative trainings are conducted to form a network model. Test data are inputted into the network model. The output data obtained by feature extraction are matched with the actual data, and the result is classified by the competitive output model.

The 3D data of the HSIs reach hundreds, and a numerical approximation is considered continuous. The curve of each pixel possesses a unique spectral plot that differs from those of other categories, and this plot is difficult to distinguish by the human eye. However, the CNN exhibits a better performance than several human visual aspects. Therefore, this study uses spectral labels to improve the classification performance of the CNN models on the HSIs.

3. Proposed CNN-Based HSI Classification Model

3.1. BN. The SGD is used to train NN in CNN training. This method is simple and effective, but the model parameters should be carefully adjusted. In particular, the learning rate and initialization parameters of the model are added in the optimization. This step significantly reduces the speed of the CNN training. The entire network should conform to the new data distribution when the distribution of the input data at the network layer changes, thereby resulting in the decreased training speed of the network and saturated gradient. If the nonlinear input data distribution is ensured to be stable, then the probability of nonlinear saturation problem will be minimal and can accelerate the training of the network. Therefore, the BN algorithm is proposed to eliminate this phenomenon and expedite the training of the network.

In the BN algorithm, for each hidden layer of neurons, the input distribution, the value interval of which is closer to the limit saturation region through nonlinear function mapping, is forced back to the normal distribution of the comparison standard, with a mean of 0 and a variance of 1. Thus, the input value of the nonlinear transformation function falls into the region, which is sensitive to the input data, to avoid a gradient disappearance problem. The most mature technique in the early DNN normalization operation is whitening; however, whitening the input of each layer would result in excessive computational costs and computational time and not all differential. Thus, BN is used in two simplified ways.

The first step indicates that zero mean and variance normalization for scalar features are used instead of whitening. Simultaneously, the input and output of the layer are normalized. The following formula is used for NNs with d-dimensional input $x = (x^{(1)}, \dots, x^{(d)})$ to normalize each dimension, where the predicted value and variance can be calculated from the corresponding batches.

$$\widehat{x}^{(k)} = \frac{x^{(k)} - E\left[x^{(k)}\right]}{\sqrt{\mathrm{var}\left[x^{(k)}\right]}}. \tag{4}$$

The primitive activation value x that corresponds to the neuron is converted by subtracting the corresponding

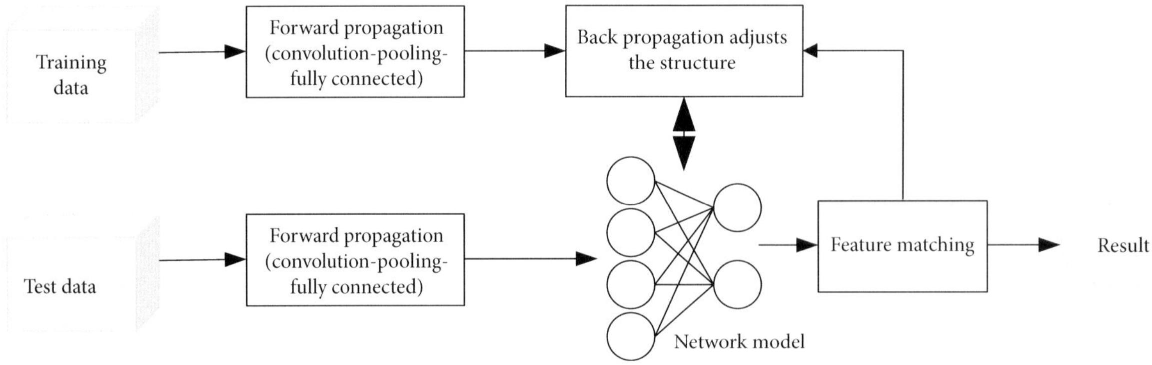

Figure 2: Traditional CNN processing flow.

batch mean E(x) and dividing it by the variance Var(x). If the normalization for input of NN layer is rather simple, then the characterization capability of the layer is reduced. Thus, the BN introduces a pair of parameters $\gamma^{(k)}$ and $\beta^{(k)}$ for each activation value $x^{(k)}$, which can scale and translate the normalized input as

$$y^{(k)} = \gamma^{(k)}\widehat{x}^{(k)} + \beta^{(k)}. \quad (5)$$

The two parameters are similar to the network parameters and are trained and modified in the same way. The characterization capability of the model is then restored. When $\gamma^{(k)} = \sqrt{\text{var}\left[x^{(k)}\right]}, \beta^{(k)} = E\left[x^{(k)}\right]$, it can obtain the original activation value based on the output and restore representation ability. That is, the network can restore the feature distribution to be learned by the original network.

The second step denotes that NN training is based on the entire training dataset. Theoretically, the entire training data-set can be used to normalize the activation value. However, it cannot be applied to the SGD because of a large amount of calculation of the dataset. The BN introduces min-batch in the SGD and calculates the corresponding mean and variance by using the minimum batch data. If a minimum batch data B exists, then the corresponding size is m. Only one of the activation values $x^{(k)}$ is considered because each dimension of the activation value is normalized, while the variable k is then ignored. Each B has m activation values. The BN algorithm can be obtained by Formula (5).

$$\text{Algorithm}_{\gamma,\beta} : x_{1\ldots m} \longrightarrow \widehat{x}_{1\ldots m} \longrightarrow y_{1\ldots m}. \quad (6)$$

The mean μ_B and variance σ_B^2 of the minimum batch data are defined as follows:

$$\begin{aligned} \mu_B &\longleftarrow \frac{1}{m}\sum_{i=1}^{m} x_i, \\ \sigma_B^2 &\longleftarrow \frac{1}{m}\sum_{i=1}^{m} (x_i - \mu_B)^2, \end{aligned} \quad (7)$$

in which $\widehat{x}$ and y can be obtained after the normalization.

$$\begin{aligned} \widehat{x}_i &\longleftarrow \frac{x_i - \mu_B}{\sqrt{\sigma_B^2 - \varepsilon}}, \\ y_i &\longleftarrow \gamma\widehat{x}_i + \beta \equiv \text{BN}_{\gamma,\beta}(x_i), \end{aligned} \quad (8)$$

where ε is a constant (tends to be 0) to ensure the stability of the calculation of variance.

The data computation becomes large and complex in applying the BN algorithm to the CNN if each neuron is normalized in every layer. Thus, this algorithm is based on the idea of weight sharing, and the mean and variance of the activation value are obtained for the entire map. Input x, output y, parameter W, and b are defined as

$$\begin{aligned} x, y &\in \mathbb{R}^{H\times W\times K\times T}, \\ \text{W} &\in \mathbb{R}^{K}, \\ b &\in \mathbb{R}^{K}, \end{aligned} \quad (9)$$

where H, W, K, and T represent the length and width of the input and output data, number of characteristic channels, and min-batch size, respectively. We explicitly define the input and output arrays as 4D data to process the feature mappings by batch. The output characteristic map is expressed by the following formula:

$$\begin{aligned} y_{ijkt} &= W_k \frac{x_{ijkt} - \mu_k}{\sqrt{\sigma_k^2 + \varepsilon}} + b_k, \\ \mu_k &= \frac{1}{HWT}\sum_{i=1}^{H}\sum_{j=1}^{W}\sum_{t=1}^{T} x_{ijkt}, \\ \sigma_k^2 &= \frac{1}{HWT}\sum_{i=1}^{H}\sum_{j=1}^{W}\sum_{t=1}^{T} \left(x_{ijkt} - \mu_k\right). \end{aligned} \quad (10)$$

The BN is performed to the activation value of each hidden layer of neurons, which can be regarded as an add-on operation layer. It is located after the activation

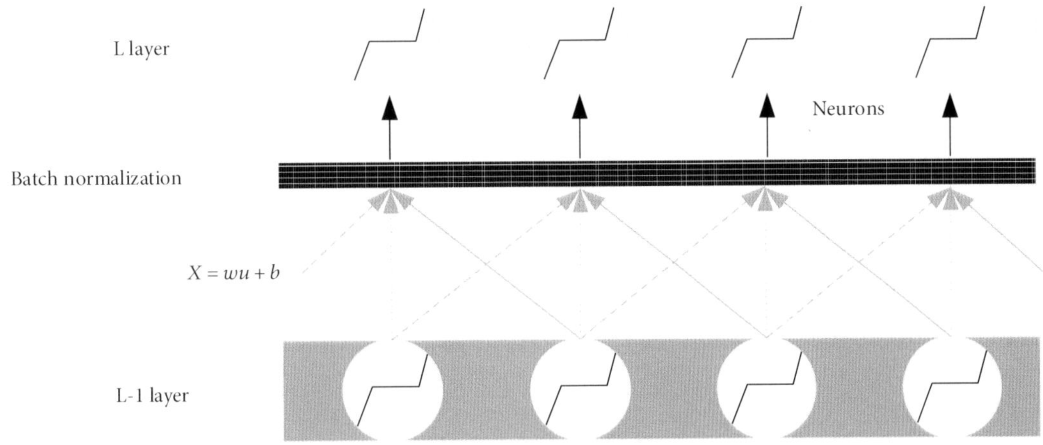

FIGURE 3: Operation layer of BN.

value $x = wu + b$ obtained and before the nonlinear activation function, as depicted in Figure 3.

The proposed BN algorithm possesses the following characteristics after the theoretical analysis and experimental verification. (1) A large initial learning rate can be selected so as to improve training speed. (2) Dropout can be removed and L2 weight attenuation coefficient can be reduced. (3) Local response normalization (LRN) can be replaced. (4) The training data can be completely disrupted (i.e., a sample is not frequently selected in each batch of training).

3.2. Softmax Loss Models. The Softmax regression model is a spread of logistic regression model in solving multiclass problems. Following the data structure of Section 4.1, it can obtain the following output of the regression function.

$$y_{ijk} = \frac{e^{x_{ijk}}}{\sum_{t=1}^{D} e^{x_{ijt}}}. \tag{11}$$

This formula is not limited by the number of feature channels and is applied to all spatial positions in a convoluted manner to translate linear predictions into categorical probabilities. The linear prediction results x_{ijk} of the kth category serve as input to the regression model, and D probability values (likelihood), which represent the possibility of the data belonging to different categories, are obtained. The Softmax regression model can be regarded as the combination of the exponential form of activation function and normalized operators. The classification loss function $\ell(x, c)$ is aimed at comparing the prediction x with the real class label c. The classification loss is defined as follows:

$$\ell(x, c) = \sum_{ijn} \omega_{ij1n} \ell\left(x_{ij:n}, c_{ij:n}\right), \tag{12}$$

where $x \in \mathbb{R}^{H\times W\times C\times N}$, $c \in \{1, \ldots, C\}^{H\times W\times 1\times N}$, and N is the size of the 3D data. The 1D vector $x_{ij:n}$ represents class C fraction, and $c_{ij:n}$ represents a real class label. Logarithmic loss function or logarithmic likelihood loss function is commonly used in logistic regression, which is based on the maximum likelihood principle. Thus, vector x represents the posterior probability $p(k) = x_k, k = 1, \ldots, C$ of the different classes. The output of the loss function is the negative logarithmic probability of the real label.

$$\begin{aligned} &\ell(x, c) = -\log x_c \cdot x \geq 0, \\ &\sum_k x_k = 1, \end{aligned} \tag{13}$$

where x is the output of the Softmax regression model and is numerically unstable. On the one hand, the score x_c should compete with other scores $x_k(k \neq c)$ to obtain a meaningful logarithmic loss. If it is not the case, then the minimization of Formula (13) can be achieved by maximizing all x_k, but the real prediction effect is that x_c is larger than $x_k(k \neq c)$. On the other hand, the Softmax regression model allows the score x to compete through the normalization factor. This study proposes the Softmax loss models, which combines the calculation module of the regression model and calculation module of logarithmic loss into a single one.

$$\ell(x, c) = -\log \frac{e^{x_c}}{\sum_{k=1}^{C} e^{x_k}} = -x_c + \log \sum_{k=1}^{C} e^{x_k}. \tag{14}$$

The combined modules result in a stable value of the output fraction x. By combining the logarithmic loss with Softmax, the loss model automatically makes the score compete $\ell(bx, c) \approx 0$, when $x_c \gg \sum_{k\neq c} x_k$. Although this model is similar to the final output result of the logarithmic loss function, the experimental results show that the Softmax loss model has the following advantages. (1) The calculation steps are minimal, while the calculated amount is small. (2) Numerical gradients are relatively stable. (3) Competitive output can improve the classification accuracy.

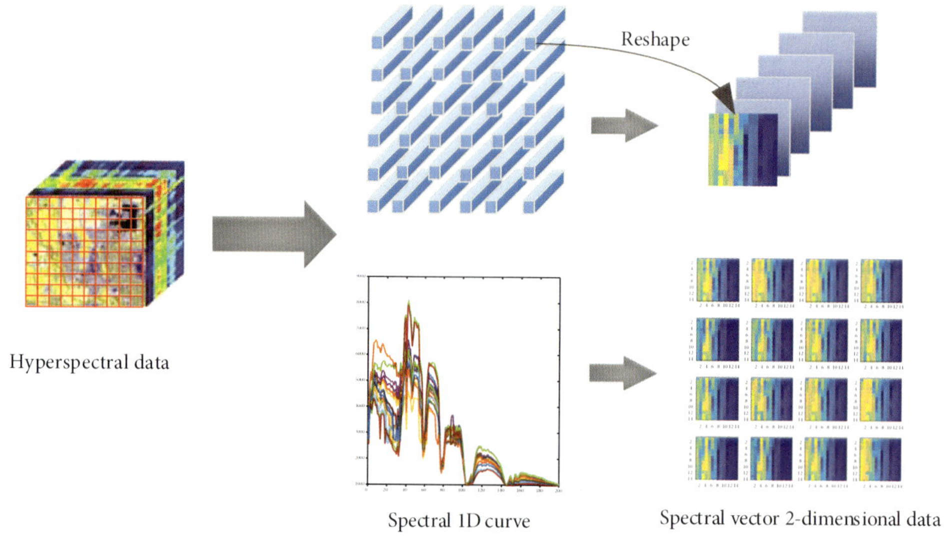

FIGURE 4: Hyperspectral image spectral vector preprocessing. Compared with the one-dimensional curve, the 2D spectrum has more obvious distinguishing features between different categories.

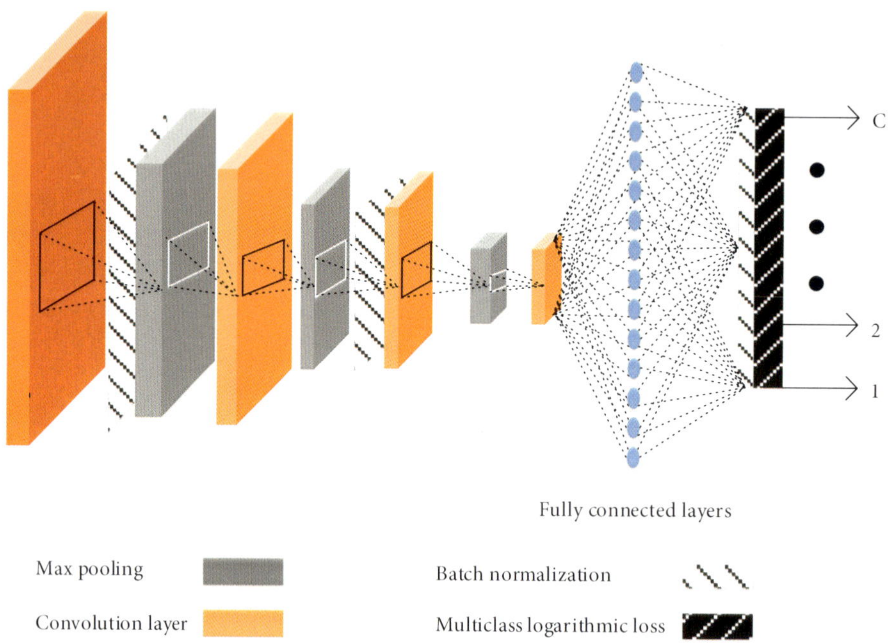

FIGURE 5: 2D-spectrum CNN structure.

3.3. Two-Dimensional Spectrum CNN Model. The hyperspectral data has a small total number of pixels, while a single pixel has rich spectral values. According to this, the 2D-spectrum method is proposed in the paper, which converts the pixel spectral vectors into two-dimensional spectral images as the input data of CNN. The convolution network can fully utilize the spatial position information between the different spectral values to improve the classification accuracy, as shown in Figure 4. The network adopts multiple convolution and max pooling layers alternately. BN layers, which regulate the data distribution and accelerate network training, are insert among the one, four, and seven layers. The Softmax loss model is then used to control the output. The 2D-spectrum CNN structure is shown in Figure 5.

4. Experimental Results

In this chapter, we first describe the dataset used in the experiment. Furthermore, the BN algorithm and Softmax loss model are used to verify the HSI classification performance of the CNN model. Finally, the proposed method is compared with other similar methods to determine the advantages and disadvantages of the 2D-spectrum CNN

TABLE 1: Hyperspectral data set features.

Dataset	Size	Spectral bands (original)	Labels	Instrument
Indian Pines	145×145	200 (224)	16	AVIRIS
Salinas	512×217	204 (224)	16	AVIRIS
KSC	512×614	176 (224)	13	NASA AVIRIS
Botswana	1476×256	145 (242)	14	NASA EO-1

(a) (b) (c)

FIGURE 6: (a) AVIRIS Infrared image, (b) Indian Pines ground-truth map, and (c) classification map.

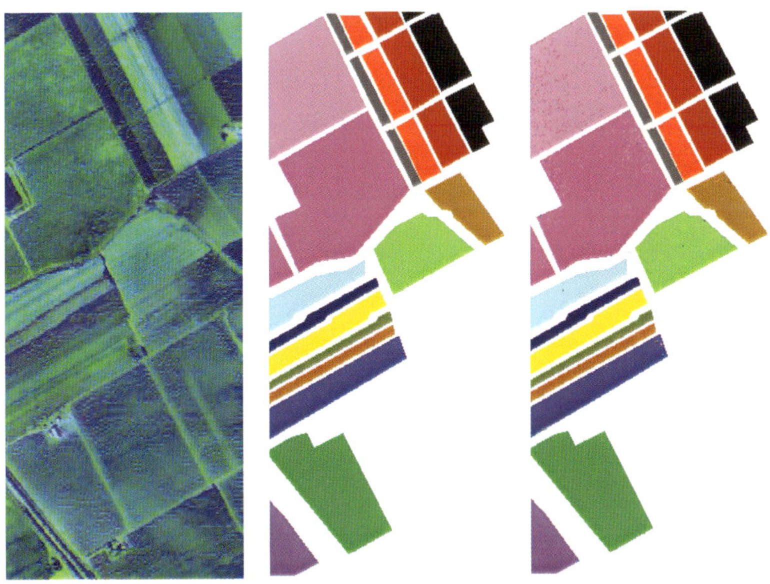
(a) (b) (c)

FIGURE 7: (a) RGB color image from random three bands, (b) Salinas ground-truth map, and (c) Salinas classification map.

model. The overall accuracy (OA) and Kappa coefficients are used as the performance measurements for each type of classification accuracy (percentage). The mathematical relationship between error and OA is

$$OA = (1 - \text{error}) \times 100\%. \quad (15)$$

Each classification result is averaged over 10 runs to avoid any deviation caused by random sampling. All of the experiments are tested on a desktop with NVIDIA GeForce GTX 1070 8G GPU, 16 GB memory, and 64-bit Windows 7 OS using MATLAB 2014b.

4.1. Hyperspectral Datasets. In the experiment, the hyperspectral data of the network model include Indian Pines, Salinas, and Kennedy Space Center (KSC) datasets and are applied to the new dataset, Botswana. Table 1 summarizes the detailed information of the four datasets.

Figures 6–8 exhibit the characteristics of the four datasets. Tables 2–5 list the number of samples trained and tested

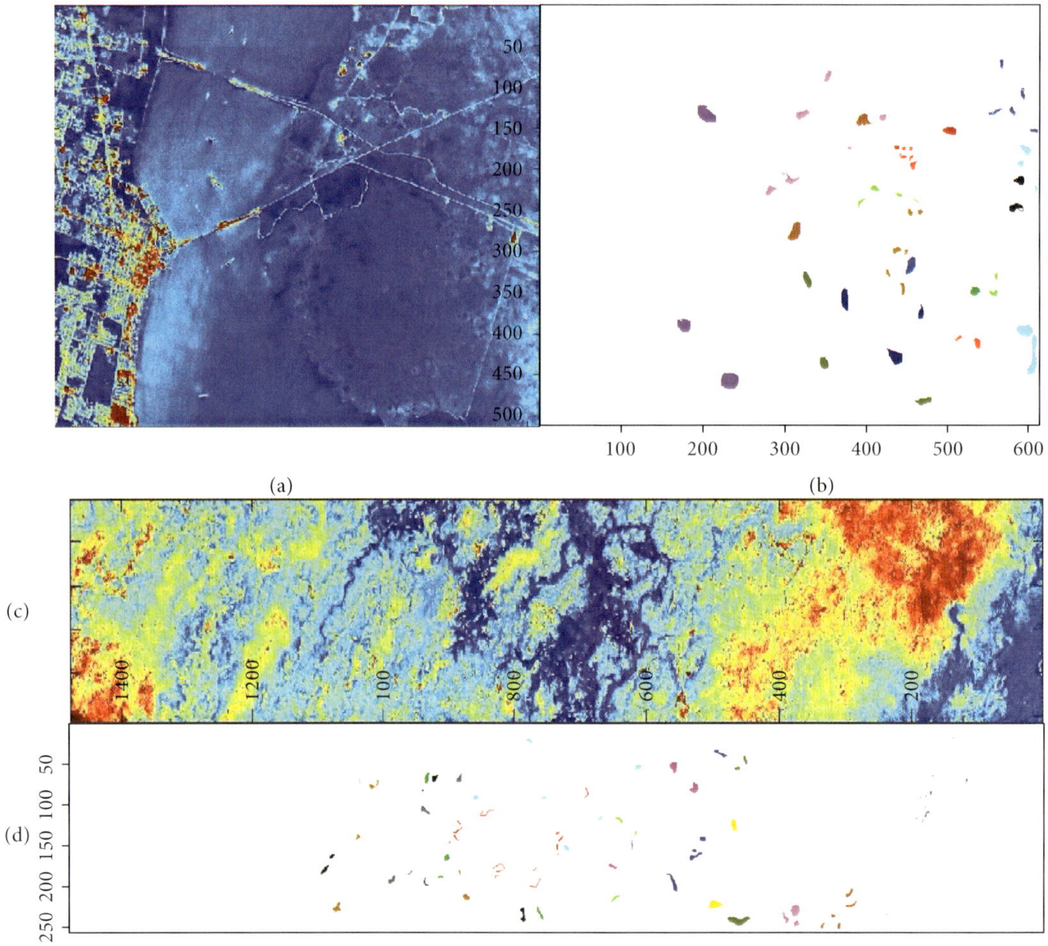

FIGURE 8: (a) KSC third band, (b) KSC ground-truth map, (c) Botswana fourth band, and (d) Botswana ground-truth map.

TABLE 2: Number of labeled training samples and testing samples of Indian Pines.

Class number	Class name	Training samples	Testing samples
1	Alfalfa	**35**	**11**
2	Corn-no till	357	1071
3	Corn-min till	207	623
4	Corn	59	178
5	Grass-pasture	120	363
6	Grass-trees	182	548
7	Grass-pasture-mowed	**21**	**7**
8	Hay-windrowed	119	359
9	Oats	**15**	**5**
10	Soybean-no till	243	729
11	Soybean-min till	613	1842
12	Soybean-clean	148	445
13	Wheat	51	154
14	Woods	316	949
15	Buildings-grass-trees-drives	96	290
16	Stone-steel-towers	**70**	**23**
Total		2652	7597

TABLE 3: Number of labeled training samples and testing samples of Salinas.

Class number	Class name	Training samples	Testing samples
1	Brocoli_green_weeds_1	503	1506
2	Brocoli_green_weeds_2	932	2794
3	Fallow	494	1482
4	Fallow_rough_plow	349	1045
5	Fallow_smooth	670	2008
6	Stubble	990	2969
7	Celery	895	2684
8	Grapes_untrained	2818	8453
9	Soil_vinyard_develop	1550	4653
10	Corn_senesced_green_weeds	820	2458
11	Lettuce_romaine_4wk	267	801
12	Lettuce_romaine_5wk	482	1445
13	Lettuce_romaine_6wk	229	687
14	Lettuce_romaine_7wk	268	802
15	Vinyard_untrained	1817	5451
16	Vinyard_vertical_trellis	452	1355
Total		13536	40593

TABLE 4: Number of labeled training samples and testing samples of KSC.

Class number	Class name	Training samples	Testing samples
1	Scrub	190	571
2	Willow swamp	60	183
3	CP hammock	64	192
4	Slash pine	63	189
5	Oak/broadleaf	40	121
6	Hardwood	58	171
7	Swamp	27	78
8	Graminoid marsh	108	323
9	Spartina marsh	130	390
10	Cattail marsh	101	303
11	Salt marsh	105	314
12	Mud flats	126	377
13	Water	232	695
Total		1304	3907

TABLE 5: Number of labeled training samples and testing samples of Botswana.

Class number	Class name	Training samples	Testing samples
1	Water	67	203
2	Hippo grass	25	76
3	Floodplain grasses 1	62	189
4	Floodplain grasses 2	53	162
5	Reeds	67	202
6	Riparian	67	202
7	Firescar	64	195
8	Island interior	50	153
9	Acacia woodlands	78	236
10	Acacia shrublands	62	186
11	Acacia grasslands	76	229
12	Short mopane	45	136
13	Mixed mopane	67	201
14	Exposed soils	23	72
Total		806	2442

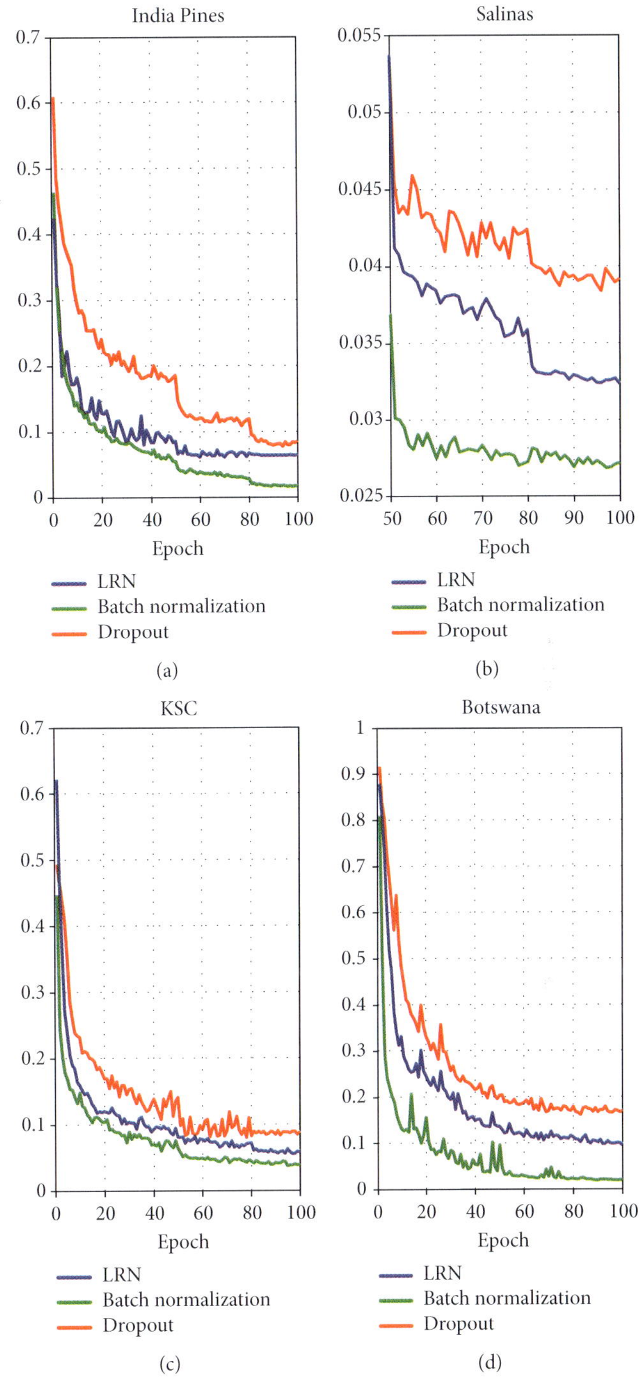

FIGURE 9: BN performance verification curve: y-axis is error.

for the corresponding dataset and are assigned according to (train) 1 : (test) 3. According to previous experiments, a considerably minimal training data will produce an underfitting phenomenon. The labels 1, 7, 9, and 16 of the Indian Pines datasets are insufficient. The above labels allocate the total amount of samples according to 3 (train) : 1 (test) to avoid underfitting and to ensure that the training sample is sufficient.

The learning rate of the CNN network model is [1e-03 5e-04 1e-05], the corresponding number of training iterations is 50, 30, and 20, and the number of samples per batch is 100.

4.2. BN Performance Verification. The BN operation layer is added in the CNN to solve the problem of modified distribution of the internal nodes caused by the change in the input data of the traditional CNN. The addition of this layer can solve the problem of training saturation and gradient disappearance, significantly increase the training speed, and improve the classification performance. According to Section 4.1, the BN operator layer can replace the traditional dropout layer and the local response normalization layer (LRN) and achieve a more satisfactory result, as presented in Figure 9 and Table 6.

TABLE 6: Performance comparison of dropout, LRN, and BN.

Dataset	Operating layer	Unit time of batch/s*	Error	Kappa
Indian pines	Dropout	0.940	0.0860	0.905
	LRN	0.898	0.0629	0.930
	B N	0.742	**0.0174**	0.978
Salinas	Dropout	0.671	0.0392	0.945
	LRN	0.360	0.0323	0.939
	B N	0.342	**0.0272**	0.972
KSC	Dropout	0.115	0.0893	0.892
	LRN	0.114	0.0592	0.922
	B N	0.099	**0.0378**	0.956
Botswana	Dropout	0.362	0.1709	0.829
	LRN	0.323	0.0970	0.886
	B N	0.283	**0.0207**	0.963

*Owing to the number of samples between the datasets and the specific parameters of the training, this entry only applies to the comparison within the same dataset.

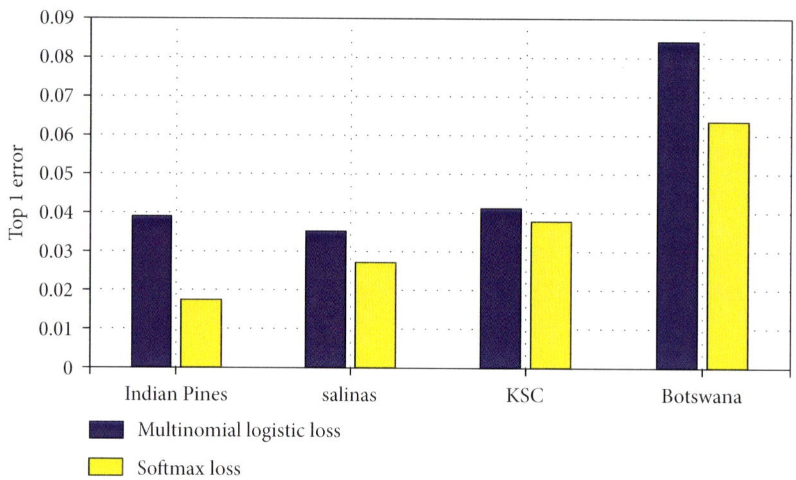

FIGURE 10: Classification accuracy of the different output models.

TABLE 7: Classification accuracy of different outputs model (error).

Output model	Performance	Indian Pines	Salinas	KSC	Botswana
Multinomial logistic loss	OA	0.0390	0.0352	0.0412	0.0842
	Kappa	0.952	0.948	0.952	0.914
Softmax loss	OA	**0.0174**	**0.0272**	**0.0378**	**0.0207**
	Kappa	0.978	0.972	0.956	0.963

Table 6 indicates that dropout and LRN are trained on the network to accelerate and prevent the overfitting phenomenon. However, the BN operation layer in training acceleration and classification accuracy are better than the two areas by data comparison, and the experimental results depict obvious differences. Figure 10 denotes the test results for error in Table 6.

4.3. Softmax Loss Models. The CNN in this study uses the Softmax loss models in the output to replace the traditional Softmax regression and multinomial logistic loss models. Table 7 compares the classification accuracy among the three models.

Figure 10 is a more intuitive histogram. The results of four hyperspectral data experiments show that the Softmax loss model has the highest classification accuracy, compared with the multinomial logistic loss model.

4.4. Comparison with Other Approaches. The proposed method in this study is compared with recent works on

TABLE 8: Comparison of CNNs with other algorithms in Indian Pines, Salinas, KSC, and Botswana.

Dataset	Performance	CNN (vector)	PPF (vector)	BASS (blocks)	2D-spectrum CNN
Indian Pines	OA	86.44	94.34	96.77	**98.26**
	Kappa	0.856	0.922	0.955	0.978
Salinas	OA	89.28	94.80	95.36	**97.28**
	Kappa	0.878	0.931	0.928	0.962
KSC	OA	88.38	93.18	94.53	**96.22**
	Kappa	0.870	0.912	0.939	**0.956**
Botswana	OA	89.45	91.57	95.42	**97.93**
	Kappa	0.885	0.908	0.939	**0.963**

Vector: one-dimensional spectral vectors. Blocks: small-area pixel division blocks.

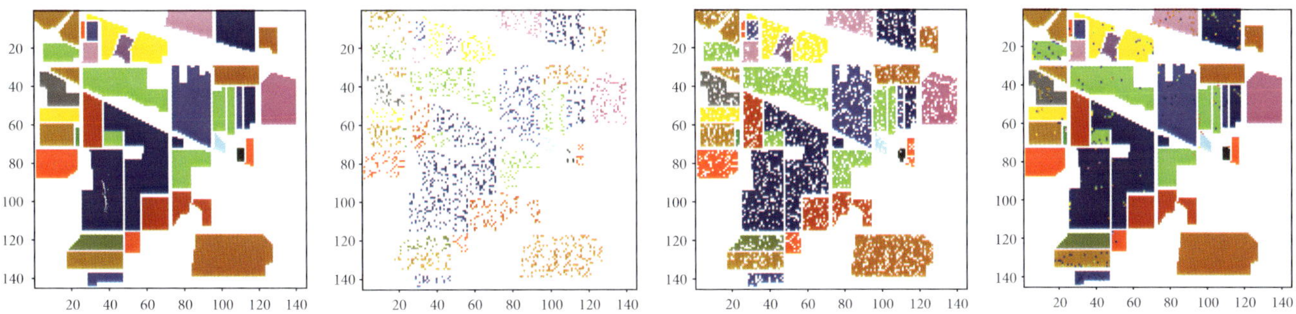

FIGURE 11: Indian Pines dataset. (a) The ground truth, (b) training samples, (c) test samples, and (d) classification maps.

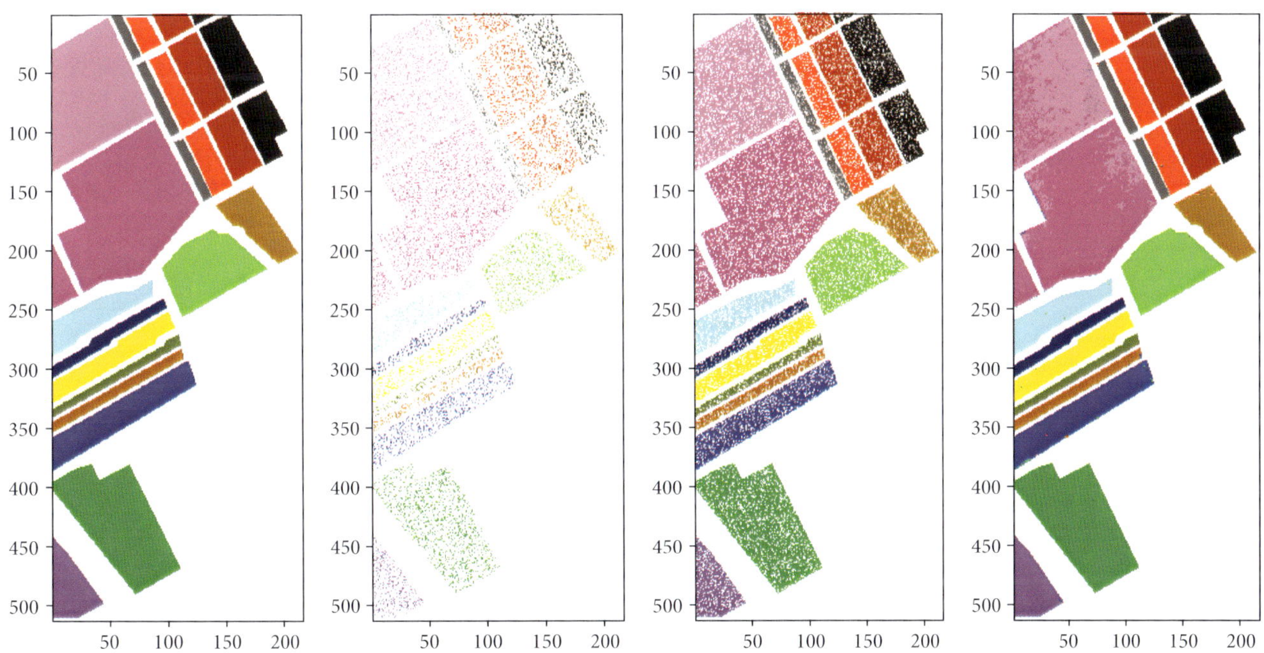

FIGURE 12: Salinas data set. (a) The ground truth, (b) training samples, (c) test samples, and (d) classification maps.

the HSI classification studies. The training accuracy evaluation indicators, namely, OA and Kappa coefficients, are important criteria for assessing network model classification performance. In Table 8, the proposed 2D-spectrum CNN model is compared with other advanced deep learning methods. The comparative data is derived from the work of

Hu et al. [27], and the CNN data based on PPFs is from the research of Li et al. [28]. The data of Band-Adaptive Spectral-Spatial (BASS) net architecture is from the paper of Santara et al. [29].

Figures 11 and 12 display the ground truth, training set, test set, and classification maps of Indian Pines and Salinas. It can be seen from the comparison of the performance that the classification performance of the 2D-spectrum CNN model is superior to other methods in different hyperspectral data in terms of overall accuracy and Kappa coefficient.

5. Conclusion

This paper proposes the 2D-spectrum CNN model that adds multilevel BN operating layers for HSI classification. The output uses a Softmax loss model for classification. In addition, many of the relevant factors include the number of iterations, learning rate, size of the input data, and size of the filter and feature graph. All of them will affect the final classification performance. The parameters obtained the optimal solution and reached the ideal effect in the experiments. The experiment results show that the CNN model proposed in this paper provides excellent performance. The training process converges quickly, thereby indicating that this method can be applied to large datasets. Furthermore, high classification accuracy can be achieved by applying sufficient iterations and number of datasets.

Conflicts of Interest

The authors declare that there are no conflicts of interest.

Authors' Contributions

Hongmin Gao and Shuo Lin contributed equally to this work and should be considered co-first authors.

Acknowledgments

This study is supported by the National Natural Science Foundation of China (no. 61701166), Projects in the National Science & Technology Pillar Program during the Twelfth Five-Year Plan Period (no. 2015BAB07B01), the Fundamental Research Funds for the Central Universities (no. 2018B16314), the China Postdoctoral Science Foundation (no. 2018M632215), National Science Foundation for Young Scientists of China (no. 51709271), and Young Elite Scientists Sponsorship Program by China Association for Science and Technology (no. 2017QNRC001).

References

[1] D. Landgrebe, "Hyperspectral image data analysis," *IEEE Signal Processing Magazine*, vol. 19, no. 1, pp. 17–28, 2002.

[2] C. J. Zhang, Y. N. Wang, X. Lu, K. N. Wang, and G. F. Wang, "Front-vehicle detection algorithm based on hypothesis and verification of improved HOG feature," *Journal of Electronic Measurement and Instrumentation*, vol. 29, no. 2, pp. 165–171, 2015.

[3] G. G. Zhang and X. H. Xu, "SAR image target recognition algorithms based on weighted texture features," *Foreign Electronic Measurement Technology*, vol. 34, no. 9, pp. 22–25, 2015.

[4] G. Licciardi, P. R. Marpu, J. Chanussot, and J. A. Benediktsson, "Linear versus nonlinear PCA for the classification of hyperspectral data based on the extended morphological profiles," *IEEE Geoscience and Remote Sensing Letters*, vol. 9, no. 3, pp. 447–451, 2012.

[5] A. Villa, J. A. Benediktsson, J. Chanussot, and C. Jutten, "Hyperspectral image classification with independent component discriminant analysis," *IEEE Transactions on Geoscience and Remote Sensing*, vol. 49, no. 12, pp. 4865–4876, 2011.

[6] T. V. Bandos, L. Bruzzone, and G. Camps-Valls, "Classification of hyperspectral images with regularized linear discriminant analysis," *IEEE Transactions on Geoscience and Remote Sensing*, vol. 47, no. 3, pp. 862–873, 2009.

[7] J. M. Keller, M. R. Gray, and J. A. Givens, "A fuzzy K-nearest neighbor algorithm," *IEEE Transactions on Systems, Man, and Cybernetics*, vol. SMC-15, no. 4, pp. 580–585, 1985.

[8] L. Jiang, C. Li, S. Wang, and L. Zhang, "Deep feature weighting for naive Bayes and its application to text classification," *Engineering Applications of Artificial Intelligence*, vol. 52, pp. 26–39, 2016.

[9] S.-B. Bai, J. Wang, G.-N. Lü, P.-G. Zhou, S.-S. Hou, and S.-N. Xu, "GIS-based logistic regression for landslide susceptibility mapping of the Zhongxian segment in the Three Gorges area, China," *Geomorphology*, vol. 115, no. 1-2, pp. 23–31, 2010.

[10] M. F. Yan, G. F. Zhao, Z. F. Liu, Q. Zhang, Z. X. Wang, and Z. Y. Du, *L-Fuzzy Sets and L-Fuzzy Support Vector Machine*, Journal of Tangshan Normal University, 2015.

[11] J. Li, J. M. Bioucas-Dias, and A. Plaza, "Spectral-spatial classification of hyperspectral data using loopy belief propagation and active learning," *IEEE Transactions on Geoscience and Remote Sensing*, vol. 51, no. 2, pp. 844–856, 2013.

[12] D. Lunga, S. Prasad, M. M. Crawford, and O. Ersoy, "Manifold-learning based feature extraction for classification of hyperspectral data: a review of advances in manifold learning," *IEEE Signal Processing Magazine*, vol. 31, no. 1, pp. 55–66, 2014.

[13] Q. Gao, W. Yang, and Q. Li, "Research on deep belief network layer tendency and its application into identifying fault images of aerial images," *Chinese Journal of Scientific Instrument*, vol. 36, no. 6, pp. 1267–1274, 2015.

[14] P. M. Atkinson and A. R. L. Tatnall, "Introduction neural networks in remote sensing," *International Journal of Remote Sensing*, vol. 18, no. 4, pp. 699–709, 1997.

[15] L. Bruzzone and D. F. Prieto, "A technique for the selection of kernel-function parameters in RBF neural networks for classification of remote-sensing images," *IEEE Transactions on Geoscience and Remote Sensing*, vol. 37, no. 2, pp. 1179–1184, 1999.

[16] F. Ratle, G. Camps-Valls, and J. Weston, "Semi-supervised neural networks for efficient hyperspectral image classification," *IEEE Transactions on Geoscience and Remote Sensing*, vol. 48, no. 5, pp. 2271–2282, 2010.

[17] G. E. Hinton, S. Osindero, and Y.-W. Teh, "A fast learning algorithm for deep belief nets," *Neural Computation*, vol. 18, no. 7, pp. 1527–1554, 2006.

[18] R. Salakhutdinov and G. E. Hinton, "Deep Boltzmann machines," in *Proceedings of the 12th International Conference on Artificial Intelligence and Statistics (AISTATS)*, pp. 448–455, Clearwater Beach, FL, USA, 2009.

[19] Y. Bengio, P. Lamblin, D. Popovici, and H. Larochelle, "Greedy layer wise training of deep networks," in *Proceedings of Neural Information Processing Systems*, pp. 153–160, Cambridge, MA, USA, 2007.

[20] Y. Chen, Z. Lin, X. Zhao, G. Wang, and Y. Gu, "Deep learning-based classification of hyperspectral data," *IEEE Journal of Selected Topics in Applied Earth Observations and Remote Sensing*, vol. 7, no. 6, pp. 2094–2107, 2014.

[21] C. Tao, H. Pan, Y. Li, and Z. Zou, "Unsupervised spectral–spatial feature learning with stacked sparse autoencoder for hyperspectral imagery classification," *IEEE Geoscience and Remote Sensing Letters*, vol. 12, no. 12, pp. 2438–2442, 2015.

[22] Y. Chen, X. Zhao, and X. Jia, "Spectral–spatial classification of hyperspectral data based on deep belief network," *IEEE Journal of Selected Topics in Applied Earth Observations and Remote Sensing*, vol. 8, no. 6, pp. 2381–2392, 2015.

[23] G. E. Hinton and R. R. Salakhutdinov, "Reducing the dimensionality of data with neural networks," *Science*, vol. 313, no. 5786, pp. 504–507, 2006.

[24] J. Yue, W. Zhao, S. Mao, and H. Liu, "Spectral–spatial classification of hyperspectral images using deep convolutional neural networks," *Remote Sensing Letters*, vol. 6, no. 6, pp. 468–477, 2015.

[25] Y. Chen, H. Jiang, C. Li, X. Jia, and P. Ghamisi, "Deep feature extraction and classification of hyperspectral images based on convolutional neural networks," *IEEE Transactions on Geoscience and Remote Sensing*, vol. 54, no. 10, pp. 6232–6251, 2016.

[26] A. Romero, C. Gatta, and G. Camps-Valls, "Unsupervised deep feature extraction for remote sensing image classification," *IEEE Transactions on Geoscience and Remote Sensing*, vol. 54, no. 3, pp. 1349–1362, 2016.

[27] W. Hu, Y. Huang, L. Wei, F. Zhang, and H. Li, "Deep convolutional neural networks for hyperspectral image classification," *Journal of Sensors*, vol. 2015, Article ID 258619, 12 pages, 2015.

[28] W. Li, G. Wu, F. Zhang, and Q. Du, "Hyperspectral image classification using deep pixel-pair features," *IEEE Transactions on Geoscience and Remote Sensing*, vol. 55, no. 2, pp. 844–853, 2017.

[29] A. Santara, K. Mani, P. Hatwar et al., "BASS net: band-adaptive spectral-spatial feature learning neural network for hyperspectral image classification," *IEEE Transactions on Geoscience and Remote Sensing*, vol. 55, no. 9, pp. 5293–5301, 2017.

[30] B. Liu, X. Yu, P. Zhang, X. Tan, A. Yu, and Z. Xue, "A semi-supervised convolutional neural network for hyperspectral image classification," *Remote Sensing Letters*, vol. 8, no. 9, pp. 839–848, 2017.

[31] N. Kruger, P. Janssen, S. Kalkan et al., "Deep hierarchies in the primate visual cortex: what can we learn for computer vision?," *IEEE Transactions on Pattern Analysis and Machine Intelligence*, vol. 35, no. 8, pp. 1847–1871, 2013.

[32] D. H. Hubel and T. N. Wiesel, "Receptive fields and functional architecture of monkey striate cortex," *The Journal of Physiology*, vol. 195, no. 1, pp. 215–243, 1968.

Weighted Voronoi Diagram-Based Simulation and Comparative Analysis of Ecosystem Service Coverage

Pengyan Zhang,[1] Wenlong Jing,[2,3,4] and Yunzeng Chen[1]

[1]*College of Environment and Planning, Henan University, Kaifeng 475004, China*
[2]*Guangzhou Institute of Geography, Guangzhou 510070, China*
[3]*Key Laboratory of Guangdong for Utilization of Remote Sensing and Geographical Information System, Guangzhou 510070, China*
[4]*Guangdong Open Laboratory of Geospatial Information Technology and Application, Guangzhou 510070, China*

Correspondence should be addressed to Wenlong Jing; jingwl@lreis.ac.cn and Yunzeng Chen; yzchen@henu.edu.cn

Academic Editor: Yichun Xie

The future of ecology lies on the promotion of eco-economy, a resource saving and environment-friendly society, and the research of ecosystem services. This article applied weighted Voronoi diagram to simulate the ecosystem service capacity in Zhongyuan Urban Agglomeration, China, using remote sensing and geographic information systems. The results show that (1) compared with the traditional Voronoi diagram, the weighted Voronoi diagram expresses more accurate on the ecosystem service range, (2) the total value of the ecosystem service in the study area was 575.7 billion yuan in 2015.The maximum value was about 253.8 billion yuan greater than the minimum value. The unit forest ecosystem has been improved after the coefficient was modified, and the value of ecosystem services reached 32,638.87 yuan/hm^2, (3) the service value of construction ecosystem has a great effect on the total value of ecosystem services; the overall pulldown effect of Kaifeng City is most considerable, and (4) the region with strong ecosystem service capacity has strong effect in annexing the nearby areas with weak ecosystem service capacity. And the distance of the annexing is inversely proportional. The closer the distance is, the stronger the annex will be.

1. Introduction

The survival of human societies is determined by the carrying capacity of their ecosystems [1, 2]. The idea that natural resources and environments are free causes ecological resource development to exceed ecosystems' basic carrying capacities. This, in turn, leads to the reduction or complete loss of ecosystem service functions [3, 4]. Now, more than two-thirds of services provided by ecosystems worldwide have already begun to decrease [5]; this trend may not be effectively reversed in the next 50 years [3, 4, 6]. Resource, ecological, and environmental crises have already seriously threatened China's development; for China to develop in a sustainable manner, scientific studies must establish macroscopic ecological security patterns and define key ecological spaces for prior protection. Therefore, exploring a scientific and effective method for delineating the range of ecosystem services is of positive significance in promoting the balanced development of ecosystem services and establishing and improving the ecosystem protection policy.

Ecosystem service functions are defined as "the natural benefits which human survival relies on and which are created and maintained by ecosystems and ecological processes" [7]. To describe ecosystem service functions more intuitively, Costanza et al. [8] conducted an assessment of ecosystem service values, which has become a key ecological research topic. Estoque and Murayama [9] found that the area of urban in the city of Baguio, the summer capital of the Philippines, expanded continuously during the study period, while the ecosystem service value dropped to 40% of that at the beginning of the study. Gómez-Baggethun and Barton [10] discovered that traditional monotonous urban planning is one-sided, and that a single socioeconomic indicator cannot assure the balanced ecological and economic development of a society. Moreover, introducing ecosystem service values has a positive impact on the development of human societies. Su et al. [11] primarily studied the Hangjiahu Plain agroecological zone, the Tianmu Mountain forest ecological zone, the Hangzhou urban ecological zone, and the Qiandao Lake Basin forest and wetland ecological zone and analyzed the effects of urbanization on regional ecosystems. Using research from around the world [12–17], Costanza et al. [6] summarized ecosystem service value assessments and further improved their evaluation. However, key research on ecosystem service values remains: for example, how should coverage of ecosystem services be delimited? How should a method of ecosystem service coverage quantification be developed?

Voronoi diagrams are partitioning that have been shown to be highly similar to some natural structures; as such, they have been studied extensively in fields including meteorology, geology, surveying and mapping, archaeology, molecular chemistry, ecology, and computer science [18, 19]. Voronoi diagrams are mainly applied to site selection and the determination of the scope of influence of different objects. Qin and Shi [20] utilized Voronoi diagrams to examine the spatial distribution of large-scale residential areas, ultimately making suggestions for planning in the region. Fan et al. [21] integrated Voronoi diagrams with spatial interaction models to investigate the Horqin Left Middle Banner, in Inner Mongolia, and proposed a method for improving this model by substituting town centrality for population size and integrating the model with weighted Voronoi diagrams. Traditional Voronoi diagrams are often used to describe the evenness of ground objects and are based on the principle of even distribution; however, they can better solve practical problems after some reasonable improvement. For example, introducing weights can compensate for the defects of traditional uniform Voronoi diagrams [22–25]. Li et al. [26] and Miao et al. [27] used Voronoi diagrams to examine parks in Shenyang and the wetland functional zone in Beijing, respectively, confirming the possibility of employing this method in ecosystem research. Nevertheless, there is still room for development in studying ecosystem services and enhancing its assessment accuracy.

To explore a suitable method of studying ecosystem service coverage areas, we introduce Voronoi diagrams and weighted Voronoi diagrams into ecosystem service value research and analyze ecosystem services coverage based on actual conditions. This approach helps quantify ecosystem service research not only at fixed points but also across regions. The Zhongyuan Urban Agglomeration is one of seven major urban agglomerations in China; it is the key development area of the Zhongyuan Economic Zone. Recent rapid social and economic development in this area has changed the types of ecosystems in the region, which in turn has led to changes in ecosystem service values in the coverage areas. In order to explore a reasonable method for estimating ecosystem service coverage areas, we will value the ecosystem services, that is, use economic methods to describe the strength of ecosystem services. Furtherly, we introduce the Voronoi diagram and the weighted Voronoi diagram to simulate and analyze the ecosystem service areas. The Voronoi diagram, the Euclidean distance map, and the rank assignment map are compared at various levels to explore the practicality of the weighted Voronoi diagram, thereby improving the spatially estimation practices of ecosystem service and providing the scope for future ecosystem service research.

2. Study Area and Data Resources

2.1. Study Area. The Zhongyuan Urban Agglomeration (shown in Figure 1) is located in central Henan Province, China; it has a total area of 58,700 square kilometers and includes nine cities: Zhengzhou, Luoyang, Kaifeng, Xinxiang, Jiaozuo, Xuchang, Luohe, Pingdingshan, and Jiyuan. This area is an important growth pole, not only of Henan Province but also of the Zhongyuan Economic Zone. In 2015, the gross national product of the agglomeration was 2.19×10^9 yuan and its population was 4.15×10^8. It has a nice natural environment with a wide distribution of varied land resources, a temperate monsoon climate, high topography in the west and north, and low topography in the east and south. The eastern part of the Zhongyuan Urban Agglomeration is located in the North China Plain, which is an essential grain production base and consists mainly of farmland. The Taihang Mountains in the west and north are important production sites of forest resources. Given the premise that ecological and economic development must be constantly promoted, it is necessary to protect regional ecologies to maintain balance in regional ecosystem services. Thus, we investigate the ecosystem service coverage of different ecosystems in the region and make suggestions for regional planning.

2.2. Data Sources. We aim to verify the applicability of weighted Voronoi diagrams to ecosystem service value research and place little emphasis on spatial and temporal variations in ecosystem service values. Weighted Voronoi diagrams of the ecosystem service value of the Zhongyuan Urban Agglomeration in 2015 are created to examine their meaning in ecosystem service research. Land use types significantly control the functions of regional ecosystem services and are commonly used as indictors of the values of regional ecosystem services [28]. We acquired land use map of the study area by interpreting Landsat OLI images in April 2015, which is in the growing seasons and the images with low cloud cover percentages were selected. We used a manual interpretation method to derive land use maps from the remote sensing images. The land use is classified into four main types: farmland, forest, water, and construction area. To evaluate the accuracy of the land use map, we selected 110 random points from the map. The Land Use Atlas in Henan Province (1 : 1 : 100000) and high-resolution satellite images from Google Earth were referred to validate the map. As a result, the overall accuracy of the land use map is 91.3%, and the Kappa coefficient is 0.88.

3. Research Methods

Based on Costanza's and Daly's [13] measurement and calculation of global ecosystem service values and the

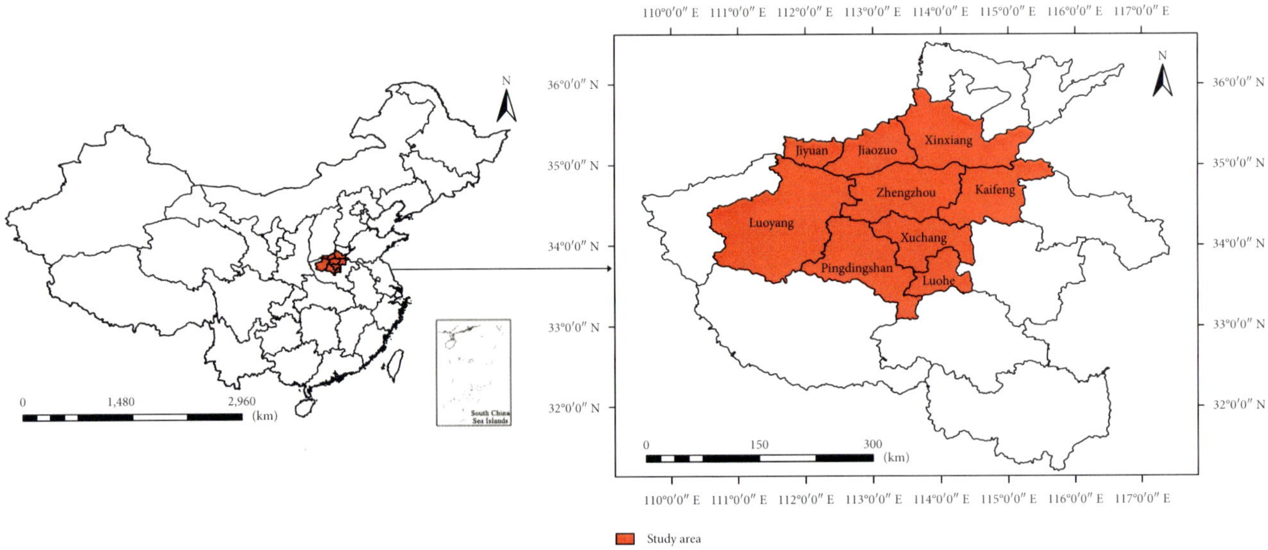

FIGURE 1: The location of the study area.

"Ecological Service Value Equivalence per Ecosystem Unit Area in China" table established by Xie at al. [29] based on actual conditions in China, this paper considers the actual condition of the Zhongyuan Urban Agglomeration to propose a novel research method, which is introduced into the correction coefficient of the study area's ecosystem service value. ArcGIS 10.1 and MATLAB are utilized to describe the spatial distribution of ecosystem service values in the Zhongyuan Urban Agglomeration. The results are compared in color gradings, traditional Voronoi diagrams, and weighted Voronoi diagrams to analyze these methods' advantages and drawbacks in ecosystem service value assessment (Figure 2) and provide a new perspective on ecosystem service value research.

3.1. The Correction Coefficient Model of Ecosystem Service Values. Based on previous research [13, 29, 30] and real-world regional conditions, the assessment coefficient of ecosystem service values (as shown in Table 1) is improved. We propose an improved model based on grain yields (1) and a correction coefficient model of ecosystem service values (2).

$$E_{\mathrm{a}} = \frac{1}{7}\left(\frac{m_{\mathrm{c}} p_{\mathrm{c}} q_{\mathrm{c}}}{M_{\mathrm{c}}}\right), \tag{1}$$

$$\mathrm{VC}_{ij} = e_{ij} E_{\mathrm{a}}, \tag{2}$$

where E_a is the value of food supply services per unit agroecosystem per unit ecosystem (yuan), m_c is the crop area (hm^2), q_c is the crop yield per unit area (kg/hm^2), p_c is the notional average price of crops (yuan/kg), M_c is the total sown area (hm^2), VC_{ij} denotes the equivalent value of ecosystem i for function j, and e_{ij} is the function of ecosystem service j in the ith ecosystem type relative to the food supply services provided by the agroecosystem. The E_a of the study area was calculated to be 1493.77 yuan/hm^2. The correction coefficient of ecosystem service value is listed in Table 1.

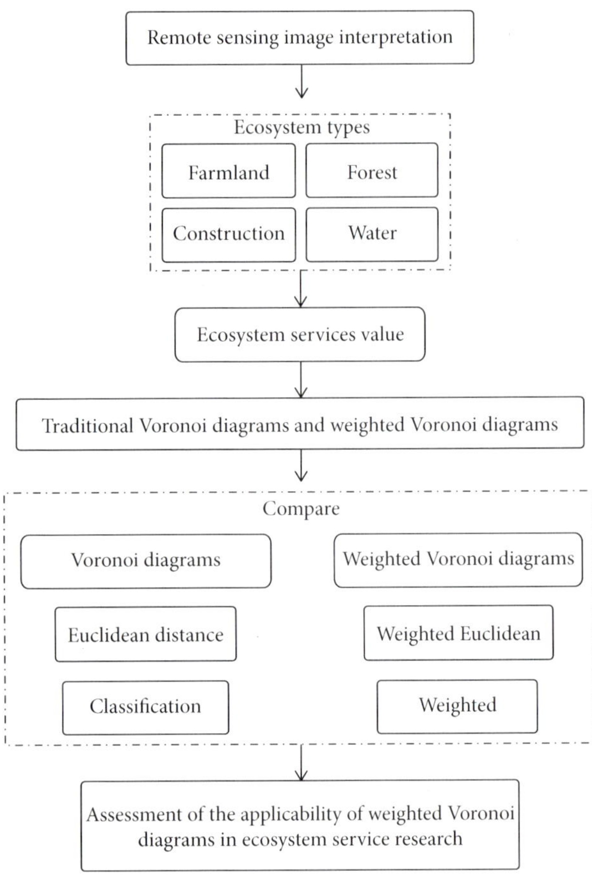

FIGURE 2: Research flowchart.

3.2. The Ecosystem Service Value (ESV) Model. The ecosystem service value (ESV) model is effective for estimating ecosystem service values based on different ecological types [30] and is expressed as illustrated in (3) and (4).

Table 1: Correction coefficient of ecosystem services values (y/hm^2·a).

Service type	Forest ecosystem	Farm ecosystem	Water ecosystem	Construction ecosystem
Gas regulation	5228.19	746.88	0.00	0.00
Climate regulation	4033.17	1329.45	687.13	0.00
Water conservation	4780.06	896.26	30443.03	−14325.25
Soil formation and protection	5825.70	2180.90	14.93	0.00
Waste disposal	1956.83	2449.78	27156.73	−4660.56
Biodiversity conservation	4869.69	1060.57	3734.42	0.00
Food production	149.37	1493.77	448.13	0.00
Raw material	3883.80	149.377	104.56	0.00
Entertainment culture	1912.02	14.9377	8290.42	0.00
Total	32638.87	10321.95	93674.31	−18985.81

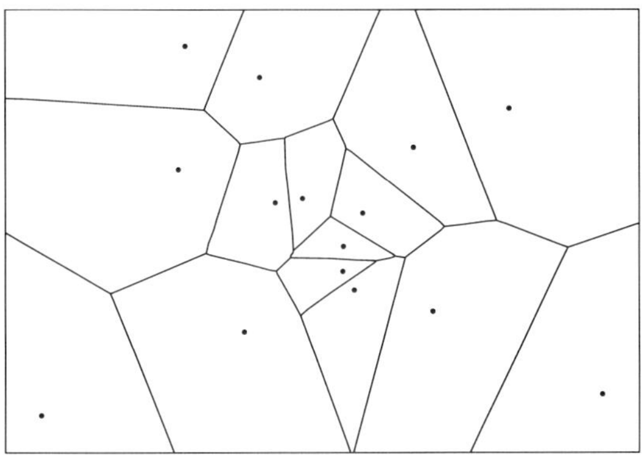

Figure 3: Traditional Voronoi diagram.

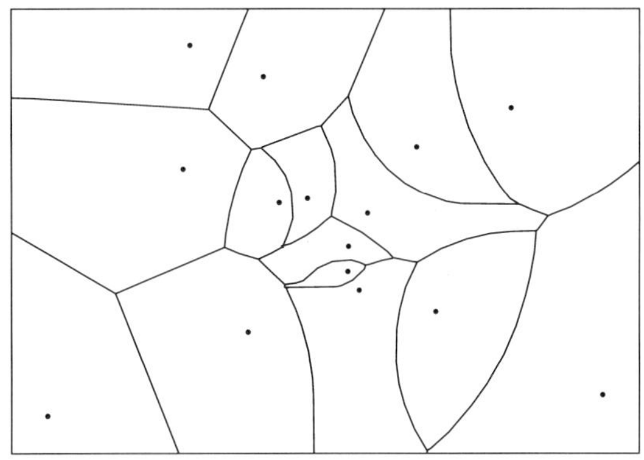

Figure 4: Weighted Voronoi diagram.

$$\mathrm{ESV}_{ij} = A_i \times \mathrm{VC}_{ij}, \tag{3}$$

$$\mathrm{ESV}_i = \sum_{j=1}^{n} \mathrm{ESV}_{ij}, \tag{4}$$

where ESV is the ecosystem service value (yuan), A_i is the area of the ith ecosystem type, and VC_{ij} denotes the equivalent coefficient of function j for ecological land i.

3.3. Voronoi Diagrams. Voronoi diagrams (abbreviated as V diagrams, as shown in Figure 3) were proposed by Ukrainian mathematician Voronoi in 1908 and were applied to high-dimensional spaces. By assuming that for a set of discrete points on Euclidean plane $P = \{p_1, p_2, \ldots, p_n\}$, where any two arbitrary points are noncoincident (i.e., $p_i \neq P_j, i \neq j, 1 \leq i, j \leq n$) and any four arbitrary points are nonconcyclic, $V(P_1) \cdots V(P_i)$ is calculated as

$$V(P_i) = \bigcap_{i \neq j} \{P | d(P, P_i) < d(P, P_j)\} \quad j = 1, 2, \ldots, n, \tag{5}$$

where $d(P, P_i)$ is the Euclidean distance between points P and P_i. Thus, the distance from each point P that belongs to $V(P_i)$ is closer to P_i than any other points P_j $(i \neq j)$ of discrete point set P.

3.4. Weighted Voronoi Diagrams. Although Voronoi diagrams can depict the basic characteristics of some natural environments, they are unable to present the condition of some relatively complex ground objects. Based on traditional Voronoi diagrams, weighted Voronoi diagrams (as shown in Figure 4) introduce weight indices and convert traditional Euclidean distances into weighted Euclidean distances [31]. Given that $P = \{P_1, P_2, \ldots, P_i\}$, nonnegative real number weights λ_i $(i = 1, 2, \ldots, n)$ are introduced; hence, the weighted distance between points P and P_i is

$$D(P, P_i) = \frac{d(P, P_i)}{\lambda_i}. \tag{6}$$

Then, $V(P_i)$ should be revised to

$$V(P_i) = \bigcap_{i \neq j} \{P | D(P, P_i) < D(P, P_j)\} \quad j = 1, 2, \ldots, n. \tag{7}$$

When $\lambda_1 = \lambda_2 = \cdots = \lambda_n$, weighted Voronoi diagrams become traditional Voronoi diagrams.

4. Results and Analysis

4.1. Ecosystem Service Value Assessment. The Zhongyuan Economic Zone has lately been an important economic development region in China; more importantly, it is also the key region for stimulating the development of the central region of China and promoting balance in China's economic development. Because the Zhongyuan Urban Agglomeration is the key region of the Zhongyuan Economic Zone,

Table 2: Comparison of different types of ecosystem service value coefficient.

Ecosystem service value coefficient	Forest ecosystem	Farm ecosystem	Water ecosystem	Construction ecosystem
Costanza [13] (yuan·hm^{-2}·a^{-1})	8035.3	761	26487.4	None
Xie et al. [29] (yuan·hm^{-2}·a^{-1})	12628.69	3547.89	44963.9	None
Modified (yuan·hm^{-2}·a^{-1})	32638.87	10321.95	93674.31	−18985.81

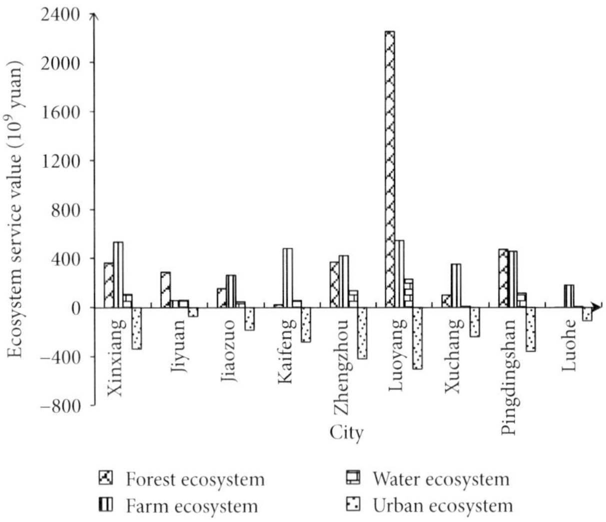

Figure 5: Ecosystem service values of nine cities in the Zhongyuan Urban Agglomeration.

its overall economic development acts as a model for the development of other regions. Therefore, proper ecosystem planning in the agglomeration and balanced ecological and economic development of the region enhances its competitiveness and encourages the development of other regions.

Before ecosystem service value assessment was completed, Costanza and Daly [13] and Xie et al. [29] performed coefficient correction of ecosystem service values according to real-world conditions in the study area. The results are listed in Table 2.

The coefficients of ecosystem service values in Table 2 and the ecosystem of different land use areas in the Zhongyuan Urban Agglomeration after satellite image interpretation were integrated to assess the ecosystem service value of each land use of the region. The results are shown in Figure 5.

The total ecosystem service value of the Zhongyuan Urban Agglomeration in 2015 was 575.7 billion yuan. The ecosystem service value of Luoyang was the highest, accounting for 43.93% of the regional total; in contrast, that of Luohe was only 1.42% of the regional total. Furthermore, the forest ecosystems of Luoyang contributed the most to the total ecosystem service value of the city, accounting for 89.09% of the total. The total ecosystem service value of Kaifeng was 28.259 billion yuan, of which the value of farm ecosystem was 48.216 billion yuan, whereas the loss in value caused by construction ecosystem expansion was up to 28.037 billion yuan. The ecosystem types combined with the administrative area could lead to the differences of ecosystem service values. Specifically, higher proportion of forest and water ecosystems in a region would increase the region's ecosystem service values and furtherly have positive influences upon the neighbor ecosystems.

4.2. Simulation of Ecosystem Service Coverage. Similar to social services, ecological services have coverage directly related to their intensity. Past ecosystem service studies could not determine the areas of influence of multiple service functions. For example, the ecosystem service function of climate regulation by forest ecosystems is affected by several factors and it is not possible to accurately assess specific service coverage. Hence, selecting an appropriate angle for measuring and calculating ecosystem service coverage can compensate for shortcomings in previous research.

Ecosystem service values quantify all kinds of ecosystem service functions and present them intuitively. The appearance of this variable allows ecosystem service coverage to be simulated and analyzed using an economic approach. The resulting Voronoi diagrams are shown in Figure 6 in a map showing a uniform distribution; however, there are considerable differences in ecosystem service values between cities. A uniform distribution cannot correctly depict the actual coverage of ecosystem services. Therefore, ecosystem service values of different regions are presented as weighted Voronoi diagrams, which partly compensate for defects in the expression of service coverage by traditional Voronoi diagrams.

The ecosystem service values of all cities in the Zhongyuan Urban Agglomeration were used as weight indices to improve the original Voronoi diagrams, as illustrated in Figure 7. These results, together with the data in Figure 5, demonstrate that the ecosystem service value of Luoyang accounts for a significant portion of the total ecosystem service value of the Zhongyuan Urban Agglomeration. Its influence is considerably greater than those of other cities. Integration of the different service boundaries of Figures 5 and 7 shows that, taking Luoyang as an example, the ecosystem service value coverage of the city expands significantly, which is highly related to the larger ecosystem service value of the city. For regions with lower service values, such as Jiyuan and Luohe, except for certain preserved coverage, areas are impacted by regions with greater ecosystem service values. That is, if the ecosystem service value is greater, the service coverage is more extensive. To an extent, these areas affect and compensate for areas with smaller ecosystem service values.

To highlight the interactions and relationships between ecosystem services of different regions, we classified ecosystem service coverage using the Jenks natural breaks classification method into five types: high-, medium to high-, medium-, medium to low-, and low-impact areas. The traditional Voronoi diagrams are compared with the weighted ones in Figure 8.

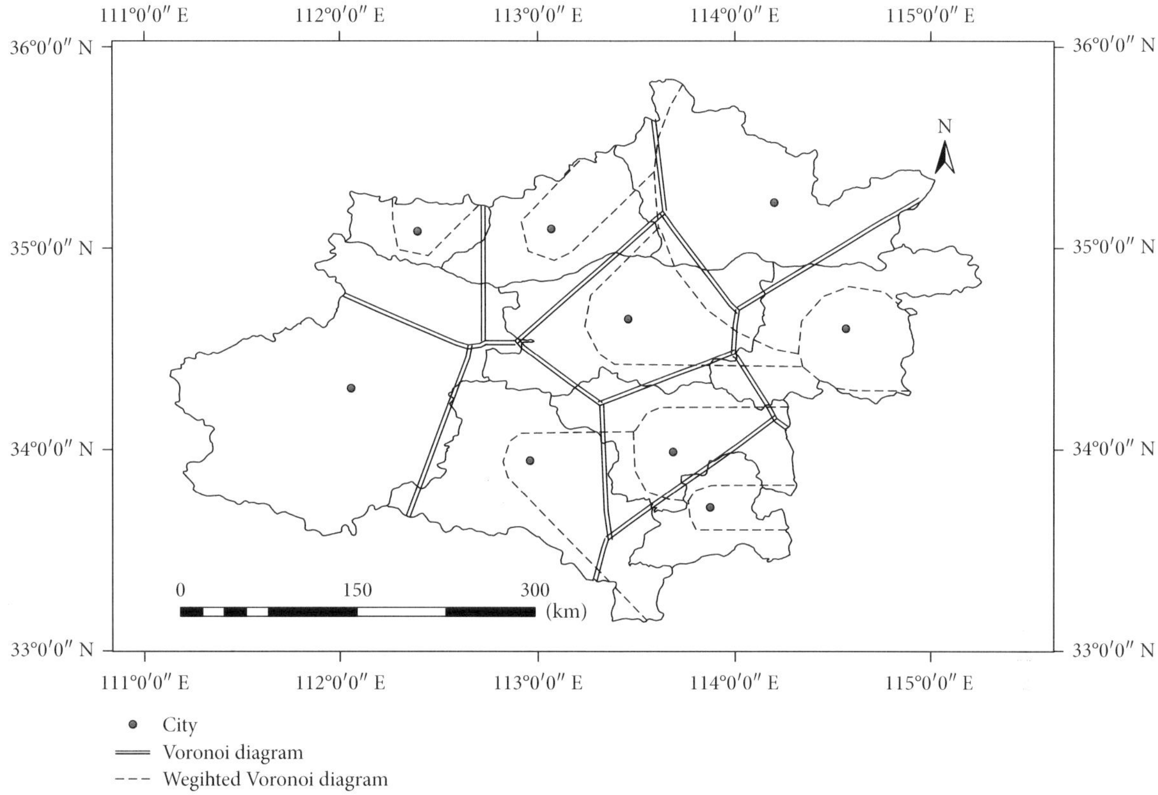

FIGURE 6: Comparison of the traditional and weighted Voronoi diagrams of the study area.

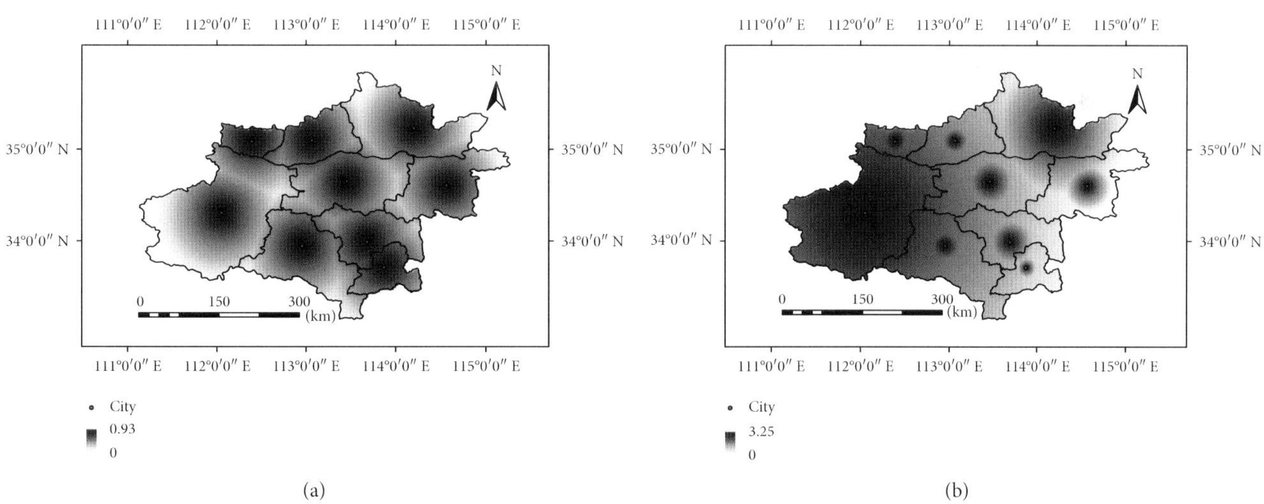

FIGURE 7: Comparison of Euclidean distances between (a) the traditional Voronoi diagrams and (b) weighted Voronoi diagrams of the study area.

As shown in Figure 8, regions may have various effects on each other, although the center of each region is mainly influenced by the region itself. In addition, regions with greater ecosystem service capabilities compensate for regions with lower capabilities, which conceal the originally insignificant ecosystem service capabilities of regions with lower service capacities. The ecosystem service interactions engulfing different regions can be monitored and controlled artificially. Rational distribution of ecosystems over the study area will considerably enhance a region's ecosystem service capability and thus its ecosystem service quality. If the stepwise progressive decreases of different kinds of ecosystem service capabilities can be successfully determined, it would be possible to more accurately evaluate the influences and coverage of ecosystem services.

From the total ecosystem service value in Figure 8, the type of ecosystem that dominates the total ecosystem services can be seen and various types of ecosystem service coverage

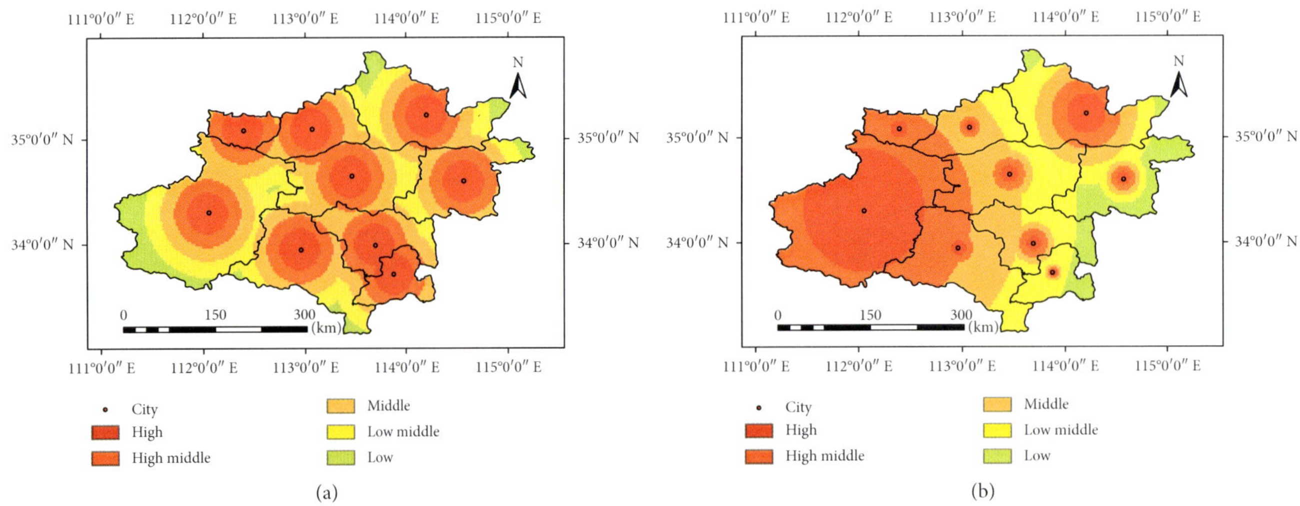

Figure 8: Comparison of ecosystem service classification between (a) traditional Voronoi diagrams and (b) weighted Voronoi diagrams of the study area.

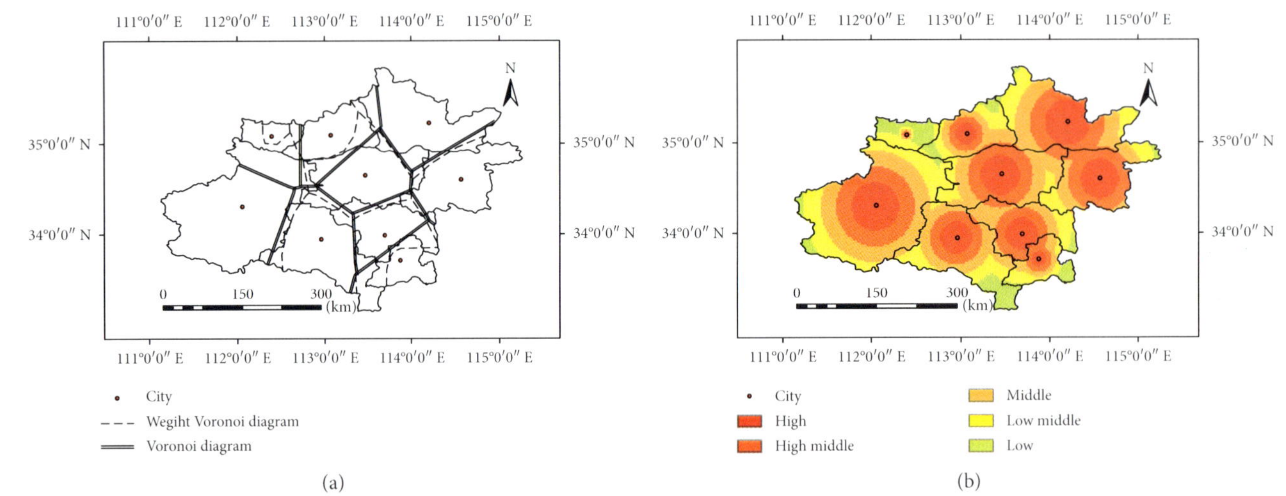

Figure 9: Service impact ranges of (a) traditional Voronoi diagram and (b) weighted Voronoi diagram in farmland ecosystem.

can be studied. Thus, the uneven distribution of land use types between regions can be obtained. We illustrate the ecosystem service capabilities and coverage of farm ecosystem, forest ecosystems, water ecosystems, and construction ecosystems of regions in the Zhongyuan Urban Agglomeration in Figures 9–12.

Excluding Jiyuan, Luohe, and Jiaozuo, the farmland ecosystem service values of the study area were 40–60 billion yuan. According to Figures 9–12, it is obvious that regions with less farmland ecosystem service capability are more considerably affected by regions with greater capabilities. For regional division, compared to traditional Voronoi diagrams, the weighted Voronoi diagrams also consider the actual spatial distribution of farmland. Additionally, compared to the traditional Voronoi diagrams, which use color grading to represent ecosystem service values, the weighted diagrams reflect the progressive reduction over space and the nonuniformity of ecosystem services, more closely representing reality. Traditional ecosystem service value evaluation only considers the ecosystem service capabilities of cities, but not the direct interactions between cities, whereas weighted Voronoi diagrams can better explain these interactions. Construction ecosystems cause negative ecosystem service values and lower overall ecosystem service capabilities. Therefore, during calculation, all construction was assigned positive values. High-impact areas are those that are more greatly affected by construction ecosystems. The ecosystem service value losses of these areas are significantly higher than those of other areas. Areas with higher levels of urbanization and more extensive construction experience stronger reduction. In the study area, there are more forest ecosystems in the west and fewer in the east. Thus, areas with higher forest ecosystem service capabilities impose greater influences on areas with lower capabilities. Luoyang has vast forest ecosystems. Strong forest ecosystem service capability is the basic characteristic of the city's ecosystem services; Luoyang's ecosystem service value is 6.91, 7.32, 15.06, 119.42, 10.15, 27.3, 27.3, and 323.29 times of those of Xinxiang, Jiyuan, Jiaozuo,

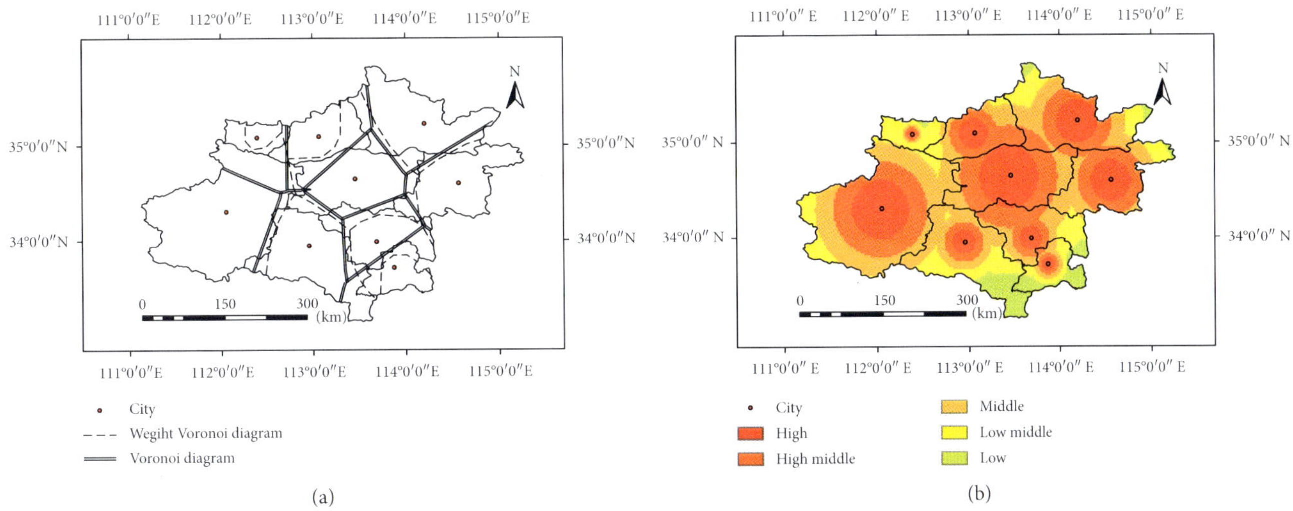

FIGURE 10: Service impact ranges of (a) traditional Voronoi diagram and (b) weighted Voronoi diagram in construction ecosystem.

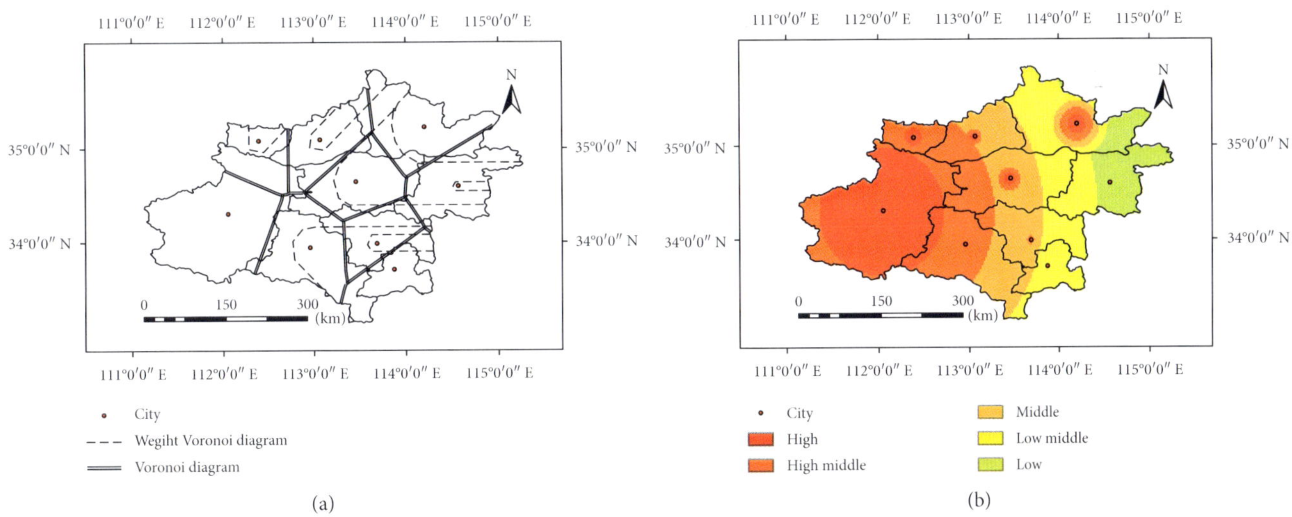

FIGURE 11: Service impact ranges of (a) traditional Voronoi diagram and (b) weighted Voronoi diagram in forest ecosystem.

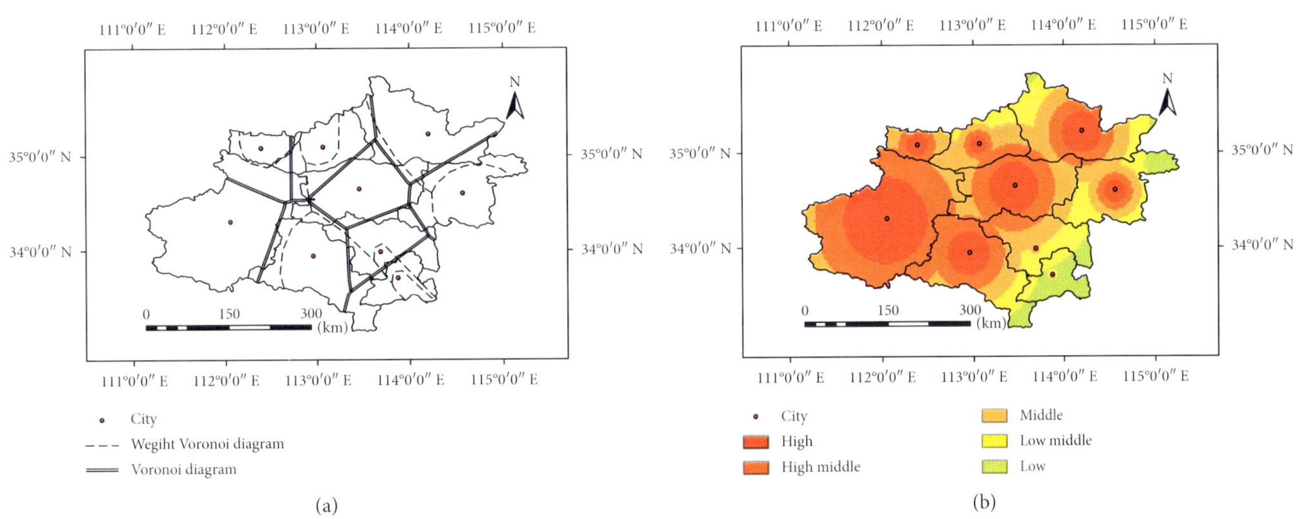

FIGURE 12: Service impact ranges of (a) traditional Voronoi diagram and (b) weighted Voronoi diagram in water ecosystem.

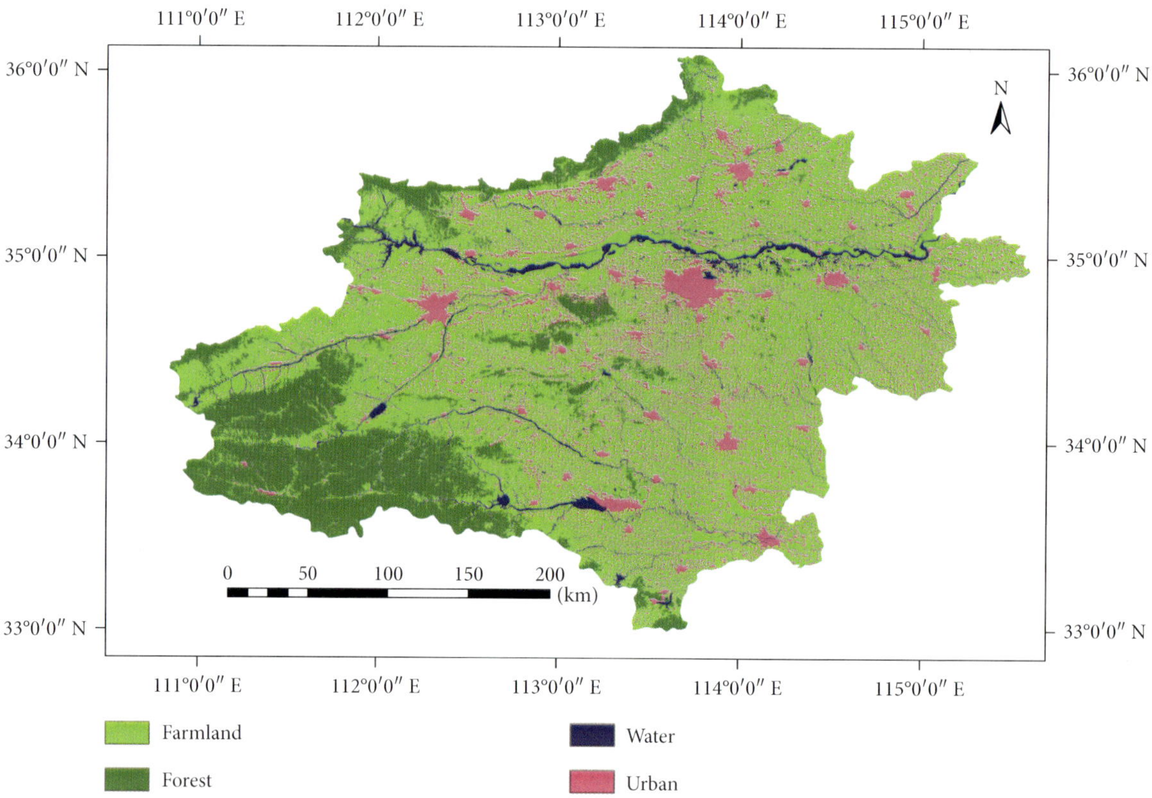

FIGURE 13: Remote sensing image interpretation data of the Zhongyuan Urban Agglomeration in 2015.

Kaifeng, Zhengzhou, Xuchang, Pingdingshan, and Luohe, respectively. Engulfing effects and compensation due to the forest ecosystem services of Luoyang gradually weaken with distance. However, the low-impact forest ecosystem service area of the city still has relatively significant effects on other areas. The weighted Voronoi diagrams, compared to the traditional Voronoi diagrams, have the advantage of intuitively presenting ecosystem service capabilities and their coverage over space. The coverage of water ecosystem services and the actual distribution of water ecosystems are relatively strongly correlated. The Yellow River runs through the central and the central-north regions of the Zhongyuan Urban Agglomeration. Rivers in Luoyang are widely distributed, which places Luohe in a low-impact water ecosystem service area. This distribution of high-impact areas in the northwest and low-impact areas in the northeast matches the satellite image interpretation results. Comparison between the satellite image interpretation (Figure 13) and ecosystem service coverage map reveals that high-, medium-, and low-impact ecosystem service areas are directly related to the distinct types of ecosystems in different cities, further confirming that weighted Voronoi diagrams are suitable for simulation of ecosystem service coverage and provide a new method and direction for these studies.

5. Discussion and Conclusion

5.1. Discussion. This study employs cities as central points for investigating ecosystem service coverage. This method considers an entire region and thereby presents the spatial distribution of ecosystem services, providing a new perspective for ecosystem service coverage research. Nevertheless, this method neglects nonuniform ecosystem distributions, ecosystem distribution within the region, and the interactions between ecosystems, which reduce the accuracy of ecosystem service coverage research. Thus, in future studies, better results can be obtained if fixed point studies can be conducted for different land use types.

Ecosystem service capabilities were classified using the Jenks natural breaks classification method. Although this method can partly reveal the basic features of research topics and data, it neglects the gradual reduction in service distance of service providers and the acceptance capabilities of service. Proper investigation of the gradual decrease of different ecosystem services over distance, and the acceptance rates of different service providers, allows the new method to be better applied to the actual production activities.

5.2. Conclusions. The total ecosystem service value of the Zhongyuan Urban Agglomeration in 2015 was 575.7 billion yuan. The total ecosystem service value of Luoyang was the highest, at 267.665 billion yuan; that of Luohe was the lowest, at only 13.881 billion yuan. The differences between regions are huge. Forest ecosystems have the greatest ecosystem service value per unit area—up to 32.639 thousand yuan per hm. Meanwhile, construction ecosystem service values significantly reduce the service values of different ecosystem types. Continuous expansion of construction ecosystems has decreased the ecosystem service value of Zhengzhou by 37.012 billion yuan, with the most remarkable decline seen in Kaifeng.

Compared to traditional Voronoi diagrams, weighted Voronoi diagrams are more suitable for studying ecosystem service coverage. Comparison between the two indicates that the spatial coverage of ecosystem services presented by the weighted Voronoi diagrams better fit actual conditions. Traditional Voronoi diagrams describe uniform ecosystem services and neglect the discrepancies in actual service coverage caused by regional differences. In weighted Voronoi diagrams, regions with higher ecosystem service capabilities partly engulf and compensate for regions with lower capabilities. These engulfing and compensating actions are inversely proportional to the distance between two regions. When two ecosystems are closer to each other, their interactions are stronger. The region with lower capability is more easily engulfed by the region with higher capability, and the ecosystem services of the former are enhanced. These actions partly contribute to the spatial and temporal distribution of ecosystems in the region and promote the union of different regions, thereby enhancing overall regional ecosystem service capability.

The ecosystem service coverage of farm ecosystem, forest ecosystems, water ecosystems, and construction ecosystems was studied separately, and the results compared to actual conditions. We found that the two agree relatively well, again proving that weighted Voronoi diagrams perform well in assessing ecosystem service coverage. The service coverage of different ecosystem types was also examined. We discovered that, for larger differences in ecosystem type and size between different regions, weighted Voronoi diagrams differ more significantly than their traditional counterparts. This result demonstrates that, for two regions whose ecosystem service differences are relatively large, traditional Voronoi diagrams do not match actual service coverage.

Disclosure

The content does not represent the official position of the Chinese government and is entirely the responsibility of the authors.

Conflicts of Interest

The authors declare that they have no conflicts of interest.

Acknowledgments

This paper was supported by the National Natural Science Foundation of China (41601175 and 41771380). The Key Scientific Research Project of Henan Province (16A610001), supported by the Program for Innovative Research Team (in Science and Technology) in the University of Henan Province (16IRTSTHN012), the GDAS' Project of Science and Technology Development (2017GDASCX-0101 and 2018GDASCX-0904), the Guangdong Innovative and Entrepreneurial Research Team Program (2016ZT06D336), the Science and Technology Program of Guangdong Province (2017B010117008), and the Science and Technology Program of Guangzhou (201604016047).

References

[1] D. E. Campbell, "Emergy analysis of human carrying capacity and regional sustainability: an example using the state of Maine," *Environmental Monitoring and Assessment*, vol. 51, no. 1/2, pp. 531–569, 1998.

[2] E. S. Nakajima and E. Ortega, "Carrying capacity using emergy and a new calculation of the ecological footprint," *Ecological Indicators*, vol. 60, pp. 1200–1207, 2016.

[3] N. M. Haddad, L. A. Brudvig, J. Clobert et al., "Habitat fragmentation and its lasting impact on Earth's ecosystemse1500052," *Science Advances*, vol. 1, no. 2, 2015.

[4] N. M. Haddad, L. A. Brudvig, J. Clobert et al., "Habitat fragmentation and its lasting impact on Earth's ecosystems," *Science Advances*, vol. 1, no. 2, article e1500052, 2015.

[5] Millennium Ecosystem Assessment, *Ecosystems and Human Well-Being; Biodiversity Synthesis*, WorldResources Institute, Washington DC, 2005.

[6] R. Costanza, R. de Groot, P. Sutton et al., "Changes in the global value of ecosystem services," *Global Environmental Change*, vol. 26, no. 1, pp. 152–158, 2014.

[7] Z. Ouyang, T. Zhao, X. Wang, and H. Miao, "Ecosystem services analyses and valuation of China terrestrial surface water system," *Acta Ecologica Sinica*, vol. 24, no. 10, pp. 2091–2099, 2004.

[8] R. Costanza, R. d'Arge, R. de Groot et al., "The value of the world's ecosystem services and natural capital," *Nature*, vol. 387, no. 6630, pp. 253–260, 1997.

[9] R. C. Estoque and Y. Murayama, "Landscape pattern and ecosystem service value changes: implications for environmental sustainability planning for the rapidly urbanizing summer capital of the Philippines," *Landscape and Urban Planning*, vol. 116, pp. 60–72, 2013.

[10] E. Gómez-Baggethun and D. N. Barton, "Classifying and valuing ecosystem services for urban planning," *Ecological Economics*, vol. 86, pp. 235–245, 2013.

[11] S. Su, R. Xiao, Z. Jiang, and Y. Zhang, "Characterizing landscape pattern and ecosystem service value changes for urbanization impacts at an eco-regional scale," *Applied Geography*, vol. 34, pp. 295–305, 2012.

[12] J. F. Barlow, "A change effects all the organisms in the ecosystem," *South Dakota Journal of Medicine*, vol. 47, no. 6, p. 186, 1994.

[13] R. Costanza and H. E. Daly, "Natural capital and sustainable development," *Conservation Biology*, vol. 6, no. 1, pp. 37–46, 1992.

[14] E. Matthews, "Global vegetation and land use: new high-resolution data bases for climate studies," *Journal of Applied Meteorology*, vol. 22, no. 3, pp. 474–487, 1983.

[15] D. Pauly and V. Christensen, "Primary production required to sustain global fisheries," *Nature*, vol. 374, no. 6519, pp. 255–257, 1995.

[16] H. Jack Ruitenbeek, "Functions of nature: evaluation of nature in environmental planning management and decision making," *Ecological Economics*, vol. 14, no. 3, pp. 211–213, 1995.

[17] J. H. Ryther, "Photosynthesis and fish production in the sea," *Science*, vol. 166, no. 3901, pp. 72–76, 1969.

[18] J. Liu, X. Sun, and S. Liu, "Weighted node network Voronoi diagram and its application to optimization of chain stores layout," *International Journal of Machine Learning and Cybernetics*, vol. 7, no. 4, pp. 679–688, 2016.

[19] A. Okabe, B. Boots, and K. Sugihara, "Nearest neighbourhood operations with generalized Voronoi diagrams: a review," *International Journal of Geographical Information Science*, vol. 8, no. 1, pp. 43–71, 1994.

[20] Y. Qin and X. Y. Shi, "Optimizing spatial distribution of residential areas based on Voronoi diagram," *Science of Surveying and Mapping*, vol. 1, pp. 136–138, 2012.

[21] Q. Fan, H. X. Zhang, Y. H. Li, and J. X. Wang, "Quantitative analysis and urban system planning of county area based on spatial-interaction model: a case in Horqin Left Middle Banner," *Inner Mongolia, Scientia Geographica Sinica*, vol. 34, no. 5, pp. 601–607, 2014.

[22] C. Xy, L. Jp, L. Chen, and H. Yj, "Intraspecific and interspecific competition analysis of Cunninghamia lanceolata ecological forest based on weighted Voronoi diagram," *Chinese Journal of Ecology*, vol. 35, no. 9, pp. 2553–2561, 2016.

[23] J. Y. Gu, Y. G. Zhou, and L. M. Guo, "Urban influence regions and spatial relation of Beijing-Tianjin Hebei region based on weighted Voronoi diagram and breakpoint theory," *Geography and Geo-Information Science*, vol. 30, no. 1, pp. 65–69, 2014.

[24] X. W. Qin, S. Q. Zhang, X. F. Li, X. D. Na, X. Pan, and H. Yu, "Assessment of red-crowned crane's habitat suitability based on weights-of-evidence," *Acta Ecologica Sinica*, vol. 29, no. 3, pp. 1074–1082, 2009.

[25] S. Xie, X. Feng, J. Wang, and W. Lu, "Radiation domain of commercial centers in Nanjing based on analysis of road network weighted Voronoi diagram," *Acta Geographica Sinica*, vol. 64, no. 12, pp. 1467–1476, 2009.

[26] X. M. Li, C. F. Liu, and W. Wu, "Recreational pressure of urban parks in Shenyang," *Chinese Journal of Ecology*, vol. 28, no. 5, pp. 992–998, 2009.

[27] L. L. Miao, W. G. Jiang, S. D. Wang, and L. Zhu, "Comprehensive assessments and zoning of ecological service functions for Beijing wetland based on RS and GIS," *Remote Sensing for Land and Resources*, vol. 25, no. 3, pp. 102–108, 2013.

[28] P. H. Verburg, K. H. Erb, O. Mertz, and G. Espindola, "Land system science: between global challenges and local realities," *Current Opinion in Environmental Sustainability*, vol. 5, no. 5, pp. 433–437, 2013.

[29] G. D. Xie, L. Zhen, C. X. Lu, Y. Xiao, and C. Chen, "Expert knowledge based valuation method of ecosystem services in China," *Journal of Natural Resources*, vol. 23, no. 5, pp. 911–919, 2008.

[30] F. Zhou, S. Y. Chen, and M. F. Wu, "Land utility structure changing trend and its response to ecological service functions—a case study in Zhanjiang City," *Journal of Safety and Environment*, vol. 7, no. 5, pp. 76–79, 2007.

[31] S. M. Feng, "Study on service coverage of distribution center based on Voronoi diagram," in *2009 International Conference on Information Engineering and Computer Science*, vol. 29no. 3, pp. 142–144, Wuhan, China, December 2009.

9

Ocean Wave Information Retrieval using Simulated Compact Polarized SAR from Radarsat-2

Xiaochen Wang,[1,2,3] Yun Shao,[1,2,3,4] Lu She,[2,3] Wei Tian,[1,2,3] Kun Li,[1,2,3] and Long Liu[1,2,3]

[1]*Laboratory of Target Microwave Properties, Deqing Academy of Satellite Applications, Huzhou 313200, China*
[2]*Institute of Remote Sensing and Digital Earth, Chinese Academy of Sciences, 1 Beichen Road, Beijing 100101, China*
[3]*University of Chinese Academy of Sciences, 19A Yuquan Road, Beijing 100049, China*
[4]*State Key Laboratory of Remote Sensing Science, Institute of Remote Sensing and Digital Earth, Chinese Academy of Sciences, 1 Beichen Road, Beijing 100101, China*

Correspondence should be addressed to Lu She; shelu_whu@163.com

Academic Editor: Biswajeet Pradhan

The main objective of this paper is to demonstrate the capability of compact polarized (CP) synthetic aperture radar (SAR) to retrieve ocean wave field parameters. Souyris' and Nord's algorithms are used to carry out the reconstruction of CP SAR pseudo quad-polarized data for the ocean surface under both the circular transmit linear receive (CTLR) and $\pi/4$ mode. The results show that, for the CP reconstruction, Nord's algorithm has a better convergence ability than Souyris'. In addition, the investigation of the reconstruction accuracy shows that the CTLR mode is superior to the $\pi/4$ mode, in terms of ocean surface reconstruction. It is, therefore, concluded that the reconstructed parameters of CP CTLR mode data by Nord's algorithm adapt to retrieve ocean wave information. The ocean wave slope spectrum and other main wave parameters are also calculated from reconstructed CP data and compared with measurements from in situ National Data Buoy Center (NDBC) matched-up buoys. Comparison of CP SAR-based wave field information with buoy outputs also shows good agreement in the case of dominate wave height, wave direction, and wave period, with biases of 0.36 m, 17.96°, and 0.88 s, respectively.

1. Introduction

Ocean waves are the most important ocean surface features in the fields of ocean physics and coastal engineering. Ocean surface parameters, such as wave height, wave direction, and wave length, are important for sea-ice monitoring, ship detection, marine pollution forecasting, and in other areas [1, 2]. The wave slope spectrum, which provides information about the 2-D directional distribution of the wave energy density, is important to ocean wave information retrieval. In recent years, remote sensing as an efficient tool for earth surface monitoring gradually plays a key role in ocean information retrieval. Of all remote sensing sensors, SAR systems are the active satellite sensors with a resolution high enough to provide information about the ocean wave spectrum and so these systems play a key role in the measurement of ocean surface waves [3–5]. In contrast to in situ buoys, SAR can operate in all weather conditions or in areas far from the shore and can measure ocean waves across the globe continuously [6]. In addition to improving the understanding of the properties of the ocean, SAR observations can be used to develop the forecast accuracy of the operational wave model. It is, therefore, important to investigate the capability of SAR to carry out ocean surface monitoring, especially in terms of ocean wave retrieval.

In recent decades, satellites such as ERS-1, ERS-2, Envisat, Radarsat-1/2, Sentinel-1 A/B, and Chinese GF-3 have carried on many SAR systems for use in ocean surface monitoring. Several wave retrieval algorithms have also been developed for ocean information retrieval. For example, K.

Hasselmann and S. Hasselmann [7, 8] proposed the Max Planck Institute (MPI) algorithm based on the closed nonlinear transformation relation between the ocean spectrum and the SAR image spectrum; the output from the wave model (WAM) is input as a first-guess spectrum in this method. Mastenbroek and de Valk [9] developed a semiparametric retrieval algorithm (SPRA) based on Hasselmann's nonlinear theory [7] by separating the wind-driven wave and swell—the wind-wave information is provided by synchronous scatterometer measurements and the swell information is derived from the rest of the SAR signal. Wang [10] proposed a semiempirical algorithm which has the objective to estimate the wave height from Envisat ASAR wave mode imagery without any prior knowledge. Shao et al. [11] develop a semiempirical function for significant wave height (Hs) and mean wave period (Tmw) retrieval from C-band VV polarization Sentinel-1 SAR. With the development of polarization technology, quad-polarized SAR has also been applied in ocean surface monitoring. Zhang et al. [12] present a synergistic method to retrieve both ocean surface wave and wind fields from spaceborne quad-polarization SAR, with consideration to the nonlinear mapping relationship between ocean wave spectra and SAR image spectra, in order to synergistically retrieve wind fields and wave directional spectra. Schuler and Lee [13, 14] proposed an algorithm for retrieving ocean wave spectra using quad-polarized SAR imagery and also investigated the results using L- and P-band SAR data. However, this method is only suitable for the L and P bands. By modifying the polarization orientation angle, He et al. [15, 16] developed a new retrieval algorithm so that it could be applied to C-band quad-polarized SAR data. He et al.'s method was also validated using Radarsat-2 quad-polarized SAR data by Zhang et al. [17, 18].

In addition, although the quad-polarized SAR can provide abundant polarization information to develop the process of ocean information retrieval, the limited swath width has also become the disadvantage of quad-polarized SAR for large-scale monitoring. For overcoming the limitation of CP SAR, a new type of SAR sensor called compact p(CP) SAR has been invented, which provides a new way to measure ocean waves and its potential for use in ocean surface monitoring has gradually become clear [19]. Compared with quad-polarized SAR, CP SAR operates at lower power but still has the capability to reconstruct quad-polarized data, thus overcoming the disadvantages of the narrow swath and high power of quad-polarized SAR and the insufficient polarized information provided by single- or dual-polarized SAR [20, 21]. CP SAR can, in fact, be interpreted as a compromise between dual-polarized and quad-polarized SAR so that swath width is increased and the operating power is reduced at the same time. To date, India's RISAT-1, launched in April 2012, and the Japanese ALOS-2 SAR satellite, launched in May 2014, have carried CP SAR. In the future, the planned Radarsat-2 constellation of satellites will also support CP SAR. Although CP SAR is carried on existing SAR platforms, there is still no well-calibrated data that can be used in research, especially in the case of ocean wave measurements. It is, therefore, important to study the potential of CP SAR for the retrieval of ocean wave information. In a previous study, Nunziata et al. [22] investigated sea-surface scattering with a polarized model and exploited the hybrid polarized (HP) mode to interpret ocean surface polarization parameters. Yin et al. [23] proposed a new extended X-Bragg model and also used C-band quad-polarized SAR data acquired by both SIR-C/X-SAR and Radarsat-2 to emulate the HP and $\pi/4$ CP modes. Geldsetzer et al. [24] proposed that the dependence between incidence angle, wind speed, and wind direction of CP SAR parameters are presented for open water. Selected 20 CP parameters are related to C-band geophysical model function (CMOD) output. CP parameters are simulated from polarized Radarsat-2 data and emulate data available on the pending Radarsat Constellation Mission (RCM). Li et al. [25] present a new empirical reconstruction algorithm for the modified N based on a data set of more than 2000 Radarsat-2 quad-polarization images and collocated buoy observations. It is proven that cross-polarization facilitates wind speed retrieval and improves the accuracy of the results, especially with respect to high wind speeds. In summary, the capability of CP SAR to reconstruct pseudo quad-pol data and ocean wind retrieval has been investigated in detailed; however, its potential for ocean surface reconstruction has not been discussed yet, let alone wave information retrieval.

In this paper, two quad-polarized SAR images are used to reconstruct the CP pseudo quad-polarized data and demonstrate the potential of CP SAR for ocean wave information retrieval. The data set used in this paper is illustrated in Section 2. In Section 3, we briefly introduced the CP theory and the ocean wave retrieval method. In Section 4, we investigate two CP algorithms used for ocean surface reconstruction and evaluate the optimum CP mode for ocean surface reconstruction. The FP and CP ocean wave retrieval results are compared with NDBC buoy data. Conclusions are given in Section 5.

2. Data Sets

2.1. SAR Data. Two scenes of Radarsat-2 quad-polarized image are used in this study. Both of two image acquisition parameters are listed in Table 1, and the study areas are shown in Figures 1(a) and 1(b). HH-, HV-, and VV-polarized single look complex image (SLC) SAR images are also shown in Figures 1(a) and 1(b).

2.2. Buoy Data. In order to validate the CP SAR wave retrieval results, NDBC buoy data available from the National Oceanic and Atmospheric Administration (NOAA) were collected along with the SAR imagery. The main wave parameters were extracted from the original measurements and are listed in Table 2. Among all the wave measurement data, the dominant wave height (DWH), the dominant wave period (DWP), and dominant wave direction (DWD) were used for validating the retrieval results.

Because of limited collection of NDBC buoy, only one buoy was collocated with SAR imagery of scene 2. So the imagery of scene 1 was performed to CP reconstruction and ocean slope spectrum retrieval methods, and scene 2 was added to validate the method.

TABLE 1: Radar parameters of C-band Radarsat-2 quad-polarized data used for this study.

Scene ID	Image file name	Observation time	Incidence angle	Scene position	
				Latitude	Longitude
1	RS2-SLC-FQ3-ASC-31-Mar-2008_18.10-PDS_00058910	31 Mar. 2008 18:10	20.8°–22.8°	35.96°N	5.65°W
2	RS2-SLC-FQ9-ASC-09-Apr-2008_02.01-PDS_00058900	09 Apr. 2008 02:01	28.0°–29.8°	37.82°N	122.46°W

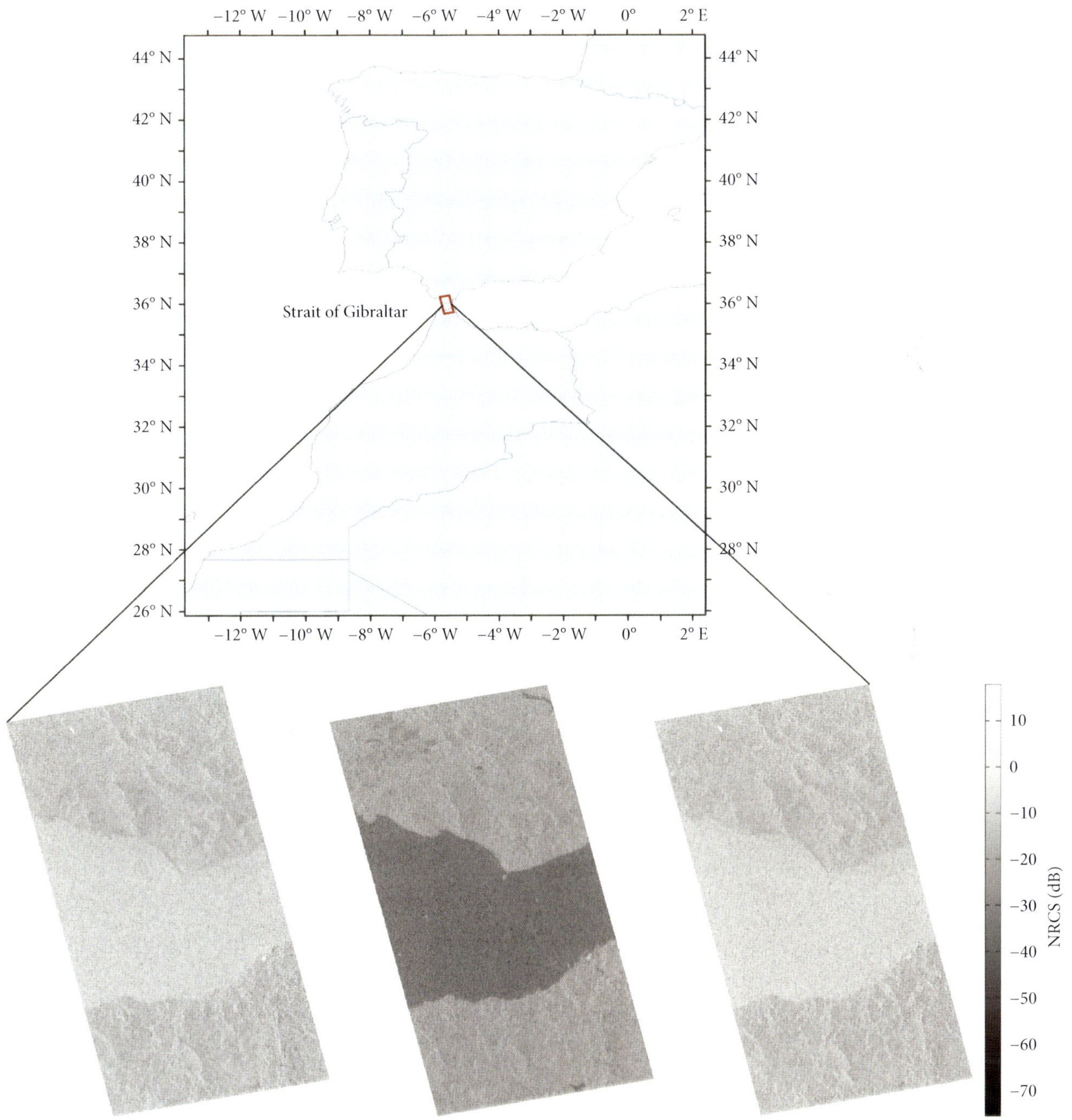

(A) HH (B) HV (VH) (C) VV

(a)

FIGURE 1: Continued.

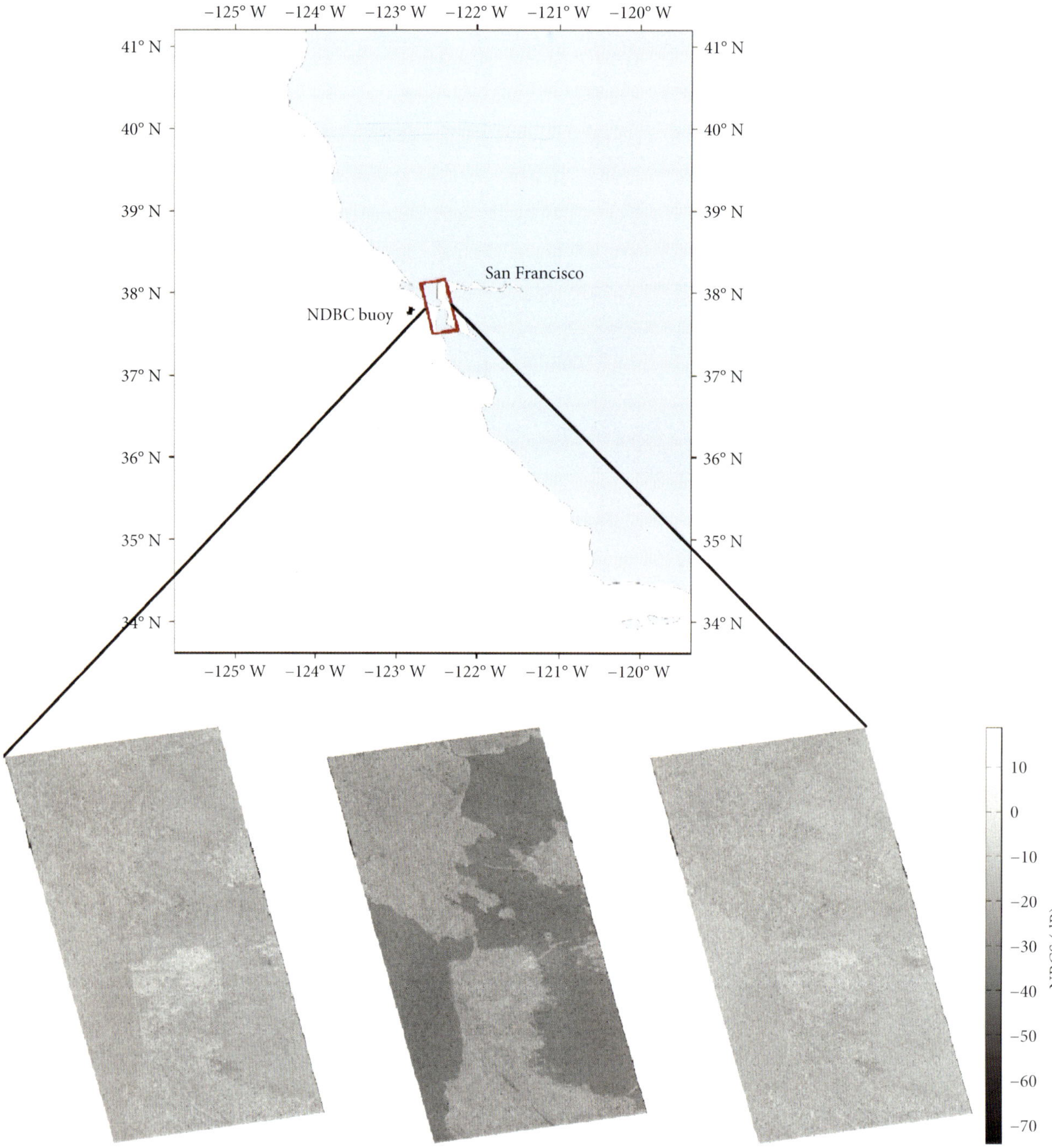

(A) HH (B) HV (VH) (C) VV

(b)

FIGURE 1: (a) The study area of scene 1: the coast near Strait of Gibraltar. The Radarsat-2 quad-polarized SLC SAR image was acquired on 2008/4/1. (A–C) correspond to the HH, HV (VH), and VV polarization channels, respectively. (b) The study area of scene 2: the coast near San Francisco. The Radarsat-2 quad-polarized SLC SAR image was acquired on 2008/4/9. The radar image and NDBC buoys are also marked. (A–C) correspond to the HH, HV (VH), and VV polarization channels, respectively.

3. Theory and Method

3.1. CP Reconstruction. CP theory is a newly developed method that can produce pseudo quad-polarized data from dual-polarized SAR data. The polarized information can be reconstructed in a satisfied precision with the decrease of sensor operation power [26, 27]. The CP has three main modes: the CTLR, dual circular polarized (DCP), and $\pi/4$ modes.

3.1.1. CTLR Mode. In the CTLR mode, the microwaves are transmitted in left-handed or right-handed circular

TABLE 2: Standard meteorological record from an NDBC in situ buoy.

Buoy ID	46237
Date	2008/4/9 01:51
Location	37.786 N 122.634 W
Water depth (m)	16.8
Dominant wave height (m)	2.18
Dominant wave period (s)	14.29
Average wave period (s)	7.03
Mean wave direction (°)	246

polarization and received in HH and VV polarization. The scattering vector $\overrightarrow{\mathbf{k}}_{\text{CTLR}}$ is expressed as

$$\overrightarrow{\mathbf{k}}_{\text{CTLR}} = \frac{[S_{\text{hh}} - iS_{\text{hv}} - iS_{\text{vv}} + S_{\text{hv}}]^T}{2}. \tag{1}$$

The relevant Hermitian covariance matrix is

$$\begin{aligned}\mathbf{C}_{\text{CTLR}} &= \left\langle \overrightarrow{\mathbf{k}}_{\text{CTLR}} \overrightarrow{\mathbf{k}}^{*T}_{\text{CTLR}} \right\rangle \\ &= \frac{1}{2}\begin{bmatrix} \langle |\mathbf{S}_{\text{hh}}|^2 \rangle & \langle i(\mathbf{S}_{\text{hh}}\mathbf{S}^*_{\text{vv}}) \rangle \\ \langle -i(\mathbf{S}_{\text{vv}}\mathbf{S}^*_{\text{hh}}) \rangle & \langle |\mathbf{S}_{\text{vv}}|^2 \rangle \end{bmatrix} \\ &+ \frac{1}{2}\langle |\mathbf{S}_{\text{hv}}|^2 \rangle \begin{bmatrix} 1 & -i \\ i & 1 \end{bmatrix} \\ &+ \frac{1}{2}\begin{bmatrix} -2\Im(\langle \mathbf{S}_{\text{hh}}\mathbf{S}^*_{\text{hv}} \rangle) & \langle \mathbf{S}_{\text{hh}}\mathbf{S}^*_{\text{hv}} \rangle + \langle \mathbf{S}_{\text{hv}}\mathbf{S}^*_{\text{vv}} \rangle \\ \langle \mathbf{S}^*_{\text{hh}}\mathbf{S}_{\text{hv}} \rangle + \langle \mathbf{S}_{\text{vv}}\mathbf{S}^*_{\text{hv}} \rangle & 2\Im(\langle \mathbf{S}_{\text{vv}}\mathbf{S}^*_{\text{hv}} \rangle) \end{bmatrix},\end{aligned} \tag{2}$$

where T stands for matrix transpose operations, $*$ stands for matrix conjugate operations, $\langle\rangle$ stands for average operations, $\mathscr{R}$ is the real part of complex number, $\mathbf{S}_{\text{pq}}$ is polarization scattering matrix and p, q are horizontal or vertical polarized channel, i is the imaginary unit, and $\Im$ is the imaginary part of complex number. It is also noticeable that this matrix can be simplified based on the assumption that the reflection is symmetrical.

$$\begin{aligned}\mathbf{C}_{\text{CTLR}} &= \left\langle \overrightarrow{\mathbf{k}}_{\text{CTLR}} \overrightarrow{\mathbf{k}}^{*T}_{\text{CTLR}} \right\rangle \\ &= \frac{1}{2}\begin{bmatrix} \langle |\mathbf{S}_{\text{hh}}|^2 \rangle & \langle i(\mathbf{S}_{\text{hh}}\mathbf{S}^*_{\text{vv}}) \rangle \\ \langle -i(\mathbf{S}_{\text{vv}}\mathbf{S}^*_{\text{hh}}) \rangle & \langle |\mathbf{S}_{\text{vv}}|^2 \rangle \end{bmatrix} \\ &+ \frac{1}{2}\langle |\mathbf{S}_{\text{hv}}|^2 \rangle \begin{bmatrix} 1 & -i \\ i & 1 \end{bmatrix}.\end{aligned} \tag{3}$$

In order to reconstruct a pseudo quad-polarized covariance matrix for the CP CTLR mode, $\langle |\mathbf{S}_{\text{hv}}|^2 \rangle$ also needs to be iteratively solved. This covariance matrix can be expressed as

$$[\mathbf{C}_{\text{CTLR}}]_{\text{pseudo-quad}} = \begin{bmatrix} C_{11} - \langle |\mathbf{S}_{\text{hv}}|^2 \rangle & 0 & iC_{12} + \langle |\mathbf{S}_{\text{hv}}|^2 \rangle \\ 0 & 2\langle |\mathbf{S}_{\text{hv}}|^2 \rangle & 0 \\ \left(iC_{12} + \langle |\mathbf{S}_{\text{hv}}|^2 \rangle\right)^* & 0 & C_{22} - \langle |\mathbf{S}_{\text{hv}}|^2 \rangle \end{bmatrix}. \tag{4}$$

It is noticed that C_{ij} in (4) is element from CP 2×2 covariance matrix.

3.1.2. DCP Mode. In DCP mode, the microwaves are transmitted in right-handed circular polarization and received in left-handed or right-handed circular polarization. The scattering vector $\overrightarrow{\mathbf{k}}_{\text{DCP}}$ is expressed as

$$\overrightarrow{\mathbf{k}}_{\text{DCP}} = \frac{[\mathbf{S}_{\text{hh}} - \mathbf{S}_{\text{vv}} + i2\mathbf{S}_{\text{hv}}\ i(\mathbf{S}_{\text{hh}} + \mathbf{S}_{\text{vv}})]^T}{2}. \tag{5}$$

The relevant Hermitian covariance matrix is

$$\begin{aligned}\mathbf{C}_{\text{DCP}} &= \left\langle \overrightarrow{\mathbf{k}}_{\text{DCP}} \overrightarrow{\mathbf{k}}^{*T}_{\text{DCP}} \right\rangle \\ &= \frac{1}{4}\begin{bmatrix} \langle |\mathbf{S}_{\text{hh}} - \mathbf{S}_{\text{vv}}|^2 \rangle & \langle -i(\mathbf{S}_{\text{hh}} - \mathbf{S}_{\text{vv}})(\mathbf{S}_{\text{hh}} + \mathbf{S}_{\text{vv}})^* \rangle \\ \langle -i(\mathbf{S}_{\text{hh}} + \mathbf{S}_{\text{vv}})(\mathbf{S}_{\text{hh}} - \mathbf{S}_{\text{vv}})^* \rangle & \langle |\mathbf{S}_{\text{hh}} + \mathbf{S}_{\text{vv}}|^2 \rangle \end{bmatrix} \\ &+ \frac{1}{4}\begin{bmatrix} 4\langle |\mathbf{S}_{\text{hv}}|^2 \rangle & 0 \\ 0 & 0 \end{bmatrix} \\ &+ \frac{1}{4}\begin{bmatrix} 4\Im(\langle(\mathbf{S}_{\text{hh}} - \mathbf{S}_{\text{vv}}) \cdot \mathbf{S}^*_{\text{hv}}\rangle) & 2\langle(\mathbf{S}_{\text{hh}} + \mathbf{S}_{\text{vv}})^*\rangle \\ 2\langle(\mathbf{S}_{\text{hh}} + \mathbf{S}_{\text{vv}}) \cdot \mathbf{S}^*_{\text{hv}}\rangle & 0 \end{bmatrix}.\end{aligned} \tag{6}$$

It should be noted that the last term can be set to zero based on the assumption of reflectional symmetry and that the covariance matrix can be simplified to

$$\begin{aligned}\mathbf{C}_{\text{DCP}} &= \left\langle \overrightarrow{\mathbf{k}}_{\text{DCP}} \overrightarrow{\mathbf{k}}^{*T}_{\text{DCP}} \right\rangle \\ &= \frac{1}{4}\begin{bmatrix} \langle |\mathbf{S}_{\text{hh}} - \mathbf{S}_{\text{vv}}|^2 \rangle & \langle -i(\mathbf{S}_{\text{hh}} - \mathbf{S}_{\text{vv}})(\mathbf{S}_{\text{hh}} + \mathbf{S}_{\text{vv}})^* \rangle \\ \langle -i(\mathbf{S}_{\text{hh}} + \mathbf{S}_{\text{vv}})(\mathbf{S}_{\text{hh}} - \mathbf{S}_{\text{vv}})^* \rangle & \langle |\mathbf{S}_{\text{hh}} + \mathbf{S}_{\text{vv}}|^2 \rangle \end{bmatrix} \\ &+ \frac{1}{4}\begin{bmatrix} 4\langle |\mathbf{S}_{\text{hv}}|^2 \rangle & 0 \\ 0 & 0 \end{bmatrix}.\end{aligned} \tag{7}$$

In order to reconstruct a pseudo quad-polarized covariance matrix for the CP DCP mode, $\langle |\mathbf{S}_{\text{hv}}|^2 \rangle$ needs to be iteratively solved. This covariance matrix can be expressed as [28]

$$[\mathbf{C}_{\mathrm{DCP}}]_{\text{pseudo-quad}} = \begin{bmatrix} C_{11} - \langle |\mathbf{S}_{\mathrm{hv}}|^2 \rangle & 0 & C_{12} + \langle |\mathbf{S}_{\mathrm{hv}}|^2 \rangle \\ 0 & 2\langle |\mathbf{S}_{\mathrm{hv}}|^2 \rangle & 0 \\ (C_{12} + \langle |\mathbf{S}_{\mathrm{hv}}|^2 \rangle)^* & 0 & C_{22} - \langle |\mathbf{S}_{\mathrm{hv}}|^2 \rangle \end{bmatrix}. \tag{8}$$

It is noticed that C_{ij} in (8) is element from CP 2×2 covariance matrix.

3.1.3. $\pi/4$ Mode. In $\pi/4$ mode, the microwaves are transmitted in linear polarization with a 45° inclination and received in HH and VV polarization. The scattering vector $\overrightarrow{\mathbf{k}}_{\pi/4}$ is expressed as

$$\overrightarrow{\mathbf{k}}_{\pi/4} = \frac{[\mathbf{S}_{\mathrm{hh}} + \mathbf{S}_{\mathrm{hv}}\ \mathbf{S}_{\mathrm{vv}} + \mathbf{S}_{\mathrm{hv}}]^T}{\sqrt{2}}. \tag{9}$$

The relevant Hermitian covariance matrix is

$$\begin{aligned} \mathbf{C}_{\pi/4} &= \left\langle \overrightarrow{\mathbf{k}}_{\pi/4} \overrightarrow{\mathbf{k}}_{\pi/4}^{*T} \right\rangle \\ &= \frac{1}{2} \begin{bmatrix} \langle |\mathbf{S}_{\mathrm{hh}}|^2 \rangle & \langle \mathbf{S}_{\mathrm{hh}} \mathbf{S}_{\mathrm{vv}}^* \rangle \\ \langle \mathbf{S}_{\mathrm{vv}} \mathbf{S}_{\mathrm{hh}}^* \rangle & \langle |\mathbf{S}_{\mathrm{vv}}|^2 \rangle \end{bmatrix} + \frac{1}{2} \langle |\mathbf{S}_{\mathrm{hv}}|^2 \rangle \begin{bmatrix} 1 & 1 \\ 1 & 1 \end{bmatrix} \\ &+ \frac{1}{2} \begin{bmatrix} 2\mathscr{R}(\langle \mathbf{S}_{\mathrm{hh}} \mathbf{S}_{\mathrm{hv}}^* \rangle) & \langle \mathbf{S}_{\mathrm{hh}} \mathbf{S}_{\mathrm{hv}}^* \rangle + \langle \mathbf{S}_{\mathrm{hv}} \mathbf{S}_{\mathrm{vv}}^* \rangle \\ \langle \mathbf{S}_{\mathrm{hh}}^* \mathbf{S}_{\mathrm{hv}} \rangle + \langle \mathbf{S}_{\mathrm{vv}} \mathbf{S}_{\mathrm{hv}}^* \rangle & 2\mathscr{R}(\langle \mathbf{S}_{\mathrm{vv}} \mathbf{S}_{\mathrm{hv}}^* \rangle) \end{bmatrix}. \end{aligned} \tag{10}$$

It should be noted that the last term can be set to zero based on the assumption of reflectional symmetry and that the covariance matrix can be simplified to

$$\begin{aligned} \mathbf{C}_{\pi/4} &= \left\langle \overrightarrow{\mathbf{k}}_{\pi/4} \overrightarrow{\mathbf{k}}_{\pi/4}^{*T} \right\rangle = \frac{1}{2} \begin{bmatrix} \langle |\mathbf{S}_{\mathrm{hh}}|^2 \rangle & \langle \mathbf{S}_{\mathrm{hh}} \mathbf{S}_{\mathrm{vv}}^* \rangle \\ \langle \mathbf{S}_{\mathrm{vv}} \mathbf{S}_{\mathrm{hh}}^* \rangle & \langle |\mathbf{S}_{\mathrm{vv}}|^2 \rangle \end{bmatrix} \\ &+ \frac{1}{2} \langle |\mathbf{S}_{\mathrm{hv}}|^2 \rangle \begin{bmatrix} 1 & 1 \\ 1 & 1 \end{bmatrix}. \end{aligned} \tag{11}$$

In order to reconstruct the pseudo quad-polarized covariance matrix for the CP $\pi/4$ mode, $\langle |\mathbf{S}_{\mathrm{hv}}|^2 \rangle$ needs to be iteratively solved. The pseudo quad-polarized covariance matrix for the $\pi/4$ mode can be expressed as [28]

$$[\mathbf{C}_{\pi/4}]_{\text{pseudo-quad}} = \begin{bmatrix} C_{11} - \langle |\mathbf{S}_{\mathrm{hv}}|^2 \rangle & 0 & C_{12} - \langle |\mathbf{S}_{\mathrm{hv}}|^2 \rangle \\ 0 & 2\langle |\mathbf{S}_{\mathrm{hv}}|^2 \rangle & 0 \\ (C_{12} - \langle |\mathbf{S}_{\mathrm{hv}}|^2 \rangle)^* & 0 & C_{22} - \langle |\mathbf{S}_{\mathrm{hv}}|^2 \rangle \end{bmatrix}. \tag{12}$$

It is noticed that C_{ij} in (12) is element from CP 2×2 covariance matrix.

It should also be noted that the scattering vectors $\overrightarrow{\mathbf{k}}_{\mathrm{CTLR}}$ and $\overrightarrow{\mathbf{k}}_{\mathrm{DCP}}$ are related by

$$\overrightarrow{\mathbf{k}}_{\mathrm{DCP}} = \frac{1}{\sqrt{2}} \begin{bmatrix} 1 & -i \\ 1 & i \end{bmatrix} \overrightarrow{\mathbf{k}}_{\mathrm{CTLR}}. \tag{13}$$

The polarization information for the CTLR and DCP modes will be similar because of the above linear relationship, which means that the CP reconstruction result for these two modes will be the same [29, 30].

3.2. Wave Spectrum Retrieval Algorithm. For the linearly polarized backscattering cross section, the quad-polarized covariance matrix is used for the ocean surface wave polarization signatures, as follows [31]:

$$\mathbf{C} = \begin{bmatrix} C_{11} & C_{12} & C_{13} \\ C_{21} & C_{22} & C_{23} \\ C_{31} & C_{32} & C_{33} \end{bmatrix} = \begin{bmatrix} |\mathbf{S}_{\mathrm{hh}}|^2 & \sqrt{2}(\mathbf{S}_{\mathrm{hh}} \mathbf{S}_{\mathrm{hv}}^*) & \mathbf{S}_{\mathrm{hh}} \mathbf{S}_{\mathrm{vv}}^* \\ \sqrt{2}(\mathbf{S}_{\mathrm{hv}} \mathbf{S}_{\mathrm{hh}}^*) & 2|\mathbf{S}_{\mathrm{hv}}|^2 & \sqrt{2}(\mathbf{S}_{\mathrm{hv}} \mathbf{S}_{\mathrm{vv}}^*) \\ \mathbf{S}_{\mathrm{vv}} \mathbf{S}_{\mathrm{hh}}^* & \sqrt{2}(\mathbf{S}_{\mathrm{vv}} \mathbf{S}_{\mathrm{hv}}^*) & |\mathbf{S}_{\mathrm{vv}}|^2 \end{bmatrix}. \tag{14}$$

It is noticed that C_{ij} in (14) is element from FP 3×3 covariance matrix. Because the maximum contribution to the wave slope is obtained using only linear polarization, here, we set the ellipticity to zero and so the linearly polarized backscattering cross section at any polarization orientation angle can be derived as [14, 15]:

$$\begin{aligned} \sigma(0, \varphi) &= \frac{1}{4}(\sigma_{\mathrm{hh}} + \sigma_{\mathrm{vv}})\left[1 + \cos^2(2\varphi)\right] \\ &+ \frac{1}{2}(\sigma_{\mathrm{hh}} - \sigma_{\mathrm{vv}}) \cos 2\varphi + \sigma_{\mathrm{hv}} \\ &+ \frac{1}{2}\mathfrak{R}(\sigma_{\mathrm{hhvv}}) \sin^2(2\varphi), \end{aligned} \tag{15}$$

where σ_{hh}, σ_{hv}, and σ_{vv} are the linearly polarized backscattering cross sections for the horizontal, cross, and vertical polarizations, respectively. σ_{hhvv} is the correlation between the horizontal and vertical polarizations, which can be directly derived from the covariance matrix.

In the linear modulation theory of ocean waves, the ocean surface height can be regarded as a superposition of sine waves and the radar backscattering cross section expressed as [16, 32, 33]

$$\xi(r, t) = \sum_k \xi_k \exp\left[i(k \cdot \mathbf{r} - \omega t)\right] + \mathrm{cc}, \tag{16}$$

$$\sigma_{\mathrm{pp}}(r, t) = \overline{\sigma_{\mathrm{pp}}}\left(1 + \left\{\sum_k T_{\mathrm{PPK}}^{\mathrm{R}} \xi_k \exp\left[i(k \cdot \mathbf{r} - \omega t)\right] + \mathrm{cc}\right\}\right), \tag{17}$$

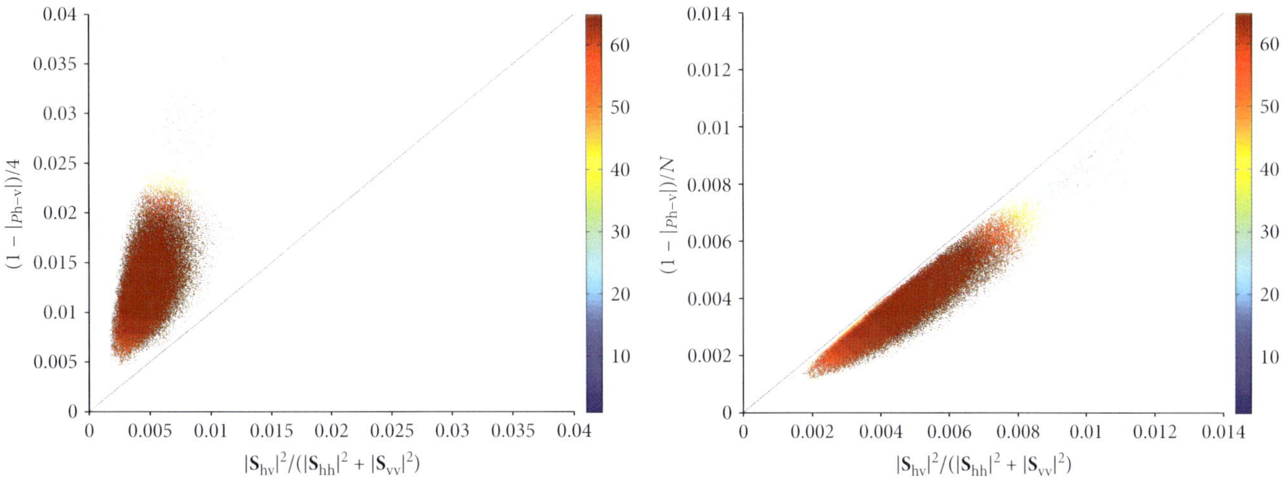

(a) Souyris' algorithm

(b) Nord's algorithm

FIGURE 2: Scatter plot between $|\mathbf{S}_{hv}|^2/(|\mathbf{S}_{hh}|^2 + |\mathbf{S}_{vv}|^2)$ and $(1-\rho_{h-v})/N$. The value of N in (a) is 4 whereas in (b) it is calculated as $|\mathbf{S}_{hh} - \mathbf{S}_{vv}|^2/|\mathbf{S}_{hv}|^2$. The convergence capability is clearly illustrated.

where ξ is the ocean surface height, k is the wavenumber, $\mathbf{r}$ is the wave space vector, ω is the wave angular frequency, and cc is the complex conjugate of the series. T^{R}_{PPK} is the modulation transfer function for SAR, which includes the tilt modulation, hydrodynamic modulation, polarization angle modulation, range and azimuthal direction modulation, and velocity bunching modulation. In all modulation functions, only the tilt and polarization orientation angle modulation depend on the radar polarization [16]. Moreover, the wave slope induced by the polarization orientation angle modulation can be neglected for horizontal or vertical polarization [16]. By (18), the terms that are not sensitive to the polarization vanish and the equation can be simplified to [16]

$$\frac{\Delta\sigma_{vv}}{\sigma_{vv}} - \frac{\Delta\sigma_{hh}}{\sigma_{hh}} = \sum_k \left(T^t_{vv} - T^t_{hh}\right)\xi_k \exp\left[i(k\cdot\mathbf{r} - \omega t)\right] + c_1,$$
$$\frac{\Delta\sigma_{\varphi\varphi}}{\sigma_{\varphi\varphi}} - \frac{\Delta\sigma_{vv}}{\sigma_{vv}} = \sum_k \left(T^t_{\varphi\varphi} - T^t_{vv} + T^p_{\varphi\varphi}\right)\xi_k \exp\left[i(k\cdot\mathbf{r} - \omega t)\right] + c_2, \tag{18}$$

where c_1 and c_2 are the complex conjugates of series, respectively; T^t_{vv} and T^t_{hh} are given by [16]

$$T^t_{vv} = ik_x \frac{4 - 0.5\left(1 - \sin^2\theta\right)}{\tan\theta\left(1 - \sin^2\theta\right)}, \tag{19}$$

$$T^t_{hh} = ik_x \frac{4 - 0.5\left(1 + \sin^2\theta\right)}{\tan\theta\left(1 + \sin^2\theta\right)}, \tag{20}$$

where k_x is the radar incidence wavenumber and θ is the incidence angle. $T^t_{\varphi\varphi}$ and $T^p_{\varphi\varphi}$ are given by (16) in [16]. By consolidating the equations, the straightforward algebraic calculation can be summarized as [16]

$$\frac{\Delta\sigma_{vv}}{\sigma_{vv}} - \frac{\Delta\sigma_{hh}}{\sigma_{hh}} = -\frac{8\tan\theta}{1+\sin^2\theta}\frac{\delta\xi}{\delta x},$$
$$\frac{\Delta\sigma_{\varphi\varphi}}{\sigma_{\varphi\varphi}} - \frac{\Delta\sigma_{vv}}{\sigma_{vv}} = A\frac{\delta\xi}{\delta x} + B\frac{\delta\xi}{\delta y}, \tag{21}$$

where $\delta\xi/\delta x$ and $\delta\xi/\delta y$ are the range and azimuthal wave slope, respectively. Parameters A and B are also given by (8) [16]. With an appropriate choice of φ, the range wave slope and azimuthal wave slope can be directly obtained. The wave slope spectrum and the other wave parameters can be estimated from wave slope.

4. Results and Discussion

4.1. Comparision of Reconstruction Algorithms. Besides the reflectional symmetry hypothesis, the relation between the copolarized and cross-polarized responses is the main condition limiting the solution of the CP reconstruction equations [21, 34].

$$\frac{|\mathbf{S}_{hv}|^2}{|\mathbf{S}_{hh}|^2 + |\mathbf{S}_{vv}|^2} \approx \frac{1-\rho_{h-v}}{N}. \tag{22}$$

Although both Souyris et al.'s [21] and Nord et al.'s [34] algorithms are used to reconstruct pseudo quad-polarized data, they differ in their definitions of N. Souyris et al. use 4 as the value of N, while Nord et al. calculate the value as

$$N = \frac{|\mathbf{S}_{hh} - \mathbf{S}_{vv}|^2}{|\mathbf{S}_{hv}|^2}. \tag{23}$$

We, therefore, used the two reconstruction algorithms to calculate the slope rate of curve between $|\mathbf{S}_{hv}|^2/|\mathbf{S}_{hh}|^2 + \mathbf{S}_{vv}{}^2$ and $(1-\rho_{h-v}/N)$ (Figure 2). It is important to note that N in Figure 2(a) is equal to 4 while in Figure 2(b)N is given

Table 3: Convergence capability for CP reconstruction as found using Souyris' and Nord's algorithms.

Image ID	Total number of pixels	Number of nonconvergent pixels			
		CTLR_Souyris	$\pi/4$_Souyris	CTLR_Nord	$\pi/4$_Nord
1	6,220,800	7651	11,341	3962	1612
2	4,176,000	173	344	41	81

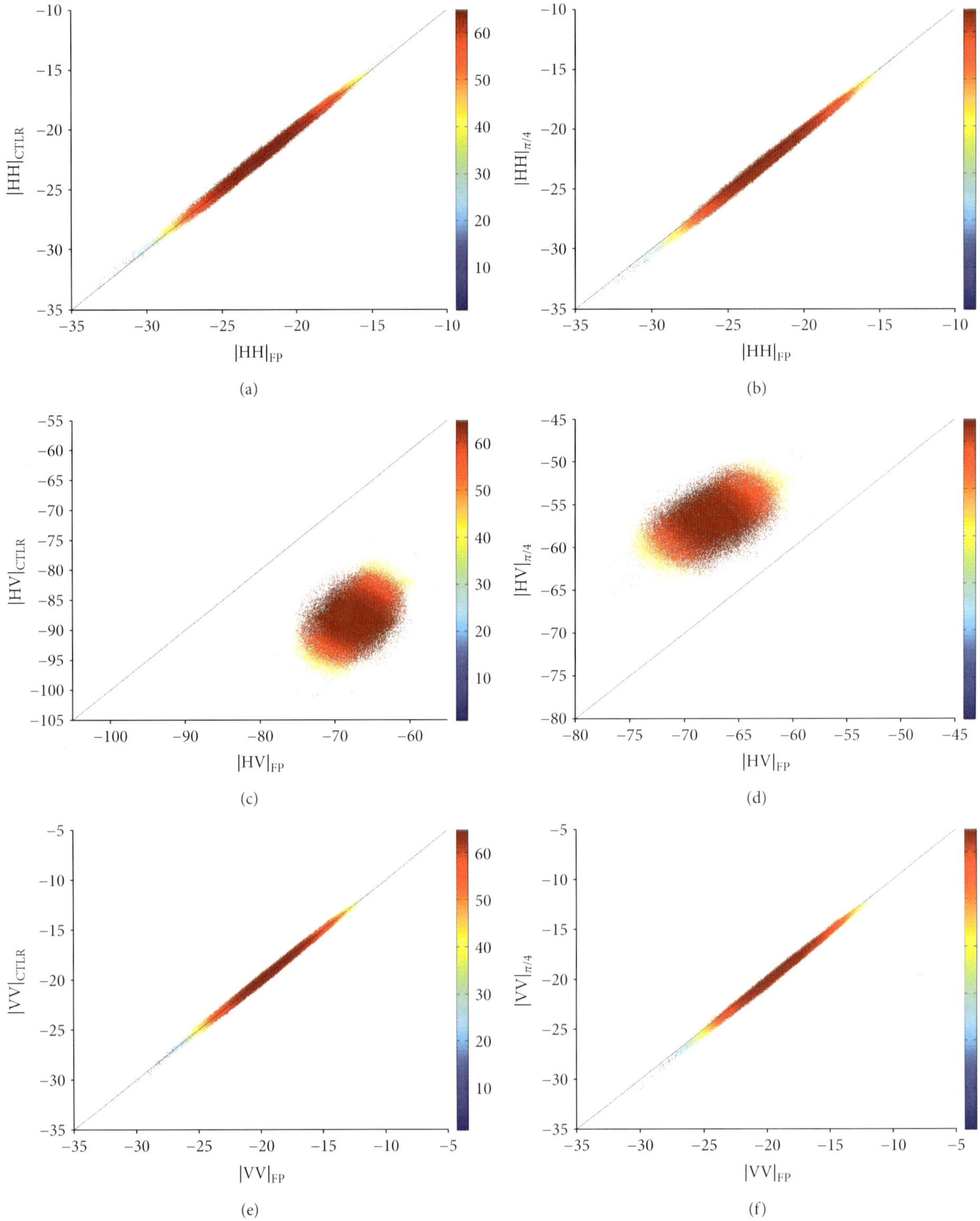

Figure 3: Reconstruction performance for HH, HV, and VV channels: scatter plots of FP and CP. (a), (c), and (e) correspond to the CTLR mode; (b), (d), and (f) correspond to the $\pi/4$ mode.

TABLE 4: Reconstruction accuracy of different polarized parameters for CTLR and $\pi/4$ modes.

Mode	σ_{hh}	σ_{hv}	σ_{vv}	ρ	l_1	l_2	l_3
CTLR	−0.0118	0.8658	−0.0050	0.0184	0.0014	−0.3394	0.8521
$\pi/4$	0.0205	−2.0466	0.0183	0.0202	0.0289	−0.3300	−0.3577

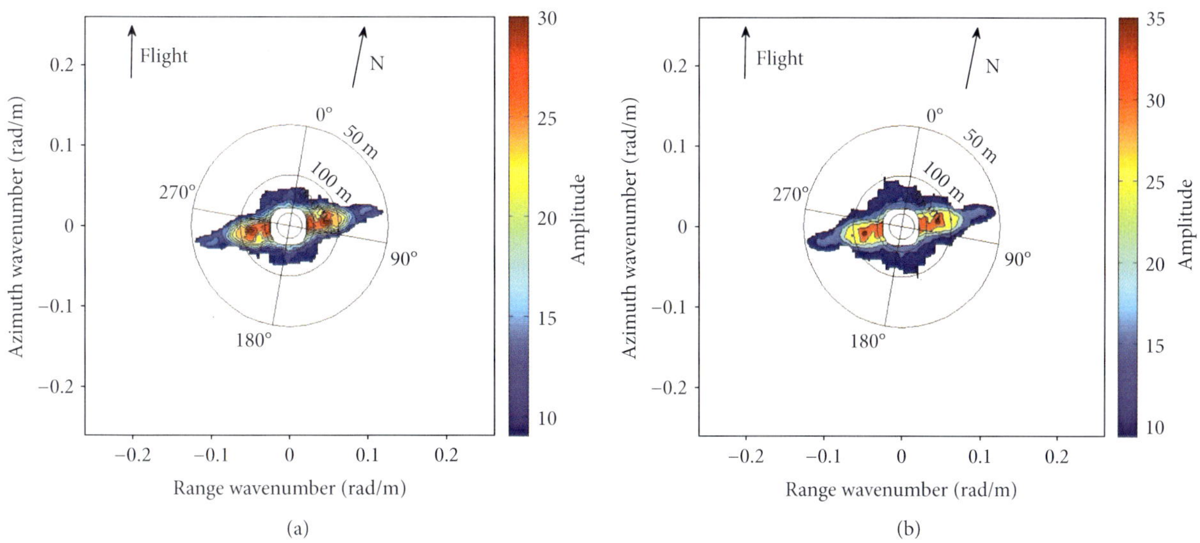

FIGURE 4: Ocean wave slope spectra derived from (a) FP data and (b) CP (CTLR mode) data (b). The color bar represents the wave energy amplitude.

FIGURE 5: The positions of subimages used for ocean information retrieval in scene 2, 1–4 represent four 512 × 512 pixel size subimage.

by $|\mathbf{S}_{hh} - \mathbf{S}_{vv}|^2/|\mathbf{S}_{hv}|^2$, the value of which was calculated from FP data. We found that the convergence of Figure 2(a) was much inferior to Figure 2(b); the correlation coefficients, r, were equal to 0.53 and 0.93, respectively. To determine the optimum algorithm, we also calculated statistics for the nonconvergence of pixels during the reconstruction process. The results for this are shown in Table 3. The number of nonconvergent pixels found when using the two subimages of the ocean surface was much lower for Nord's algorithm than for Souyris' algorithm. Nord's algorithm is, therefore, more suitable for reconstructing pseudo quad-polarized data of the ocean surface.

4.2. Evaluation of CP Reconstruction Accuracy. CP SAR has three main modes: CTLR mode, DCP mode, and $\pi/4$ mode. Each CP mode has different orientation and ellipticity angles, which describe the transmitted and received wave status. Because of the linear relationship between the DCP and CTLR modes, these two modes have similar polarization information; therefore, we discuss only the $\pi/4$ mode and the CTLR mode in this paper. Figure 3 shows scatter plots detailing how well the derived pseudo quad-pol results fit the original quad-pol ones. The figure shows that the values of σ_{hh} and σ_{vv} between FP and CP are concentrated along the one-to-one line, which means that for σ_{hh} and σ_{vv}, the reconstruction of the pseudo quad-polarized data was good. However, the reconstruction of σ_{hv} shows a slight underestimation in the CTLR mode and an overestimation in the $\pi/4$ mode, which means that the reconstruction of σ_{hv} was not as good as for σ_{hh} and σ_{vv}. The reason for the low reconstruction accuracy for σ_{hv} may be related to the weak volume scattering at the ocean surface. It has been previously documented that the CP reconstruction algorithms are generally applied to fields and forests, where the total scattering contains a strong volume scattering component. However, the cross-polarization information is not directly used in our paper, Therefore, the inferior reconstruction performance of σ_{hv} will not make any influence on retrieval results.

To examine the reconstruction performance in more detail, we also selected 7 types of polarization parameters, including copolarization backscattering coefficientσ_{hh}, σ_{vv}, cross-polarization backscattering coefficientσ_{hv}, copolarization correlation coefficientρ, and coherence matrix eigenvaluesl_1,l_2, andl_3, for each parameter, calculating the reconstruction accuracy for the CTLR and $\pi/4$ modes. The

TABLE 5: Comparison of ocean parameters derived from FP quad-pol and CP CTLR mode pseudo quad-pol data with NDBC buoy measurements.

Parameters	Mode	Zone	Retrieved value	Buoy	Error	Average error
Dominant wave height (m)	Fp	1	1.99	2.18	0.19	0.14
		2	2.35		0.17	
		3	2.34		0.16	
		4	2.14		0.04	
	Cp	1	2.37		0.19	0.36
		2	2.69		0.51	
		3	2.62		0.44	
		4	2.38		0.20	
Wave direction (°)	Fp	1	259.1	246	19.25	17.12
		2	264.57		19.38	
		3	265.06		18.02	
		4	263.73		15.19	
	Cp	1	265.25		13.1	17.96
		2	265.38		18.57	
		3	264.02		19.06	
		4	261.19		17.73	
Wave period (s)	Fp	1	14.41	14.29	0.12	0.61
		2	13.74		0.54	
		3	12.72		1.56	
		4	14.52		0.23	
	Cp	1	14.01		0.27	0.88
		2	13.31		0.97	
		3	12.58		1.70	
		4	14.88		0.59	

reconstruction accuracy $(\text{Pol}_{\text{cp}} - \text{Pol}_{\text{fp}})/\text{Pol}_{\text{fp}}$ was used to describe the amount of deviation in the CP reconstruction. The reconstruction accuracies for σ_{hh}, σ_{hv}, σ_{vv}, ρ, l_1, l_2, and l_3 are listed in Table 4. We used the descriptions "high," "medium," and "low" to describe the deviations, where "high" was defined as less than 10%, "medium" corresponded to 10%–30%, and "low" was defined as larger than 30%. Table 4 shows that, for both the CTLR and $\pi/4$ modes, the reconstruction accuracy for the parameters σ_{hh}, σ_{vv}, ρ, and l_1 was high, with the values for the CTLR mode being superior to those for the $\pi/4$ mode. We, therefore, decided to use the CRTL mode for the subsequent ocean surface parameter reconstruction and ocean wave slope spectrum retrieval.

4.3. Retrieval of Ocean Wave Slope Spectra. In this section, we describe how Nord's algorithm was used to reconstruct the pseudo quad-pol data in the CTLR mode. The retrieved ocean wave slope spectra based on the reconstructed pseudo quad-polarized data are shown in Figure 4. Compared with the ocean wave slope spectrum derived from the FP data, the results for the CP CTLR mode obtained in this study are satisfied. It is also clear that the wave slope spectrum derived from FP presents more details compared to that derived from the CP data. However, the two wave slope spectra are similar on the whole.

In order to make a quantitative assessment of the wave retrieval results, the main wave parameters of dominant wave height, wave direction, and wave period were also extracted from the wave slope spectra and compared with the NDBC buoy data. The expressions of main wave parameters are

$$\begin{aligned} H_{\text{d}} &= \tan\left(S_{\text{rms}}\right)\frac{\lambda_{\text{d}}}{2}, \\ \text{WD}_{\text{d}} &= \tan\left(\frac{\lambda_{\text{d\_az}}}{\lambda_{\text{d\_r}}}\right), \\ \text{WP}_{\text{d}} &= \frac{2\pi}{\sqrt{k_{\text{d}} g \tan h(k_{\text{d}} h)}}, \end{aligned} \tag{24}$$

where S_{rms} represents the root mean square of wave slope, λ_{d} represents the dominated wave length, $\lambda_{\text{d\_az}}$ and $\lambda_{\text{d\_r}}$, respectively, represent the components of λ_{d} in azimuth and range direction, k_{d} is the dominated wavenumber, g is acceleration of gravity, and h is the water depth.

Figure 5 shows four subimages' position which is performed to retrieve result statistics, and the 180° ambiguity of wave slope spectrum has been removed according to the measured wind direction of buoy. The results shown in Table 5 show that the average errors in the dominant wave height, wave direction, and wave period derived from the FP data are 0.14, 17.12, and 0.61, while those derived from the CP data are 0.36, 17.96, and 0.88, respectively. It should be noted that, although the average errors in the dominant wave height, wave direction, and wave period derived from the FP data were much larger than for the CP, the retrieved CP SAR results are also satisfactory for ocean wave measurements. It is also true that the retrieval accuracy obtained using CP SAR in this study was limited by the reconstruction of the CP pseudo quad-pol data—this was especially due to the loss in precision for σ_{hv}. In future studies, the retrieval accuracy obtained using CP SAR will need to be improved.

5. Conclusion

In this paper, we investigated the capacity and potential of CP for ocean wave measurement. A Radarsat-2 quad-polarized image was used to reconstruct CP pseudo quad-polarized data and validate the potential of this for ocean slope spectrum retrieval. The main conclusions of this study are as follows.

For ocean surface waves, Nord's reconstruction algorithm is much more effective than Souyris'. Because of its convergence capacity and iteration time, we recommend adopting Nord's algorithm for the reconstruction of CP pseudo quad-polarized data.

For both the CTLR and $\pi/4$ modes, the reconstruction accuracy for the parameters σ_{hh}, σ_{vv}, ρ, and l_1 was high whereas for σ_{hv}, l_2, and l_3 it was low. In terms of the retrieval of ocean wave slope spectra using the CP mode, the CTLR mode is superior to the $\pi/4$ mode and, therefore, for the retrieval of ocean wave slope spectra, we recommend using the CTLR mode.

The potential of CP for ocean wave slope spectrum retrieval has been demonstrated in this paper. The values for the main wave parameters were also validated using in situ NDBC buoy data. A comparison of the CP SAR-based wave field information with the buoy outputs showed good agreement in terms of the dominant wave height, wave direction, and wave period, with biases of 0.36 m, 17.96°, and 0.88 s, respectively. Overall, the results for the retrieval of ocean wave slope spectra and ocean wave parameters using CP SAR were satisfactory. Moreover, considering only single SAR imagery is performed to wave retrieval algorithm study, the universality validation of ocean wave information retrieval from CP SAR will still need more results to support. In our future study, we will collect more quad-pol SAR to match up with buoy for plotting the wave parameter difference between FP and CP SAR precisely.

Conflicts of Interest

The authors declare that they have no conflicts of interest regarding the publication of this article.

Acknowledgments

The authors gratefully acknowledge the financial support provided by the National Key Research and Development Program of China (2016YFB0502504, 2016YFB0502500) and National Science Foundation of China under Grant nos. 41301500, 41431174, 61471358, and 41401427. This article is partly sponsored by the scholarship funding from the Chinese Academy of Sciences. The authors would also like to acknowledge Professor Zhang from the First Institute of Oceanography, State Oceanic Administration, for supporting research on CP theory. The authors would also like to acknowledge Dr. Bian and Dr. Liu from Institute of Remote Sensing and Digital Earth for polishing the original manuscript.

References

[1] J. Chen, Z. Lee, C. Hu, and J. Wei, "Improving satellite data products for open oceans with a scheme to correct the residual errors in remote sensing reflectance," *Journal of Geophysical Research Oceans*, vol. 121, no. 6, pp. 3866–3886, 2016.

[2] J. Chen, X. He, B. Zhou, and D. Pan, "Deriving colored dissolved organic matter absorption coefficient from ocean color with a neural quasi-analytical algorithm," *Journal of Geophysical Research Oceans*, vol. 122, no. 11, pp. 8543–8556, 2017.

[3] S. K. Chaturvedi, P. Shanmugam, C. S. Yang, and U. Guven, "Detection of ocean wave parameters using synthetic aperture radar (SAR) data," *Journal of Navigation*, vol. 66, no. 2, pp. 283–293, 2013.

[4] D. Kim, "Wind retrieval from X-band SAR image using numerical ocean scattering model," *Korean Journal of Remote Sensing*, vol. 25, no. 3, pp. 243–253, 2009.

[5] M. Sugimoto, N. Shiroto, and K. Ouchi, "Estimation of ocean wave height using polarization ratio of synthetic aperture radar data," in *2011 IEEE International Geoscience and Remote Sensing Symposium (IGARSS)*, pp. 2821–2824, Vancouver, BC, Canada, July 2011.

[6] P. W. Vachon, H. E. Krogstad, and J. S. Paterson, "Airborne and spaceborne synthetic aperture radar observations of ocean waves," *Atmosphere-Ocean*, vol. 32, no. 1, pp. 83–112, 1994.

[7] K. Hasselmann and S. Hasselmann, "On the nonlinear mapping of an ocean wave spectrum into a synthetic aperture radar image spectrum and its inversion," *Journal of Geophysical Research Oceans*, vol. 96, no. C6, pp. 10713–10729, 1991.

[8] S. Hasselmann and K. Hasselmann, "Computations and parameterizations of the nonlinear energy transfer in a gravity-wave spectrum. Part I: a new method for efficient computations of the exact nonlinear transfer integral," *Journal of Physical Oceanography*, vol. 15, no. 11, pp. 1369–1377, 1985.

[9] C. Mastenbroek and C. F. de Valk, "A semiparametric algorithm to retrieve ocean wave spectra from synthetic aperture radar," *Journal of Geophysical Research Oceans*, vol. 105, no. C2, pp. 3497–3516, 2000.

[10] H. Wang, J. Zhu, J. Yang, and C. Shi, "A semiempirical algorithm for SAR wave height retrieval and its validation using Envisat ASAR wave mode data," *Acta Oceanologica Sinica*, vol. 31, no. 3, pp. 59–66, 2012.

[11] W. Shao, Z. Zhang, X. Li, and H. Li, "Ocean wave parameters retrieval from Sentinel-1 SAR imagery," *Remote Sensing*, vol. 8, no. 9, 2016.

[12] B. Zhang, X. Li, W. Perrie, and Y. He, "Synergistic measurements of ocean winds and waves from SAR," *Journal of Geophysical Research Oceans*, vol. 120, no. 9, pp. 6164–6184, 2015.

[13] D. L. SCHULER and J. S. LEE, "A microwave technique to improve the measurement of directional ocean wave spectra,"

International Journal of Remote Sensing, vol. 16, no. 2, pp. 199–215, 1995.

[14] D. L. Schuler, J. S. Lee, D. Kasilingam, and E. Pottier, "Measurement of ocean surface slopes and wave spectra using polarimetric SAR image data," *Remote Sensing of Environment*, vol. 91, no. 2, pp. 198–211, 2004.

[15] Y. He, H. Shen, and W. Perrie, "Remote sensing of ocean waves by polarimetric SAR," *Journal of Atmospheric and Oceanic Technology*, vol. 23, no. 12, pp. 1768–1773, 2006.

[16] Y. He, W. Perrie, T. Xie, and Q. Zou, "Ocean wave spectra from a linear polarimetric SAR," *IEEE Transactions on Geoscience and Remote Sensing*, vol. 42, no. 11, pp. 2623–2631, 2004.

[17] B. Zhang, W. Perrie, and Y. He, "Validation of RADARSAT-2 fully polarimetric SAR measurements of ocean surface waves," *Journal of Geophysical Research Oceans*, vol. 115, no. C6, pp. 302–315, 2010.

[18] Y. He, B. Zhang, and W. Perrie, "Validation of RADARSAT-2 polarimetric SAR measurements of ocean waves," in *2009 IEEE International Geoscience and Remote Sensing Symposium*, pp. III-168–III-171, Cape Town, South Africa, July 2009.

[19] R. L. Paes, F. Nunziata, and M. Migliaccio, "On the capability of hybrid-polarity features to observe metallic targets at sea," *IEEE Journal of Oceanic Engineering*, vol. 41, no. 2, pp. 346–361, 2016.

[20] R. K. Raney, "A perspective on compact polarimetry," *IEEE Geoscience and Remote Sensing Letters*, vol. 160, pp. 12–18, 2011.

[21] J.-C. Souyris, P. Imbo, R. Fjortoft, S. Mingot, and J.-S. Lee, "Compact polarimetry based on symmetry properties of geophysical media: the $\pi/4$ mode," *IEEE Transactions on Geoscience and Remote Sensing*, vol. 43, no. 3, pp. 634–646, 2005.

[22] F. Nunziata, M. Migliaccio, and X. Li, "Sea oil slick observation using hybrid-polarity SAR architecture," *IEEE Journal of Oceanic Engineering*, vol. 40, no. 2, pp. 426–440, 2015.

[23] J. Yin, J. Yang, Z. S. Zhou, and J. Song, "The extended Bragg scattering model-based method for ship and oil-spill observation using compact polarimetric SAR," *IEEE Journal of Selected Topics in Applied Earth Observations and Remote Sensing*, vol. 8, no. 8, pp. 3760–3772, 2015.

[24] T. Geldsetzer, F. Charbonneau, M. Arkett, and T. Zagon, "Ocean wind study using simulated RCM compact-polarimetry SAR," *Canadian Journal of Remote Sensing*, vol. 41, no. 5, pp. 418–430, 2015.

[25] H. Li, J. Wu, W. Perrie, and Y. He, "Wind speed retrieval from hybrid-pol compact polarization synthetic aperture radar images," *IEEE Journal of Oceanic Engineering*, no. 99, pp. 1–12, 2017.

[26] R. Shirvany, M. Chabert, and J. Y. Tourneret, "Ship and oil-spill detection using the degree of polarization in linear and hybrid/compact dual-pol SAR," *IEEE Journal of Selected Topics in Applied Earth Observations and Remote Sensing*, vol. 5, no. 3, pp. 885–892, 2012.

[27] R. Touzi and F. Charbonneau, "Requirements on the calibration of hybrid-compact SAR," in *2014 IEEE Geoscience and Remote Sensing Symposium*, pp. 1109–1112, Quebec City, QC, Canada, July 2014.

[28] S. Boularbah, M. Ouarzeddine, and A. Belhadj-Aissa, "Investigation of the capability of the compact polarimetry mode to reconstruct full polarimetry mode using RADARSAT2 data," *Advanced Electromagnetics*, vol. 1, no. 1, p. 19, 2012.

[29] X. Zhang, J. Zhang, M. Liu, and J. Meng, "Assessment of C-band compact polarimetry SAR for sea ice classification," *Acta Oceanologica Sinica*, vol. 35, no. 5, pp. 79–88, 2016.

[30] P. C. Dubois-Fernandez, J. C. Souyris, S. Angelliaume, and F. Garestier, "The compact polarimetry alternative for spaceborne SAR at low frequency," *IEEE Transactions on Geoscience and Remote Sensing*, vol. 46, no. 10, pp. 3208–3222, 2008.

[31] J. J. van Zyl, H. A. Zebker, and C. Elachi, "Imaging radar polarization signatures: theory and observation," *Radio Science*, vol. 22, no. 4, pp. 529–543, 1987.

[32] J.-S. Lee, D. L. Schuler, and T. L. Ainsworth, "Polarimetric SAR data compensation for terrain azimuth slope variation," *IEEE Transactions on Geoscience and Remote Sensing*, vol. 38, no. 5, pp. 2153–2163, 2000.

[33] J.-S. Lee, D. L. Schuler, T. L. Ainsworth, E. Krogager, D. Kasilingam, and W.-M. Boerner, "On the estimation of radar polarization orientation shifts induced by terrain slopes," *IEEE Transactions on Geoscience and Remote Sensing*, vol. 40, no. 1, pp. 30–41, 2002.

[34] M. E. Nord, T. L. Ainsworth, J. S. Lee, and N. J. S. Stacy, "Comparison of compact polarimetric synthetic aperture radar modes," *IEEE Transactions on Geoscience and Remote Sensing*, vol. 47, no. 1, pp. 174–188, 2009.

10

COMS-Based Retrieval of Daily Actual Evapotranspiration over Korea

Na-Yeon Park,[1] Jae-Dong Jang,[1] Youngmi Kim,[2] Eun-Ha Sohn,[1] Mi-Lim Ou,[3] Jun-Dong Park,[1] Minji Seo,[4] Kyeong-Sang Lee,[4] and Kyung-Soo Han[4]

[1] *National Meteorological Satellite Center, Korea Meteorological Administration, Jincheon-gun 27803, Republic of Korea*
[2] *National Institute of Meteorological Sciences, Korea Meteorological Administration, Seogwipo-si 63568, Republic of Korea*
[3] *Climate Change Monitoring Bureau, Korea Meteorological Administration, Seoul 07062, Republic of Korea*
[4] *Department of Spatial Information Engineering, Pukyong National University, Busan 48513, Republic of Korea*

Correspondence should be addressed to Kyung-Soo Han; kyung-soo.han@pknu.ac.kr

Academic Editor: Saro Lee

Evapotranspiration (ET) from the land surface is an important hydrometeorological factor in the exchange of energy between the atmosphere and land surface. The accurate quantification for management of water resources and understanding of climate change are crucial, requiring continuous temporal and spatial monitoring. The objective of this study is to apply and estimate daily actual ET using semiempirical method, B-method, which is based on surface energy balance over heterogeneous area, Korea. To estimate daily ET, we used geostationary meteorological satellite data (Communication, Ocean and Meteorological Satellite, COMS) and polar-orbiting satellite data (Système Pour l'Observation de la Terre, SPOT). Estimated daily ET using only satellite data was relatively accurate and reflects land surface characteristics. It had high periodicity and spatial resolution over a wide area on clear-sky days. The daily ET was overestimated by about 1 mm/day at the two flux tower measurements sites, but the simulated seasonal variation and pattern were in good agreement with flux tower measurements. In the mixed forest, the root-mean-square error (RMSE) was 0.94 mm/day and the bias was 1.05 mm/day, while, in the rice paddy, RMSE was 1.12 mm/day and bias was 1.21 mm/day.

1. Introduction

Evapotranspiration (ET) from land surfaces, which drives latent heat flux (expressed in mm/day), is the sum of water loss from plant transpiration and water evaporation from land surfaces. An ET of 1 mm/day is equivalent to a loss of 10,000 liters per hectare per day [1]. It is a phenomenon in which liquid water changes to vapor, resulting in the transfer of water and energy between land surfaces and the atmosphere. As such, it is one of the most important components in the water cycle, representing total water consumed by plants and evaporated from water bodies and nonvegetated surfaces. A recent estimate of global ET components was obtained using a multimodel output from the Global Soil Wetness Project (GSWP) 2, which found that 48% of mean global ET came from transpiration, 36% from soil evaporation, and 16% from canopy ET [2]. ET is an essential variable for understanding surface energy balance (SEB), the global water cycle, and water and vegetation dynamics in terrestrial ecosystems [3–8].

ET plays a critical role in the global water cycle. On a global basis, the mean land surface ET accounts for approximately 60% of mean precipitation [9, 10]. At regional scales, ET consumes > 90% of annual rainfall, and agricultural regions are of crucial importance for water resource management in arid and semiarid climates [11, 12], highlighting the importance of accurate ET estimates. There are several methods for in situ measurement of ET including pan evaporation (large/small), atmometers, evapotranspirometers tanks, lysimeters, Bowen ratio [13] based on meteorological data, the eddy covariance method using flux tower measurements, and scintillometers [14]. Although in situ measuring methods provide accurate values at point scales, they offer limited spatiotemporal representation and are complex, labor intensive, and expensive, particularly in forests where ET retrieval is more difficult due to the evident experimental limitations [15].

Satellite remote-sensing method provides a wide range of coverage, frequent updates, and consistent quality. Several methods have been developed to estimate ET from remote-sensing data including empirical and statistical methods with remotely sensed vegetation indices [16–18], physical models that calculate ET as the residual of SEB using remotely sensed thermal infrared (IR) data [19–25], and other physical models such as the Penman-Monteith equations [26], the revised Resistance-Surface energy balance and the Penman-Monteith (RS-PM) equations [27–29] and Priestley-Taylor equations [30–32]. On a global or continental scales, the spatial ET is provided by the European Organization for the Exploitation of Meteorological Satellites (EUMETSAT) Satellite Application Facility (SAF) on Land Surface Analysis (LSA) and the National Aeronautics and Space Administration (NASA). EUMETSAT LSA SAF has developed ET products over large areas (Europe, Africa, and South America) using the spinning enhanced visible and IR imager onboard Meteosat Second Generation (MSG) [33]. It provides 30 min and daily ET estimates with a 3 km spatial resolution at nadir. It is based on a simplified Soil-Vegetation-Atmosphere Transfer (SVAT) model that uses a combination of remote-sensing and meteorological data from the European Center for Medium-Range Weather Forecast (ECMWF) as inputs. The range of LSA SAF daily ET values is 0–10 mm/day with 20% uncertainty, and the product is highly correlated (85–95%) with ECMWF products. Due to limited validation in dry or very dry conditions (mainly in Africa), care should be taken when using the product [34]. NASA has developed ET products over global terrestrial regions using the MODerate-resolution Imaging Spectrometer (MODIS) on the Aqua and Terra satellites. It provides 8-day, monthly, and yearly ET with a 1 km resolution. It is based on Penman-Monteith equations, which use a combination of remote-sensing data and meteorological data from NASA's Global Modeling and Assimilation Office (GMAO, v4.0.0) as inputs. MODIS ET has been validated using 46 AmeriFlux towers [27].

The semiempirical method using remote-sensing data, well-known approaches, proposed by [35], which is based on the SEB, uses the linear relationship between the sensible heat flux and the difference between surface and air temperatures near the land surface. The net integrated daily ET at the surface can be obtained from surface and air temperatures measured near the time of the local maximum ("net integrated daily radiation"), which is considered a semiempirical method. The B-method is determined either by fitting the linear least squares to the data or by simulations based on a SVAT [36–38] or boundary layer model. Many studies using the B-method have shown that the error of estimated daily ET is about ±1 mm/day; *B* coefficients depend on surface roughness, wind speed, and atmospheric stability [39, 40].

Remote-sensing based ET estimation using B-method has been reported in many previous studies. This method can be used to estimate ET at the continental scale (e.g., for drought detection, water-reserve estimations for irrigation [41], and rainfall monitoring in Sahel regions [42, 43]); at the local scale (e.g., grasslands [12], wheat fields and meadows [41], and sugar beets and wheat stump [39]); and at the regional scale (e.g., barley fields [10 × 10 km] in a semiarid region of Spain using the Advanced Very High Resolution Radiometer sensor onboard the National Oceanic Atmosphere Administration (NOAA) satellite [44], in a northern boreal forest region of Finland using the MODIS sensor [15], and in western Québec, Canada (approximately 224,000 km^2), using NOAA visible and GOES thermal IR satellite data [45]). B-method was designed to simplify surface energy budget through the semiempirical way, which determine *B*-coefficient for specific land type. It means that this method has a limitation to obtain the spatially gridded information over wide area covering various land types. It could be solved through *B*-coefficient generation model adapting over various land types, which is firstly proposed for northeast Asia region, in this study.

The objective of this study is (1) to estimate the daily actual ET using semiempirical method, B-method, considering various type of land cover and surface roughness in heterogeneous regions, Korea, by using geostationary meteorological satellite data (Communication, Ocean and Meteorological Satellite, COMS) and polar-orbiting satellite data (Système Pour l'Observation de la Terre, SPOT) and (2) to evaluate estimated daily actual ET derived from B-method using flux tower ET measurements and compare it to MODIS averaged 8-day ET (MOD16).

2. Theory and Data

2.1. Theory. The simplified semiempirical regression method used in this study was first proposed by [35] and is based on the land-SEB equation using thermal IR data, from which land surface temperature can be extracted. The energy arriving at the surface is equal to the energy leaving the surface during the same period. All of the fluxes of energy should be considered when deriving an SEB equation, which can be expressed as follows:

$$\mathrm{Rn} = \mathrm{LE} + H + G, \tag{1}$$

where Rn is the net radiation (W/m^2), LE is the latent heat flux (energy consumed by ET; W/m^2), H is the sensible heat flux (W/m^2), and G is the ground heat flux (W/m^2). Many studies estimating daily ET have assumed that ground heat flux is a negligible part of the daily energy balance, and therefore set G to 0 because the magnitude of daily or 10-day soil heat flux is relatively small compared to net radiation [46]. Therefore, LE is calculated as the remaining difference between Rn and H:

$$\mathrm{ET} \approx \mathrm{LE} = \mathrm{Rn} - H. \tag{2}$$

Although the net radiation can be derived through remote-sensing data, the sensible heat flux is complex and cannot be easily obtained. The numerous meteorological variables involved (e.g., wind speed, relative humidity, and atmosphere pressure) make it difficult to calculate ET. Therefore, in this study, we used the simplified method proposed by [35] to estimate daily actual ET. The method is based on the linear relationship between the sensible heat flux and the difference between instantaneous surface and air temperatures measured near midday (about 13:00–14:00 local time) over diverse surfaces with variable vegetation cover [3]. The simplified regression method can be expressed as follows:

$$\mathrm{ET}_d - \mathrm{Rn}_d = -B_d \cdot \left(T_{S_{\mathrm{mid}}} - T_{a_{\mathrm{mid}}}\right), \quad (3)$$

where ET_d is the daily actual ET (mm/day), Rn_d is the daily net radiation (mm/day), B_d is a semiempirical coefficient (mm/day/K), $T_{S_{\mathrm{mid}}}$ is the land surface temperature (Ts) (K) of midday, and $T_{a_{\mathrm{mid}}}$ is the near-surface air temperature (Ta) (K) of midday.

2.2. Study Area and Data

2.2.1. Study Area. The study area covered 220,847 km^2 (1,100 km from north to south) of the Korea, adjacent to the western Pacific Ocean (33–43°N, 122–132°E; Figure 1). The climate is considered temperate, with four distinct seasons. On average, the warmest month is August, while January is the coldest. The humidity is highest in July, reaching 80–90%, and lowest in January and April (30–50%). The study area had various types of land cover, including water, evergreen coniferous forests (0.06%), deciduous broadleaf forests (4.07%), mixed forests (56.96%), savannas (0.06%), grasslands (0.74%), permanent wetlands (0.01%), croplands/natural vegetation mosaics (36.05%), developed urban areas (2.01%), snow/ice (0.003%), and barren or sparsely vegetated areas (0.032%), as identified by the MODIS International Geosphere Biosphere Programme (IGBP) land cover map from 2012. There were mainly croplands in the western region and mixed forests in the eastern region.

2.2.2. Data. Figure 2 shows the components of the surface energy balance (i.e., net radiation [Rn], air temperature [Ta], and *B*-coefficients) for derived daily actual ET algorithm. We used the geostationary orbit satellite (COMS) and polar orbit satellite-based input variables for estimate daily actual ET algorithms.

(1) COMS Data. COMS, the first Korean geostationary meteorological satellite, was launched successfully on 27 June 2012 and has been operating at 128.2°E since April 1, 2012. The meteorological imager (MI) onboard COMS has five channels, which include one visible channel (1 km; 0.55–0.8 μm) and four IR channels (4 km, shortwave IR [SWIR]; 3.5–4.0 μm, water vapor; 6.5–7.0 μm, IR1; 10.3–11.3 μm, IR2; 11.5–12.5 μm) with 15 min acquisition intervals over the extended northern hemisphere. Although its high temporal resolution and ability to provide 96 daily acquisitions are advantageous, its spatial resolution at nadir (4 km) is low. The COMS Level 1 product (IR1, IR2, and visible) and COMS Level 2 product (land surface temperature [LST] and insolation [INS]) were used to estimate daily ET. Land surface temperatures were estimated using the split-window technique with a spatial resolution of 4 km. Compared to MODIS LST product, the COMS LST product has a root-mean-square error (RMSE) of 3.12 K and a bias of −1.65 K. It used the mean values 13:00–14:00 local time. COMS INS product was estimated using the physical model [47] for clear and cloudy skies with a spatial resolution of 4 km using daily mean values. Compared to pyranometer measurements, the COMS INS product has an RMSE of 60–140 W/m^2.

(2) SPOT Data. It is critical to understand the effects of vegetation when estimating ET. Transpiration is the dominant component of ET across a variety of ecohydrological systems [2]. Because COMS/MI has no near-IR (NIR) channels and cannot account for the impact of vegetation, we considered these variables using the Normalized Difference Vegetation Index (NDVI) from SPOT satellites, which were launched by the French Centre National d'Etudes Spatiales to observe vegetation. The SPOT satellite is a sun-synchronous polar-orbiting satellite with an inclination of 98.7° at an altitude of 820 km. The VGT sensor onboard SPOT is a multispectral instrument with a wide swath width (2,280 km) and a 1 km spatial resolution. The sensor has four spectral bands: blue (0.43–0.47 μm), red (0.61–0.68 μm), NIR (0.78–0.89 μm), and SWIR (1.58–1.74 μm). In this study, the surface reflectance (i.e., surface albedo) was calculated using the SPOT/VGT 10-day synthesis (D10) product corrected Bidirectional Reflectance Distribution function (BRDF) in the red and NIR channels. The BRDF is a critical tool for reliably reducing the anisotropy associated with surface scattering effects, one of the main fluctuating error sources of relative solar-target-sensor geometry [48]. Albedo was determined using monthly mean data from 2001 to 2011. NDVI varies seasonally according to the type of vegetation. To consider of vegetation status over Korea, the NDVI was calibrated using multiple polynomial regression [49]. In addition, geo- and atmosphere-corrected data, as well as S10 data, were used. The S10 data were synthesized in three 10-day sections per month, from days 1–10, days 11–20, and day 21 to the last day, using the maximum value composite method. We also used the Normalized Difference Water Index (NDWI), using NIR and SWIR channels as proposed by [50], to estimate the satellite-based air temperature and account for the effects of drying and wetting conditions of the surface.

(3) Ancillary Data. To derive daily ET, surface roughness length (Z_0), the MODIS land cover map, and a digital elevation model (DEM) were used. Z_0 is a critical parameter for estimating aerodynamic resistance between the land surface and atmosphere, as well as daily ET, because fluxes of mass and energy in and out of canopies are dependent on turbulent flow. There are many limitations when measuring Z_0 using satellite data; therefore, it is obtained using indirect methods. In Europe, these data are provided by land cover. In this study, Z_0 of the forest region was estimated using Ecoclimap-1 DB [51, 52], which is a class-based input for global and regional climate models produced in Europe that provides ground information for the same land cover and climate at a 1 km spatial resolution. In addition, NDVI-based Z_0 [53] was used in nonforested areas. The DEM used the GTOPO30 data provided by the EROS data center of the United States Geological Survey. GTOPO30 is part of the World Geodetic System 1984 and has a resolution of 30 arcs (about 1 km or 0.0083333°).

(4) Data Reprocessing and Matchup Data Base (MDB). To reprocess the data for estimating the COMS ET, the remote-sensing data (i.e., air temperature, surface temperature, INS,

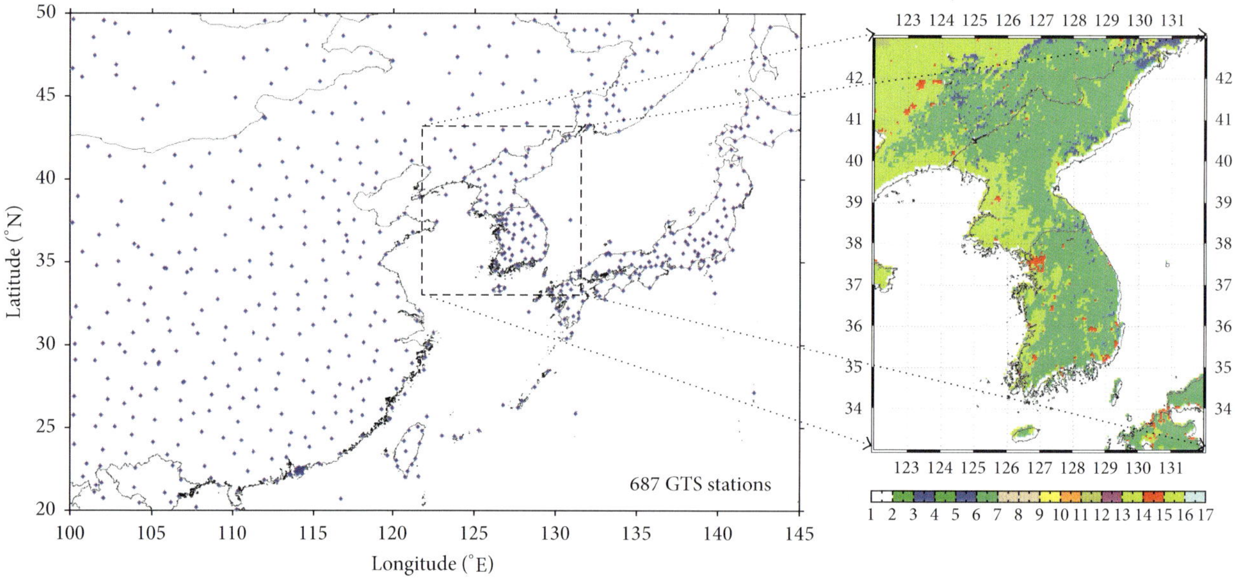

FIGURE 1: Location of Korea in this study area. The blue points represent the Global Telecommunication System (GTS) over North-East Asia regions. The right figure represents MODIS land cover types map over Korea (1: water bodies; 2: evergreen needleleaf forest; 3: evergreen broadleaf forest; 4: deciduous needleleaf forest; 5: deciduous broadleaf forest; 6: mixed forest; 7: closed shrublands; 8: open shrublands; 9: woody savannas; 10: savannas; 11: grasslands; 12: permanent wetlands; 13: croplands; 14: urban and built-up; 15: cropland/natural vegetation mosaics; 16: snow/ice; 17: barren or sparsely vegetated).

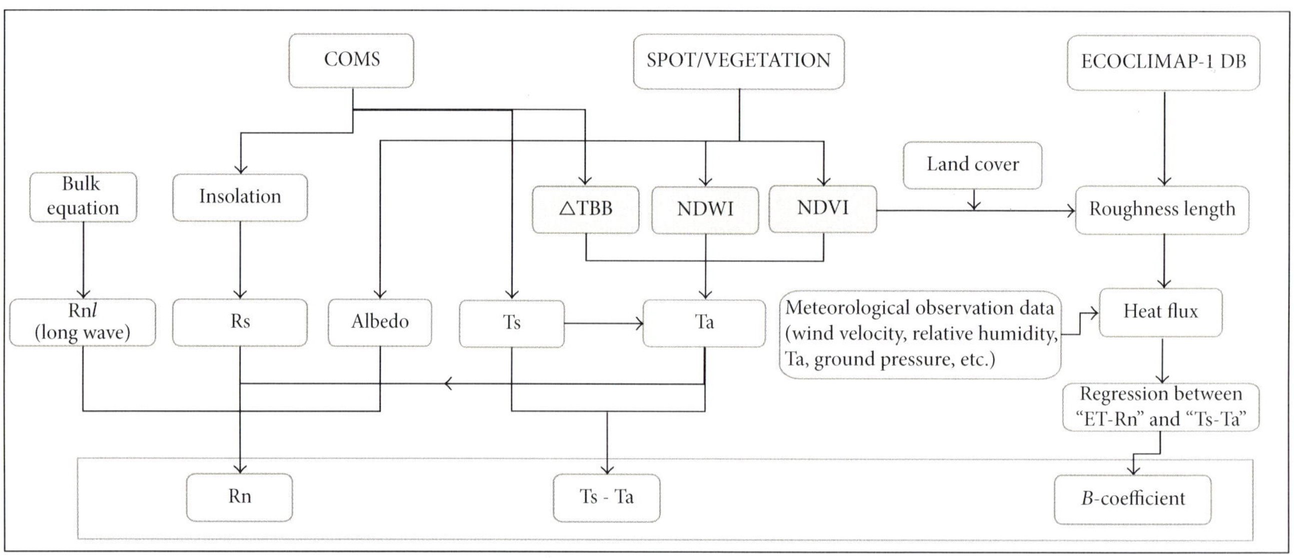

FIGURE 2: The geostationary orbit satellite (COMS), polar orbit satellite (SPOT), and ancillary data (ECOCLIMAP-1 DB) based input variables for estimate daily actual evapotranspiration (ET) algorithms.

albedo, outgoing radiation, vegetation index, land cover map, Z_0 map, and DEM) and meteorological data (i.e., temperature, wind speed, relative humidity, and atmospheric pressure) must have coincident temporal and spatial resolutions (1 km). All of the data were standardized to a Plate Carrée projection (Geographic/WGS84 projection) with a 1 km spatial resolution in the course of 1 year (April 2011 to March 2012) in Korea (33–43°N, 122–132°E). Meteorological observations were collected from a total of 93 observatories (manned observation stations) in Korea. The satellite and auxiliary data were standardized with the meteorological observations by averaging the nearest time and distance (within 1 km) around the observation station. The data were analyzed by linear regression of the difference between ET and Rn and the difference between the surface (Ts) and air temperatures (Ta) observed from 13:00 to 14:00.

(5) Validation Data. To evaluate the accuracy of the COMS daily ET, we used flux tower measurement data and MODIS ET (MOD16) [28]. The flux tower measurements were

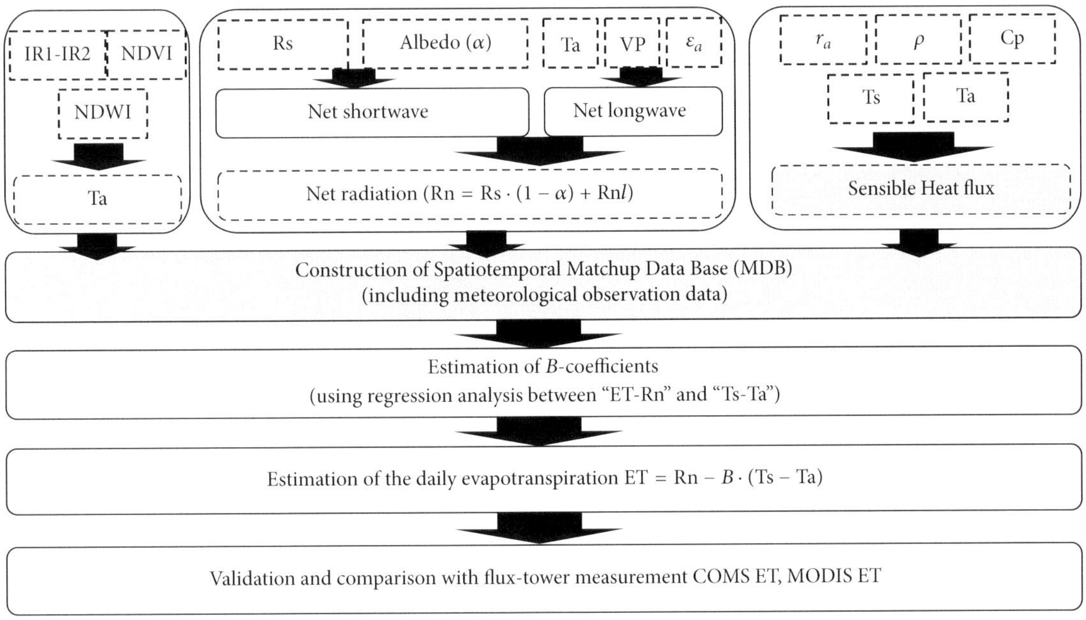

FIGURE 3: Flowchart of the methodology for estimating the Communication, Ocean and Meteorological Satellite (COMS) daily evapotranspiration (ET).

obtained by converting the latent heat flux observed every 30 min from the tower using the eddy covariance method to measure directly the vertical fluxes of heat, water vapor, and CO_2 between the surface and atmosphere. This was transformed as follows:

$$ET = \frac{LE}{\lambda}, \quad (4)$$

where λ is the latent heat of vaporization (J/kg). This assumes ideal conditions in which the ground is horizontally uniform and flat, there is no sink or source, and the average vertical flux due to turbulence is defined as $F = \overline{\omega' c'}$ (where ω' is vertical wind speed and c' is physical quantity). Wind velocity was measured in the x, y, and z directions and CO_2, H_2O, and temperature were measured at a rate of 10 Hz to understand the amount of water movement over time. Table 1 summarizes the information about the flux towers used in this study. The Seolmacheon flux tower (SMK) was in a mixed forest in the middle part of the river basin at an elevation of 293 m. The average height of the trees was 15 m. More than 90% of the forest is composed of 20–40-year-old coniferous and broadleaf trees. Their ability to replenish water is low due to the shallow topography, and the rocks and gravel are scattered on mountain slopes [54]. The Cheongmicheon flux tower (CFK) was in a rice paddy. The surrounding area was homogeneous and the terrain flat. The 500 m radius around the tower mostly consisted of rice, and the soil was dominated by sandy loam. The MODIS ET is based on the Penman-Monteith equation, which considers aerodynamic resistance and surface resistance and uses daily meteorological reanalysis data (air pressure, air temperature, humidity, radiation, etc.) and remotely sensed vegetation data (Leaf Area Index, the Fraction of Absorbed Photosynthetically Active Radiation, albedo, and land cover) from MODIS. This algorithm estimates ET and accounts for canopy conductivity, environmental factors (e.g., minimum temperature and difference in saturated water vapor pressure), and evaporation in the soil. The MODIS ET was accumulated at 8-day, monthly, yearly, and 1 km resolutions.

3. Methodology

Figure 3 shows the flowchart for estimating daily ET. First, we estimated the satellite-based air temperature (Section 3.1), daily Rn using net shortwave and longwave radiation (Section 3.2), and sensible heat flux (Section 3.3). The spatial and temporal resolutions of each input variable from the satellite data (e.g., air temperature, daily Rn, and land surface temperature) and meteorological measurements (e.g., humidity, temperature, and wind speed) were standardized using the matchup database (MDB). Then the *B*-coefficients were estimated by considering the relationship between sensible heat flux and the difference between land surface and air temperatures based on the MDB (Section 3.4). Finally, daily ET was estimated using the *B*-coefficients (Section 3.5).

3.1. Estimation of Daily Air Temperature (Ta). Near-surface air temperature (Ta) is a critical variable for estimating daily ET. Remote-sensing data can be used to estimate Rn and land surface temperature; however, it is difficult to estimate temperature using satellite data. Generally, Ta values are obtained from meteorological stations 2 m above the ground, although this can be difficult in mountainous terrain. Due to the limited distribution of meteorological stations, we used an algorithm for satellite-based midday air temperature retrieval over northeast Asia [55]. This algorithm uses COMS Level 1B brightness temperature the difference between the IR 1

TABLE 1: Flux tower measurement information.

Flux Site	Location/position	Vegetation Type	Elevation	Annual mean precipitation (mm)	Annual mean Air temperature (°C)	Canopy Height (m)
SMK (KoFlux Seolmacheon Site)	Paju, South Korea/37°56′19″N, 126°57′17″E	Mixed Forest	293 m above sea level	1332 mm	11.5°C	15 m
CFK (KoFlux Cheongmicheon Site)	Yeuju, South Korea/37°9′35″N, 127°39′10″E	Rice paddy	141 m	1170 mm	11.5°C	1 m

and IR 2 channels, land surface temperature, SPOT/VGT NDWI, and NDVI. Validated to ground measurements, these estimates have an R^2 value of 0.9401, a RMSE of 2.89 K, and a bias of 0.49 K for 13,542 matchup cases points.

3.2. Estimation of Daily Net Radiation (Rn). Daily Rn is one of the important parameters of total heat energy for estimating daily ET. It can be estimated using the sum of the difference between the incoming shortwave (Rs) and reflected outgoing shortwave solar radiation (0.15–5 μm), and the difference between the downwelling atmospheric and the surface emitted and reflected longwave radiation (3–100 μm). The daily Rn can be expressed as follows:

$$\text{Rn} = \text{Rn}_s + \text{Rn}_l = \text{Rs} \cdot (1 - \alpha) + \text{Rn}_l, \quad (5)$$

where Rn_s is net shortwave radiation (W/m^2), Rn_l is net longwave radiation (W/m^2), Rs is incoming solar radiation (W/m^2), and α is surface albedo (dimensionless). Rn_s is the difference between incoming and reflected solar radiation. Rs (i.e., surface solar radiation [insolation]) was estimated based on a physical model [48]. The reflectance of the surface (i.e., surface albedo) was calculated according to a previous study [56] to derive the narrowband-to-broadband albedo as the mean value of normalized reflectance in visible and NIR channels obtained from SPOT/VGT satellites. Rn_l is the difference between outgoing longwave radiation from upwelling surface radiation and incident downwelling atmospheric radiation. It is calculated with the Stefa-Boltzmann law and Brunt equation [57] using a combined $\varepsilon_{\text{bulk}}$ between the air and land surface with maximum and minimum air temperatures:

$$\text{Rn}_l = L\downarrow - L\uparrow = \varepsilon_{\text{bulk}} \cdot \sigma \cdot \left(\frac{T_{\max}{}^4 + T_{\min}{}^4}{2} \right), \quad (6)$$

where $\varepsilon_{\text{bulk}}$ is the net emissivity, σ is the Stefan-Boltzmann constant (5.67×10^{-8} W/m^2/K^4), $T_{\max}$ is daily maximum temperature (K), and $T_{\min}$ is daily minimum temperature (K). $\varepsilon_{\text{bulk}}$ [58, 59] was calculated as follows:

$$\varepsilon_{\text{bulk}} = -(a \cdot \left(\frac{R_S}{R_{SO}} + b \right) \cdot (a_1 + b_1 \cdot \sqrt{e_a}), \quad (7)$$

where a and b are regression coefficients to correct for cloudiness, a_1 and b_1 are regression coefficients to correct for water in the air, e_a is actual vapor pressure (kPa), R_{so} is clear-sky radiation (W/m^2), R_s is actual solar radiation (W/m^2), and R_s/R_{so} is relative short wave radiation (W/m^2). The coefficients a, b, a_1, and b_1 can be expressed as follows [60]:

$$\begin{aligned} a &= 1.126, \\ b &= -0.07, \\ &\qquad \text{If } \frac{Rs}{Rso} > 0.7, \\ a &= 1.017, \\ b &= -0.06, \\ &\qquad \text{If } \frac{Rs}{Rso} \le 0.7, \\ a_1 &= 0.26 + 0.1 \\ &\quad \cdot \exp\left\{ - [0.0154 \cdot (30 \cdot m + N - 207)]^2 \right\}, \\ b_1 &= -0.139, \end{aligned} \quad (8)$$

where m is the number of the month (i.e., 1 for January in the northern hemisphere) and N is day of the month. The daily clear-sky solar radiation uses an exponential model and a cosine model [61] as follows:

$$\begin{aligned} Rso &= A' \cdot \exp\left[- \left(\frac{\text{JD} - C'}{B'} \right)^2 \right], \\ A' &= 31.25 + 0.001113 \cdot Z, \\ B' &= 270 - 3.008 \cdot L, \end{aligned} \quad (9)$$

where A' is the peak radiation, B' is width of the curve, C' is Julian day of the day when the maximum solar radiation occurs (usually around day 172), JD is Julian day, Z is elevation above sea level (m), and L is latitude (°).

3.3. Estimation of Sensible Heat Flux (H). The calculation of sensible heat flux (H) is the most complex component of the surface energy budget. H is the energy emitted from the surface to the atmosphere by the heat transfer processes (convection and conduction) due to temperature differences between the surface and the atmosphere [62]. It is estimated following Ohm's Law. H can be calculated by the following bulk aerodynamic resistance equation [63] as follows:

$$H = \frac{\rho \cdot C_P \cdot (T_S - T_a)}{r_a}, \tag{10}$$

where H is sensible heat flux, ρ is the air density (kg/m^3), C_P is the specific heat of air at constant pressure (J/kg/K), T_a is the air temperature at reference height (2 m), T_S is the land surface temperature, and r_a is the aerodynamic resistance for the heat transport (s/m). Estimating sensible heat flux using the slope between the surface and air temperatures is complicated and many meteorological parameters (surface/air temperature, relative humidity, wind speed, aerodynamic resistance, and surface roughness length) are required. Therefore, it is difficult to apply to a wide area. In this study, H was simplified as follows:

$$H = B_d \cdot \left(T_{S_d} - T_{a_d}\right), \tag{11}$$

where B_d is a semiempirical coefficient that varies with land characteristics. It is defined as an "exchange coefficient," which is weighted by $\mathrm{Rn}_d/\mathrm{Rn}_i$ ratio and shows the instantaneous Rn contribution at midday for daily radiation on clear days. $\mathrm{Rn}_d/\mathrm{Rn}_i$ is affected by the change in solar energy depending upon surface conditions and season. It is approximately 0.33 in barren regions [12]. B_d is calculated as follows:

$$\begin{aligned} B_d &\approx \frac{\mathrm{Rn}_d}{\mathrm{Rn}_i} \cdot \frac{\rho_m \cdot C_p}{r_a} \cdot \frac{1}{\lambda}, \\ \rho_m &= \frac{P}{\mathrm{RT}'}, \\ P &= 101.3 - 0.01055 \cdot Z, \\ T' &= \frac{T}{1 - 0.378\left(e_a/P\right)}, \end{aligned} \tag{12}$$

where λ is the latent heat of vaporization (kJ/kg), the subscripts d and i represent the values observed at daily and instantaneously at noon, respectively, P is atmospheric pressure (kPa), R is the specific gas constant for dry air (0.287 kJ/kg/K), T' is the virtual temperature (K), and e_a is the actual vapor pressure of the air (kPa). The actual vapor pressure is computed as follows:

$$e_a = e^o_{dp} = \mathrm{RH} \cdot e^o_{\mathrm{db}}, \tag{13}$$

where RH is relative humidity (%) and e^o_{db} is the saturation vapor pressure at dew point temperature (kPa). The latent heat of vaporization is the heat required to change a unit mass of water from a liquid to a vapor at a constant pressure and temperature. In thermodynamic references, it is also known as the enthalpy of vaporization [64, 65]. Latent heat of vaporization can be estimated using the following linear regression equation [66] as follows:

$$\lambda = 2501 - 2.3601 \cdot T, \tag{14}$$

where T is air temperature (°C). Aerodynamic resistance, r_a, is resistance to heat transport between the land surface and the canopy and is affected by many factors including surface roughness (vegetation height and structure), wind speed, and atmospheric stability. Various methods for calculating r_a have been developed ranging from extremely elementary (a function of wind speed only) to quite rigorous (accounting for atmospheric stability, wind speed, surface aerodynamic roughness, etc.) [67]. Based on the Monin-Obukhov Similarity Theory for the atmospheric surface layer, the aerodynamic resistance is computed by the following equation [68–71] as follows:

$$r_a = \frac{\ln\left((Z-d)/Z_0\right)\ln\left((z-d)/Z_0h\right)}{K^2 \cdot U}, \tag{15}$$

where K is the Karman constant (=0.41), Z is height of the anemometer (=10 m), U is wind velocity at height Z (m/s), Z_0 is surface roughness length for momentum transfer (m), d is the zero-plane displacement height (m), and Z_0h is surface roughness length for the transfer of heat and vapor (m). Aerodynamic resistance is affected by numerous factors (e.g., surface roughness [vegetation height, vegetation structure], wind speed, and atmospheric stability). Although it is impossible to measure broad vegetation height accurately, it can be estimated using surface roughness as follows:

$$\begin{aligned} h &= \frac{Z_0}{0.123}, \\ d &= 0.67 \cdot h, \\ Z_0h &= 0.1 \cdot Z_0, \end{aligned} \tag{16}$$

where h is vegetation height (m) and Z_0 is surface roughness length (m). In this study, Z_0 for densely vegetated areas was calculated using ECOCLIMAP-DB 1. Areas with low vegetation height were estimated using the vegetation index [53] as follows:

$$Z_0 = \exp\left(-5.5 + 5.8 \cdot \mathrm{NDVI}\right). \tag{17}$$

The B-coefficients change according to the ratio of Rn, the density of wet air, and the aerodynamic resistance (i.e., the change in surface vegetation). B-coefficients depend on surface roughness, wind speed, and atmospheric stability.

3.4. Estimation of B-Coefficients. In this study, the B-coefficients from [35] were calculated using the spatiotemporal coincidence MDB composed of the ground-based data and satellite data. The B-coefficients determined values of the slopes from the regression between ET and Rn and Ts and Ta. There was a negative correlation with Z_0. Figure 4 shows a linear relationship between ET − Rn and Ts − Ta, which demonstrates their dependence on Z_0. It also presents the variation in the slopes for Z_0. The larger the surface roughness, the greater the deviation in B values. The slopes decreased with increasing surface roughness length (Table 2). In this study, the B-coefficients were representative of the surface roughness range as follows: Z_0, 0.0207 m (0.0048–0.0597 m), 0.1044 m (0.0602–0.1695 m), 0.2110 m

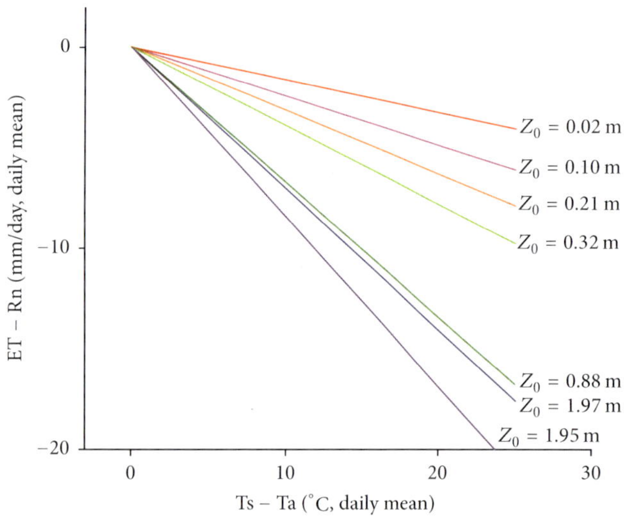

FIGURE 4: The relationship between ET − Rn and Ts − Ta using different surface roughness lengths (Z_0).

TABLE 2: Surface roughness length value according to *B*-coefficients.

Z_0 (m)	*B*-coefficient
0.0207	0.1629
0.1044	0.2449
0.2110	0.3157
0.3200	0.3905
0.8800	0.6718
0.9700	0.7029
1.9500	0.8435

(0.1710–0.2486 m), 0.3200 m (0.2540–0.4864 m), 0.8800 m (0.8600–0.9100 m), 0.9700 m (0.9200–1.000 m), and 1.9500 m (1.9200–1.9700 m). The *B*-coefficient model was calculated using representative *B* values, which were derived using an exponential function model and reflected the surface characteristics of the region:

$$B = a_1 \cdot 1 - \exp\left(-a_2 \cdot Z_0\right) + a_3, \tag{18}$$

where $a_1 = 0.8224$, $a_2 = 1.3153$, and $a_3 = 0.1381$.

3.5. Estimation of Daily ET. We developed a method for estimating daily ET using the *B*-coefficient model, which reflected vegetation and Z_0. This algorithm only used satellite data, specifically COMS satellite data:

$$\begin{aligned} \mathrm{ET}_d &= \mathrm{Rn}_d - B \cdot \left(\mathrm{Ts}_d - \mathrm{Ta}_d\right), \\ B &= 0.7705 \cdot \left(1 - \exp\left(-1.3153 \cdot Z_0\right)\right) + 0.1381. \end{aligned} \tag{19}$$

4. Results

4.1. Estimation of Daily COMS ET. The daily COMS ET was estimated under clear skies, except in the urban and desert regions. ET was estimated in Korea (33–43°N, 122–132°E) from 1 April 2011 to the present in real time. The daily COMS ET estimates only used satellite data (e.g., land surface temperature, air temperature, and solar radiation) from geostationary meteorological satellites. Therefore, these data allow for spatial and temporal monitoring of energy and water cycles over a wide area of Korea. They also reflect vegetation changes and regional characteristics of the land surface using vegetation indices and surface roughness. Figure 5 shows the spatial distribution of the COMS average monthly ET from January to December 2012. Changes in the growth and decline of vegetation from spring to winter on the Korea, which is mainly composed of forests and cultivated areas, were clearly reflected. The spatial and temporal distributions of COMS ET based on the seasonal land characteristics were also clear. In the spring (March, April, and May), the western plains area had many ET values exceeding 0–4 mm/day. Values over 3-4 mm/day increased during April and May, which was due to the freshwater cultivation of the plains area. In summer (June, July, and August), ET had a maximum of about 6 mm/day in the eastern part of the Korea. In autumn, ET was 0–4 mm/day in September, 0–3 mm/day in October, and 0–2 mm/day in November; this decrease in ET reflected the vegetation decline and harvest that occurs during these months. In winter (December, January, and February), ET was generally 0–2 mm/day because the vegetation was inactive. The COMS ET is characterized by being impacted on the surface roughness according to the land cover types.

4.2. Validation. To evaluate the accuracy of the daily COMS ET estimates, we compared them with daily flux tower ET measurements and MOD16 data via temporal and spatial analyses.

4.2.1. Validation of Estimated ET with Flux Tower Measurements ET and MODIS ET. In this study, we compared the COMS daily ET with daily flux tower ET measurements and the MOD16 data. The study period was from January to December 2012 and compared COMS data points (1 km) to the nearest the flux tower measurements in forests and cropland. Figures 6(a) and 6(b) show the results of time series comparisons of COMS daily ET, flux tower ET measurements, and MOD16. Figure 6(a) shows the results of daily ET in a mixed forest; flux tower ET was 0–4 mm/day, COMS daily ET was 0–4.5 mm/day, and MOD16 was 0–8 mm/day. COMS ET had similar seasonal patterns to flux tower ET measurements. Conversely, MOD16 was estimated to be 4–8 mm/day from May to August, which was higher than the flux tower measurements. Figure 6(b) shows the results of daily ET in a rice paddy, where flux tower ET was 0–5.5 mm/day, COMS daily ET was 0–5.5 mm/day, and MOD16 was 0–4.5 mm/day. The COMS daily ET was higher than the flux tower ET measurements by about 1 mm/day in spring (March–May) when vegetation was growing and in winter (December and January) when vegetation was inactive after harvest. Freshwater cultivation was mainly conducted in rice paddies from April to September; the ET tended to decrease after harvest time in September. Because rice paddies are filled with water during planting, the ET is somewhat higher than that of mixed forests. In addition, in August, flux tower

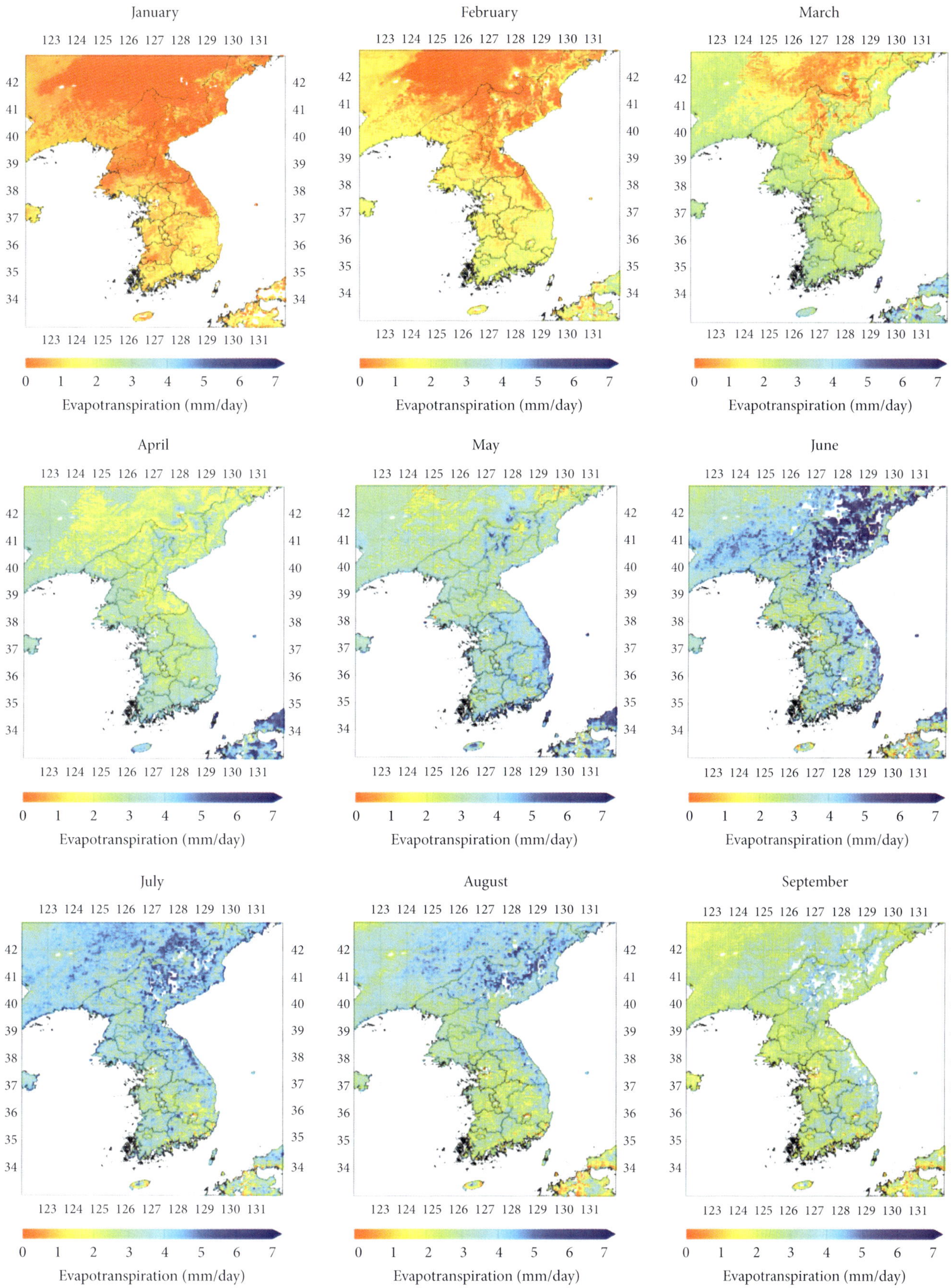

FIGURE 5: Continued.

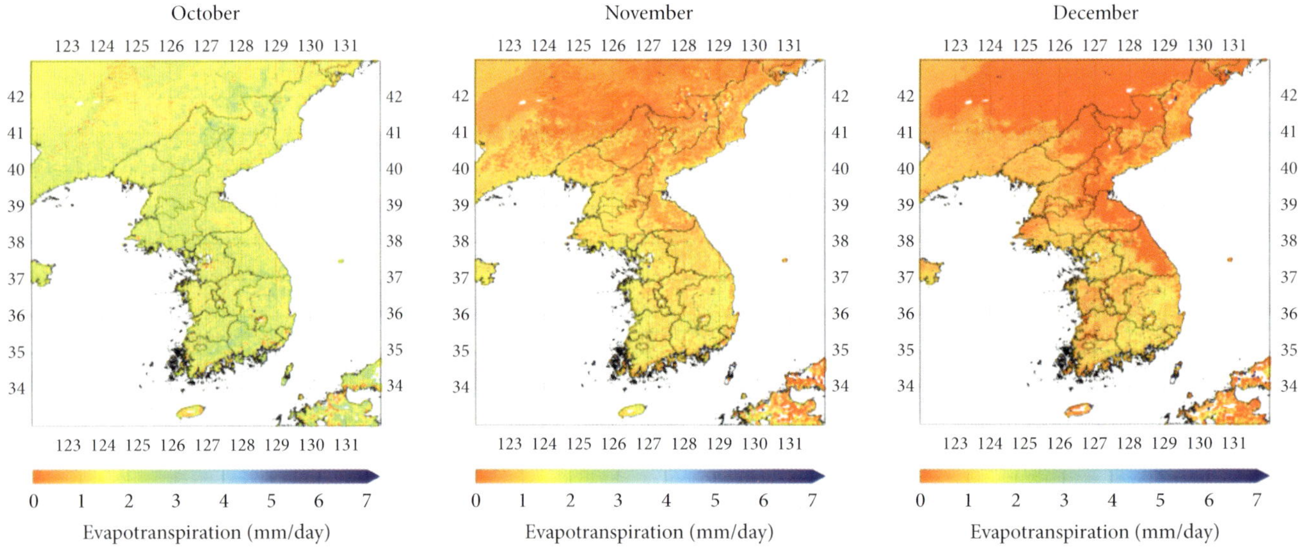

Figure 5: Spatial distribution of Communication, Ocean and Meteorological Satellite (COMS) monthly mean evapotranspiration (ET; mm/day) over Korea in 2012.

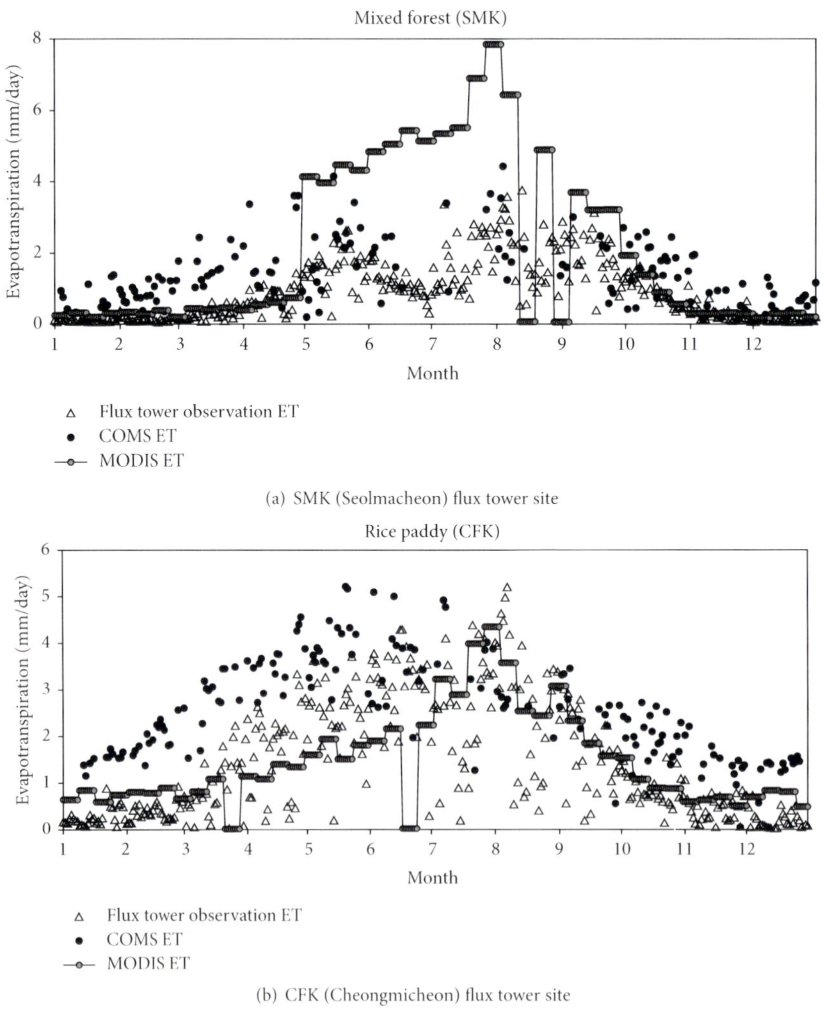

Figure 6: Time series of the Communication, Ocean and Meteorological Satellite (COMS; blue points), MODerate-resolution Imaging Spectrometer (MODIS; gray points and line), and daily flux tower (black triangle) evapotranspiration (ET) measurements during 2012 at flux tower measurements sites in (a) a mixed forest (Seolmacheon, SMK) and (b) a rice paddy (Cheongmicheon, CFK).

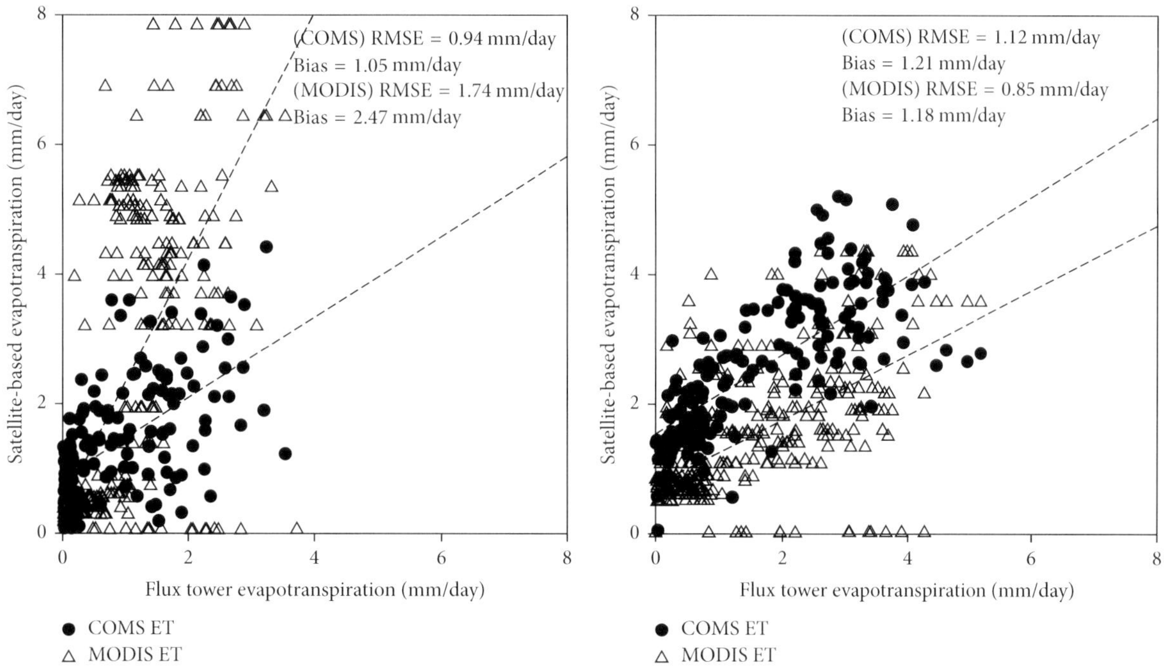

(a) SMK (Seolmacheon) flux tower site

(b) CFK (Cheongmicheon) flux tower site

FIGURE 7: Comparison between flux tower measurements of daily evapotranspiration (ET; mm/day) and Communication, Ocean and Meteorological Satellite (COMS; blue circles) ET, as well as MODerate-resolution Imaging Spectrometer (MODIS; black triangles) ET over Korea at flux tower sites in (a) a mixed forest (Seolmacheon, SMK) and (b) a rice paddy (Cheongmicheon, CFK).

ET measurements were higher than the COMS daily ET estimates. We assumed that this included a small number of observations affected by clouds, the observation error of instantaneous ET due to remaining water at the surface and being cut off by the evaporation of water by rainfall and vegetation in the summer. Figure 7 shows a scatterplot comparing the daily COMS ET and MOD16 data with flux tower ET measurements from the nearest flux tower measurement sites; the flux tower ET measurements were 0–4 mm/day, the COMS daily ET estimates were 0–5 mm/day, and MOD16 values were 0–8 mm/day. MOD16 values tended to be more than twice as high as the COMS ET estimates. The correlation of flux tower measurements with COMS ET estimates (RMSE: 0.94 mm/day, bias: 1.05 mm/day) was lower than that with MOD16 values (RMSE: 1.74 mm/day, bias: 2.47 mm/day) by 0.1. However, the RMSE and bias of COMS ET estimates were better than those of MOD16 by 0.8 mm/day and 1.42 mm/day, respectively (Table 3). Figure 7(b) shows that, in the croplands, flux tower ET was 0–5.5 mm/day, COMS daily ET was 0–5.5 mm/day, and MOD16 was 0–4.5 mm/day. MOD16 values were slightly lower than the flux tower measurements in agricultural areas with active water supplies because of the lack of regional characteristics. In these areas, the correlation of flux tower measurements with COMS ET estimates (RMSE: 1.12 mm/day, bias: 1.21 mm/day) was higher than that with MOD16 values (RMSE: 0.85 mm/day, bias: 1.18 mm/day) by 0.12. However, the RMSE and bias of COMS ET estimates were worse than those of MOD16 by 0.27 mm/day and 0.02 mm/day, respectively (Table 4)

TABLE 3: Comparisons between daily flux tower evapotranspiration (ET) measurements and Communication, Ocean and Meteorological Satellite (COMS) ET estimates, as well as MODerate-resolution Imaging Spectrometer (MODIS) ET estimates in a mixed forest (SMK) from January to December 2012.

	R	RMSE (mm/day)	Bias (mm/day)	Standard error (mm/day)	Number of observations
COMS ET	0.598	0.941	1.05	0.751	125
MODIS ET	0.696	1.741	2.47	1.673	300

4.2.2. Comparisons of Estimated ET with MODIS ET. Figure 8 shows the spatial distribution of the monthly average COMS and MODIS ET estimates in Korea. Figure 9 shows the probability density function (PDF) of monthly average COMS and MODIS ET estimates in January (winter), April (spring), July (summer), and October (autumn). COMS ET had significant changes in seasonal ET due to vegetation growth and decline throughout Korea (Figure 8(a)). It had a normal distribution function, which ranged from 0 to 5.5 mm/day over the entire season. In January, the estimated COMS ET was 0–2 mm/day. In April, when vegetation was growing, the ET range of 2–4 mm/day accounted for more than 50% of total values. In July, when rainfall and temperature increased, the mode was about 4 mm/day with range of 0–5.5 mm/day. In October,

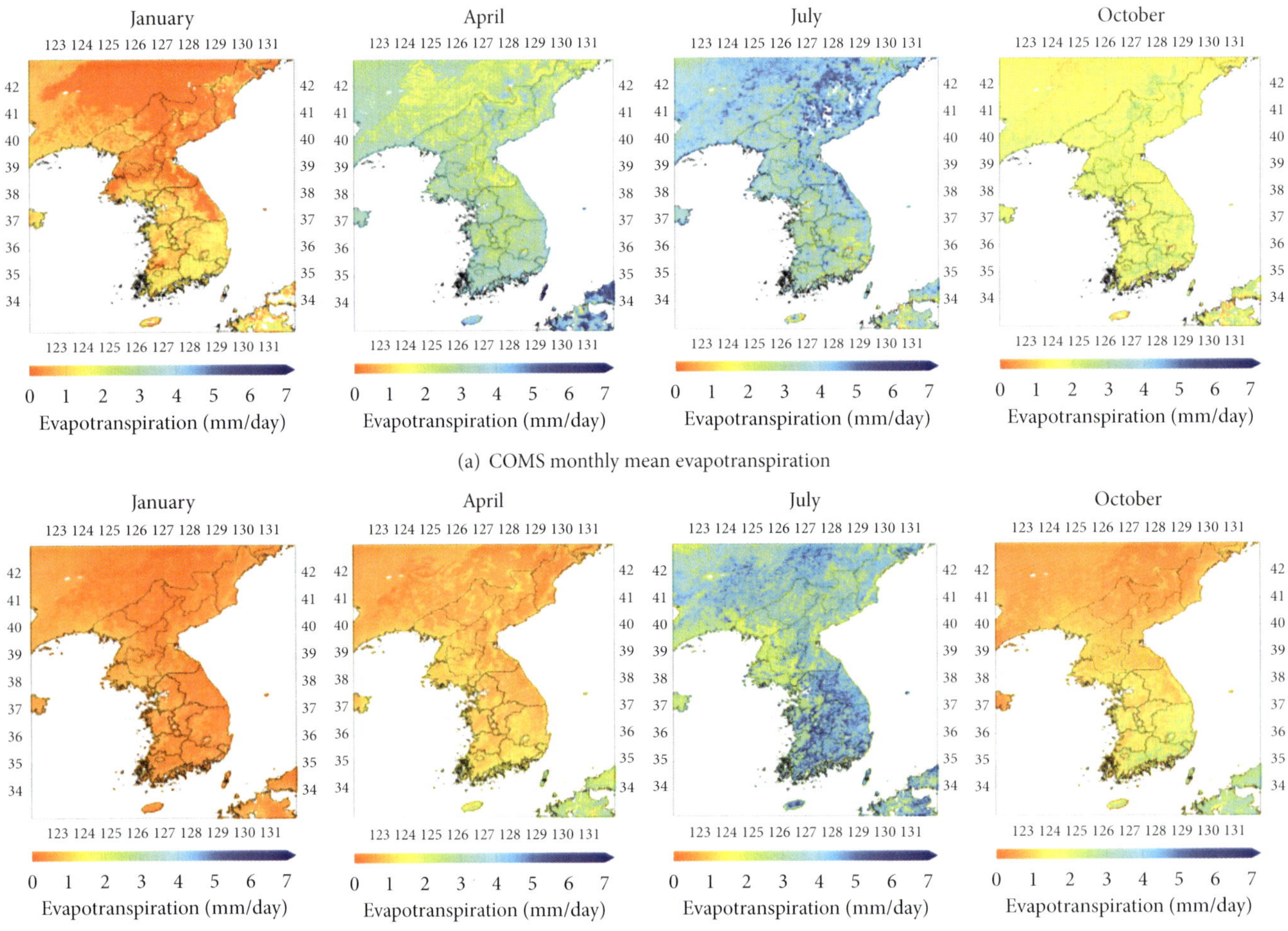

(a) COMS monthly mean evapotranspiration

(b) MODIS monthly mean evapotranspiration

Figure 8: (a) Communication, Ocean and Meteorological Satellite (COMS) monthly mean actual evapotranspiration (ET) and (b) MODerate-resolution Imaging Spectrometer (MODIS) monthly mean ET. Seasonally representative months were selected for this figure: January (winter), April (spring), July (summer), and October (autumn).

Table 4: Comparisons between daily flux tower evapotranspiration (ET) measurements and Communication, Ocean and Meteorological Satellite (COMS) ET estimates, as well as MODerate-resolution Imaging Spectrometer (MODIS) ET estimates in a rice paddy (CFK) from January to December 2012.

	R	RMSE (mm/day)	Bias (mm/day)	Standard error (mm/day)	Number of observation
COMS ET	0.737	1.122	1.205	0.722	132
MODIS ET	0.620	0.853	1.182	0.7943	301

when vegetation declined, the ET was 0–3 mm/day and the > 80% of estimates were below 2 mm/day. MODIS ET estimates ranged from 0 to 6 mm/day (Figure 8(b)). In January, MODIS ET estimates were generally < 1 mm/day, and over 80–90% of values were within 0–2 mm/day except in July (Figure 9). In January, about 80% of values were 0.5 mm/day, which was the same in the spring. In July, values were still low, with those <1 mm/day accounting for 60% of the total. In October, when vegetation declined, the values of 0–1 mm/day accounted for about 70% of the total.

5. Summary and Concluding

Recently, the importance of ET has been recognized in many fields of research, including water management, drought management, and climate change, which require continuous temporal and spatial monitoring. The flux tower ET measurements are disadvantageous because they can be difficult to interpret due to the expected energy imbalance in inhomogeneous and complex terrain. Due to spatial discontinuity of in situ measurements, effective monitoring of terrestrial biogeochemical cycles components at the global scale can be achieved only by satellite observations. To observe ET over a constant time interval and a wide geographic area, various studies are being conducted using satellite observations. Variable parameters (e.g., insolation, land surface temperature, near-surface temperature, albedo, vegetation, and land cover map) based on satellite observation are required to

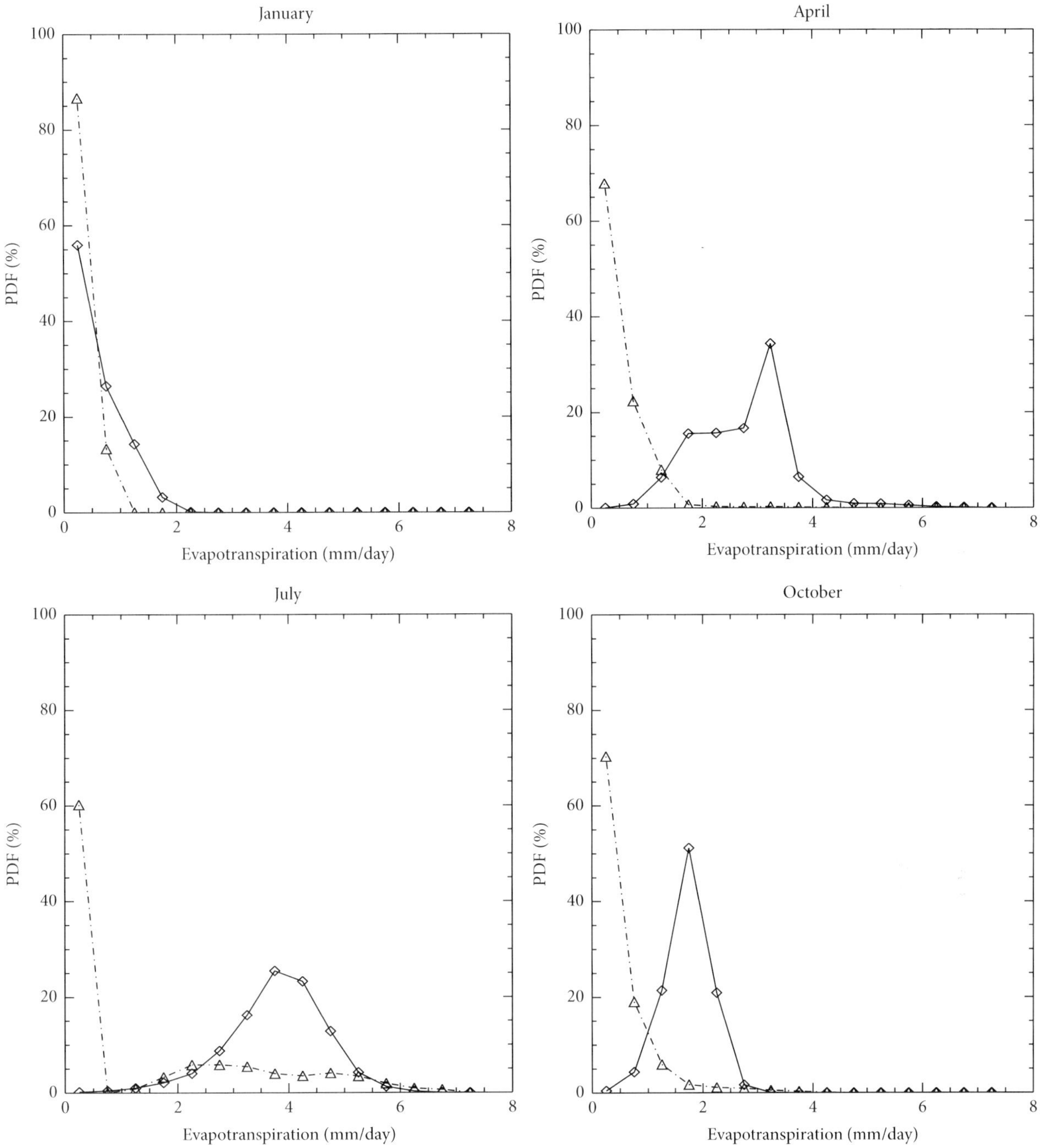

FIGURE 9: Probability density functions (PDF) of monthly mean evapotranspiration from the Communication, Ocean and Meteorological Satellite (COMS; blue line) and MODerate-resolution Imaging Spectrometer (MODIS; the black dash line) over Korea in 2012. Seasonally representative months were selected in this figure: January (winter), April (spring), July (summer), and October (autumn).

estimate the ET of large area. To observe and quantify ET continuously over Korea, we developed an algorithm of simplified daily ET using a combination of geostationary (COMS) and polar-orbiting satellite data (SPOT/VGT). The algorithm for estimating ET reflects vegetation, land surface type, and surface characteristics. The COMS daily ET had a range of 0–6 mm/day in 1 year; seasonal characteristics were also reflected. To evaluate the accuracy of our estimates, we compared them with the flux tower measurements and MODIS ET. In the mixed forest, the correlation coefficient (R) between COMS ET and flux tower measurements was about 0.6, the RMSE was 0.94 mm/day, and the bias is 1.05 mm/day, while, in the rice paddy, these values were 0.73, 1.12 mm/day, and 1.21 mm/day, respectively. Compared to the flux tower measurements, the MODIS ET had a RMSE of 2.47 mm/day in the forest and 0.85 mm/day in the cropland, which were higher than those of COMS ET. The values from MODIS ET also tended to be somewhat higher in the forested area (range: 0–8 mm/day) than those from COMS. In addition, the MODIS ET in the flooded cropland ranged from 0 to 4.5 mm/day, which was lower than the COMS ET and flux tower measurements. The spatial distribution of daily ET estimates tended to change with vegetation growth and decline. However, MODIS ET, which is an 8-day cumulative estimate, was 0–2 mm/day at the time of vegetation growth and decline (particularly in spring and autumn) and was not sensitive to regional characteristics. This suggests that the results do not fully reflect local characteristics when using GMAO model data as meteorological data.

COMS daily ET using only satellite data (e.g., air temperature, surface temperature, net short- and longwave radiation, and albedo) were relatively accurate. In addition, the ET algorithm simplified the generally complex computational process of sensible heat flux using the *B*-coefficients, which are a function of surface roughness. It had a high periodicity, a temporal resolution of 1 day, and a spatial resolution of 1 km over a wide area (except urban and desert areas) on clear-sky days. In this study, *B*-coefficient generation model adapts over various land types, which is firstly proposed for northeast Asia region. In many studies, the error of ET estimated using the methods of [35] is about ±1 mm/day, which is consistent with the error found in this study. This is sufficiently low to provide reliable information for water management [66]. However, the COMS daily ET was estimated and applied differently to the surface roughness based on land cover type. The daily ET was calculated using surface roughness, whose variability can significantly affect estimates. Various other necessary input parameters can also affect ET estimates. That is, input variables can increase the error and uncertainty of the ET values. In the future, we will conduct sensitivity analysis to assess these effects. The cycles and intensities of climate change and droughts are increasing [71]. COMS ET is an important ecological and hydrological parameter expressing the characteristics of the surface and climate through satellite observation. It provides the scientific basis for water management and policies and will play an important role in increasing the application of satellite data. Quantitative observation and monitoring using the spatial distribution of COMS ET over many years not only helps to elucidate climate change in Korea but also provides a means for predicting and investigating the mechanisms of global energy and water cycles, as well as validating hydrological and climate models [72] and numerical model inputs [73]. In the future, highly accurate ET estimates can be used for water resource management (e.g., irrigation for agriculture) and regional evaluation of drought.

Disclosure

The English language in the document has been checked by at least two native professional editors, both native speakers of English.

Conflicts of Interest

The authors declare that they have no conflicts of interest.

Acknowledgments

This study was funded by the KMA/NMSC's R&D Project, "Technical Development of Satellite Data Application for Operational Weather Service," and the Korea Meteorological Administration Research and Development Program under Grant KMIPA 2015-5041.

References

[1] C. Lozoya, C. Mendoza, A. Aguilar, A. Román, and R. Castelló, "Sensor-Based Model Driven Control Strategy for Precision Irrigation," *Journal of Sensors*, vol. 2016, Article ID 9784071, 2016.

[2] D. M. Lawrence, P. E. Thornton, K. W. Oleson, and G. B. Bonan, "The partitioning of evapotranspiration into transpiration, soil evaporation, and canopy evaporation in a GCM: impacts on land-atmosphere interaction," *Journal of Hydrometeorology*, vol. 8, no. 4, pp. 862–880, 2007.

[3] M. Jung, M. Reichstein, P. Ciais et al., "Recent decline in the global land evapotranspiration trend due to limited moisture supply," *Nature*, vol. 467, no. 7318, pp. 951–954, 2010.

[4] J. Peng, M. Borsche, Y. Liu, and A. Loew, "How representative are instantaneous evaporative fraction measurements of day-time fluxes?" *Hydrology and Earth System Sciences*, vol. 17, no. 10, pp. 3913–3919, 2013.

[5] K. C. Wang and R. E. Dickinson, "A review of global terrestrial evapotranspiration: observation, modeling, climatology, and climatic variability," *Reviews of Geophysics*, vol. 50, no. 2, 2012.

[6] A. Ershadi, M. F. McCabe, J. P. Evans, N. W. Chaney, and E. F. Wood, "Multi-site evaluation of terrestrial evaporation models using FLUXNET data," *Agricultural and Forest Meteorology*, vol. 187, pp. 46–61, 2014.

[7] A. J. Dolman, D. G. Miralles, and R. A. M. de Jeu, "Fifty years since Monteith's 1965 seminal paper: The emergence of global ecohydrology," *Ecohydrology*, vol. 7, no. 3, pp. 897–902, 2014.

[8] Q. Mu, M. Zhao, and S. W. Running, "MODIS Global Terrestrial Evapotranspiration (ET) Product (NASA MOD16A2/A3)," Algorithm Theoretical Basis Document, Collection 5, 2013.

[9] Z. Su, "The Surface Energy Balance System (SEBS) for estimation of turbulent heat fluxes," *Hydrology and Earth System Sciences*, vol. 6, no. 1, pp. 85–99, 2002.

[10] R. Rivas and V. Caselles, "A simplified equation to estimate spatial reference evaporation from remote sensing-based surface temperature and local meteorological data," *Remote Sensing of Environment*, vol. 93, no. 1-2, pp. 68–76, 2004.

[11] C. Abdelghani, J. C. B. Hoedjes, J.-C. Rodriquez et al., "Using remotely sensed data to estimate area-averaged daily surface fluxes over a semi-arid mixed agricultural land," *Agricultural and Forest Meteorology*, vol. 148, no. 3, pp. 330–342, 2008.

[12] B. Seguin and B. Itier, "Using midday surface temperature to estimate daily evaporation from satellite thermal IR data," *International Journal of Remote Sensing*, vol. 4, no. 2, pp. 371–383, 1983.

[13] D. L. Spittlehouse and T. A. Black, "Evaluation of the bowen ratio/energy balance method for determining forest evapotranspiration," *Atmosphere - Ocean*, vol. 18, no. 2, pp. 98–116, 1980.

[14] A. L. Ruhoff, A. R. Paz, W. Collischonn, L. E. Aragao, H. R. Rocha, and Y. S. Malhi, "A MODIS-Based Energy Balance to Estimate Evapotranspiration for Clear-Sky Days in Brazilian Tropical Savannas," *Remote Sensing*, vol. 4, no. 12, pp. 703–725, 2012.

[15] J. M. Sánchez, V. Caselles, R. Niclós, E. Valor, C. Coll, and T. Laurila, "Evaluation of the B-method for determining actual evapotranspiration in a boreal forest from MODIS data," *International Journal of Remote Sensing*, vol. 28, no. 6, pp. 1231–1250, 2007.

[16] E. P. Glenn, A. R. Huete, P. L. Nagler, and S. G. Nelson, "Relationship between remotely-sensed vegetation indices, canopy attributes and plant physiological processes: What vegetation indices can and cannot tell us about the landscape," *Sensors*, vol. 8, no. 4, pp. 2136–2160, 2008.

[17] E. P. Glenn, K. Morino, K. Didan et al., "Scaling sap flux measurements of grazed and ungrazed shrub communities with fine and coarse-resolution remote sensing," *Ecohydrology*, vol. 1, no. 4, pp. 316–329, 2008.

[18] P. L. Nagler, J. Cleverly, E. Glenn, D. Lampkin, A. Huete, and Z. Wan, "Predicting riparian evapotranspiration from MODIS vegetation indices and meteorological data," *Remote Sensing of Environment*, vol. 94, no. 1, pp. 17–30, 2005.

[19] W. G. M. Bastiaanssen, M. Menenti, R. A. Feddes, and A. A. M. Holtslag, "A remote sensing surface energy balance algorithm for land (SEBAL): 1. Formulation," *Journal of Hydrology*, vol. 212, no. 1–4, pp. 198–212, 1998.

[20] P. H. Gowda, J. Chävez, T. A. Howell, T. H. Marek, and L. L. New, "Surface energy balance based evapotranspiration mapping in the Texas high plains," *Sensors*, vol. 8, no. 8, pp. 5186–5201, 2008.

[21] K. Nishida, R. R. Nemani, J. M. Glassy, and S. W. Running, "Development of an evapotranspiration index from Aqua/MODIS for monitoring surface moisture status," *IEEE Transactions on Geoscience and Remote Sensing*, vol. 41, no. 2, pp. 493–501, 2003.

[22] J. Overgaard, D. Rosbjerg, and M. B. Butts, "Land-surface modelling in hydrological perspective - A review," *Biogeosciences*, vol. 3, no. 2, pp. 229–241, 2006.

[23] R. G. Allen, M. Tasumi, A. Morse et al., "Satellite-based energy balance for mapping evapotranspiration with internalized calibration (METRIC)—applications," *Journal of Irrigation and Drainage Engineering*, vol. 133, no. 4, pp. 395–406, 2007.

[24] W. Kustas and M. Anderson, "Advances in thermal infrared remote sensing for land surface modeling," *Agricultural and Forest Meteorology*, vol. 149, no. 12, pp. 2071–2081, 2009.

[25] Y. Lee and S. Kim, "The Modified SEBAL for Mapping Daily Spatial Evapotranspiration of South Korea Using Three Flux Towers and Terra MODIS Data," *Remote Sensing*, vol. 8, no. 12, article 983, 2016.

[26] A. Tsouni, C. Kontoes, D. Koutsoyiannis, P. Elias, and N. Mamassis, "Estimation of actual evapotranspiration by remote sensing: Application in Thessaly plain, Greece," *Sensors*, vol. 8, no. 6, pp. 3586–3600, 2008.

[27] H. A. Cleugh, R. Leuning, Q. Mu, and S. W. Running, "Regional evaporation estimates from flux tower and MODIS satellite data," *Remote Sensing of Environment*, vol. 106, no. 3, pp. 285–304, 2007.

[28] Q. Mu, M. Zhao, and S. W. Running, "Improvements to a MODIS global terrestrial evapotranspiration algorithm," *Remote Sensing of Environment*, vol. 115, no. 8, pp. 1781–1800, 2011.

[29] K. Jang, S. Kang, J. Kim et al., "Mapping evapotranspiration using MODIS and MM5 Four-Dimensional Data Assimilation," *Remote Sensing of Environment*, vol. 114, no. 3, pp. 657–673, 2010.

[30] C. H. Priestley and R. J. Taylor, "On the Assessment of Surface Heat Flux and Evaporation Using Large-Scale Parameters," *Monthly Weather Review*, vol. 100, no. 2, pp. 81–92, 1972.

[31] V. Venturini, S. Islam, and L. Rodriguez, "Estimation of evaporative fraction and evapotranspiration from MODIS products using a complementary based model," *Remote Sensing of Environment*, vol. 112, no. 1, pp. 132–141, 2008.

[32] J. Baik and M. Choi, "Evaluation of geostationary satellite (COMS) based Priestley-Taylor evapotranspiration," *Agricultural Water Management*, vol. 159, pp. 77–91, 2015.

[33] N. Ghilain, A. Arboleda, and F. Gellens-Meulenberghs, "Evapotranspiration modelling at large scale using near-real time MSG SEVIRI derived data," *Hydrology and Earth System Sciences*, vol. 15, no. 3, pp. 771–786, 2011.

[34] LSA-SAF: Product User Manual (PUM) Evapotranspiration (ET), "PUM_METv2.5," http://landsaf.meteo.pt/, 2015.

[35] R. D. Jackson, R. J. Reginato, and S. B. Idso, "Wheat canopy temperature: A practical tool for evaluating water requirements," *Water Resources Research*, vol. 13, no. 3, pp. 651–656, 1977.

[36] T. N. Carlson and M. J. Buffum, "On estimating total daily evapotranspiration from remote surface temperature measurements," *Remote Sensing of Environment*, vol. 29, no. 2, pp. 197–207, 1989.

[37] T. N. Carlson, W. J. Capehart, and R. R. Gillies, "A new look at the simplified method for remote sensing of daily evapotranspiration," *Remote Sensing of Environment*, vol. 54, no. 2, pp. 161–167, 1995.

[38] R. R. Gillies, W. P. Kustas, and K. S. Humes, "A verification of the 'triangle' method for obtaining surface soil water content energy fluxes from remote measurements of the normalized difference vegetation index (NDVI) and surface radiant temperature," *International Journal of Remote Sensing*, vol. 18, no. 15, pp. 3145–3166, 1997.

[39] A. Vidal and A. Perrier, "Analysis of a simplified relation for estimating daily evapotranspiration from satellite thermal IR data," *International Journal of Remote Sensing*, vol. 10, no. 8, pp. 1327–1337, 1989.

[40] J.-P. Lagouarde and K. J. McAneney, "Daily sensible heat flux estimation from a single measurement of surface temperature and maximum air temperature," *Boundary-Layer Meteorology*, vol. 59, no. 4, pp. 341–362, 1992.

[41] D. Courault, P. Clastre, J. Guinot, and B. Seguin, "Analyse des sécheresses de 1988 à 1990 en France à partir de l'analyse combinée de données satellitaires NOAA-AVHRR et d'un modèle agrométéorologique," *Agronomie*, vol. 14, no. 1, pp. 41–56, 1994.

[42] E. Assad, B. Seguin, Y. Kerr, J. P. Fréteaud, and J. P. Lagouarde, "The possible applications of Meteosat for monitoring the rain season in the Sahel zone: The case of Senegal," *Advances in Space Research*, vol. 7, no. 3, pp. 67–71, 1987.

[43] A. Rosema, "Comparison of meteosat-based rainfall and evapotranspiration mapping in the sahel region," *International Journal of Remote Sensing*, vol. 11, no. 12, pp. 2299–2309, 1990.

[44] A. Brasa Ramos, F. Martín De Santa Olalla, and V. Caselles, "Maximum and actual evapotranspiration for barley (Hordeum vulgare L.) through NOAA satellite images in Castilla-La Mancha, Spain," *Journal of Agricultural Engineering Research*, vol. 63, no. 4, pp. 283–294, 1996.

[45] K.-S. Han, A. A. Viau, and F. Anctil, "Hourly evapotranspiration derived from NOAA-AVHRR visible and GOES-IMAGER thermal infrared data," *International Journal of Remote Sensing*, vol. 31, no. 11, pp. 2817–2836, 2010.

[46] R. G. Allen, L. S. Pereira, D. Raes, and M. Smith, "Crop evapotranspiration - Guidelines for computing crop water requirements," FAO Irrigation and Drainage paper 56, FAO, Rome, vol. 300, no. 9, 1998.

[47] H. Kawamura, S. Tanahashi, and T. Takahashi, "Estimation of insolation over the Pacific Ocean off the Sanriku Coast," *Journal of Oceanography*, vol. 54, no. 5, pp. 457–464, 1998.

[48] S.-I. Kim, D.-S. Ahn, K.-S. Han, and J.-M. Yeom, "Improved Vegetation Profiles with GOCI Imagery Using Optimized BRDF Composite," *Journal of Sensors*, vol. 2016, Article ID 7165326, 2016.

[49] J. M. Yeom, K. S. Han, and Y. S. Kim, "Identification of contaminated pixels in 10-day NDVI image," in *Proceedings of the Korean Society of Remote Sensing Spring Conference*, vol. 12, pp. 113–116, 2006.

[50] B.-C. Gao, "NDWI—a normalized difference water index for remote sensing of vegetation liquid water from space," *Remote Sensing of Environment*, vol. 58, no. 3, pp. 257–266, 1996.

[51] V. Masson, J.-L. Champeaux, F. Chauvin, C. Meriguet, and R. Lacaze, "A global database of land surface parameters at 1-km resolution in meteorological and climate models," *Journal of Climate*, vol. 16, no. 9, pp. 1261–1282, 2003.

[52] J.-L. Champeaux, K.-S. Han, D. Arcos, F. Habets, and V. Masson, "Ecoclimap2: A new approach at global and European scale for ecosystems mapping and associated surface parameters database using SPOT/VEGETATION data - First results," in *Proceedings of the 2004 IEEE International Geoscience and Remote Sensing Symposium Proceedings: Science for Society: Exploring and Managing a Changing Planet. IGARSS 2004*, pp. 2046–2049, Anchorage, Alaska, USA, September 2004.

[53] R. K. Gupta, T. S. Prasad, and D. Vijayan, "Estimation of roughness length and sensible heat flux from wiFS and NOAA AVHRR data," *Advances in Space Research*, vol. 29, no. 1, pp. 33–38, 2002.

[54] H. Kwon, J. Lee, Y. Lee, J. Lee, S. Jung, and J. Kim, "Seasonal Variations of Evapotranspiration Observed in a Mixed forest in the Seolmacheon Catchment," *Korean Journal of Agricultural and Forest Meteorology*, vol. 11, no. 1, pp. 39–47, 2009.

[55] J.-H. Ryu, K.-S. Han, J. Cho et al., "Estimating midday near-surface air temperature by weighted consideration of surface and atmospheric moisture conditions using COMS and SPOT satellite data," *International Journal of Remote Sensing*, vol. 36, no. 13, pp. 3503–3518, 2015.

[56] R. W. Saunders, "The determination of broad band surface albedo from AVHRR visible and near-infrared radiances," *International Journal of Remote Sensing*, vol. 11, no. 1, pp. 49–67, 1990.

[57] D. Brunt, "Notes on radiation in the atmosphere," *Quarterly Journal of the Royal Meteorological Society*, vol. 58, no. 247, pp. 389–420, 1932.

[58] H. Bjornlund, "Water governance for sustainable irrigation: the role of civil society," in *Proceedings of the Sustainable Irrigation 2014*, pp. 211–222, Poznan, Poland, June 2014.

[59] R. D. Burman, R. H. Cuenca, and A. Weiss, "Techniques for estimating irrigation water requirements," *Advances in Irrigation*, 1983.

[60] J. L. Wright, "New evapotranspiration crop coefficients," *Journal of the Irrigation and Drainage Division*, vol. 108, no. IR2, pp. 57–74, 1982, Proceeding of the American Society of Civil Engineers.

[61] D. F. Heermann, G. J. Harrington, and K. M. Stahl, "Empirical estimation of daily clear sky solar radiation.," *Journal of Climate & Applied Meteorology*, vol. 24, no. 3, pp. 206–214, 1985.

[62] A. S. Monin and A. M. Obukhov, "Basic laws of turbulent mixing in the surface layer of the atmosphere," *Contributions of the Geophysical Institute of the Slovak Academy of Sciences*, vol. 24, no. 151, pp. 163–187, 1954.

[63] J. L. Monteith, *Principles of Environmental Physics*, Edward Amold, London, UK.

[64] H. F. Silver and J. E. Nydahl, *Introduction to Engineering Thermodynamics*, West Publishing Company, St Paul, Minn, USA, 1977.

[65] G. J. Van Wylen and R. E. Sonntag, *Fundamentals of Classical Thermodynamics*, John Wiley & Sons, Ltd, New York, NY, USA, 1985.

[66] A. R. Harris and I. M. Mason, "An extension to the split-window technique giving improved atmospheric correction and total water vapour," *International Journal of Remote Sensing*, vol. 13, no. 5, pp. 881–892, 1992.

[67] Z. Li, R. Tang, Z. Wan et al., "A review of current methodologies for regional evapotranspiration estimation from remotely sensed data," *Sensors*, vol. 9, no. 5, pp. 3801–3853, 2009.

[68] M. E. Jensen and R. G. Allen, *Evapotranspiration and Irrigation Water Requirements*, American Society of Civil Engineers, Reston, Va, USA, 1990.

[69] R. G. Allen, "A penman for all seasons," *Journal of Irrigation and Drainage Engineering*, vol. 112, no. 4, pp. 348–368, 1986.

[70] R. G. Allen, M. E. Jensen, J. L. Wright, and R. D. Burman, "Operational estimates of reference evapotranspiration," *Agronomy Journal*, vol. 81, no. 4, pp. 650–662, 1989.

[71] M. Zhao and S. W. Running, "Drought-induced reduction in global terrestrial net primary production from 2000 through 2009," *Science*, vol. 329, no. 5994, pp. 940–943, 2010.

[72] D. Michel, C. Jiménez, D. G. Miralles et al., "The WACMOS-ET project - Part 1: Tower-scale evaluation of four remote-sensing-based evapotranspiration algorithms," *Hydrology and Earth System Sciences*, vol. 20, no. 2, pp. 803–822, 2016.

[73] B. Mueller, S. I. Seneviratne, C. Jimenez et al., "Evaluation of global observations-based evapotranspiration datasets and IPCC AR4 simulations," *Geophysical Research Letters*, vol. 38, no. 6, 2011.

11

Comparison between a Machine-Learning-Based Method and a Water-Index-Based Method for Shoreline Mapping using a High-Resolution Satellite Image Acquired in Hwado Island, South Korea

Yun-Jae Choung[1] and Myung-Hee Jo[2]

[1]*Research Institute of Spatial Information Technology, Geo C&I Co. Ltd., 435 Hwarang-ro, Dong-gu, Daegu 41165, Republic of Korea*
[2]*Department of Aero-Satellite Geo-Informatics Engineering, School of Convergence and Fusion System Engineering, College of Science and Technology, Kyungpook National University, 2559 Gyeongsang-daero, Sangju 37224, Republic of Korea*

Correspondence should be addressed to Myung-Hee Jo; mhjo7312@gmail.com

Academic Editor: Eduard Llobet

Shoreline-mapping tasks using remotely sensed image sources were carried out using the machine learning techniques or using water indices derived from image sources. This research compared two different methods for mapping accurate shorelines using the high-resolution satellite image acquired in Hwado Island, South Korea. The first shoreline was generated using a water-index-based method proposed in previous research, and the second shoreline was generated using a machine-learning-based method proposed in this research. The statistical results showed that both shorelines had high accuracies in the well-identified coastal zones while the second shoreline had better accuracy than the first shoreline in the coastal zones with irregular shapes and the shaded areas not identified by the water-index-based method. Both shorelines, however, had low accuracies in the coastal zones with the shaded areas not identified by both methods.

1. Introduction

A coastal zone is defined as "the coastal waters (including the lands therein and thereunder) and the adjacent shorelands (including the waters therein and thereunder) strongly influenced by each and in proximity to the shorelines of several coastal states, and include islands, transitional and intertidal areas, salt marshes, wetlands, and beaches" [1]. Coastal erosions generally cause serious damage in the ecosystems and human lives in coastal zones [2]. A shoreline is defined as "the line along which a large body of water meets the land" [3]. A shoreline-mapping task is critical for the prevention of coastal erosion, the management of coastal zones, the preservation of coastal properties, and the description of the detailed coastal shapes [4, 5]. Historically, the shoreline-mapping tasks have been carried out using the ground-surveying methods, but due to the irregular coastal surfaces and the huge size of coastal areas, the ground-surveying method is not an efficient method for the shoreline-mapping tasks [6].

Research on shoreline mapping using the remote sensing datasets has been carried out because the utilization of such datasets is efficient for acquiring the surface and geometric information of wide coastal zones with high accuracy and without human access [4–6]. Li et al. (2001) and Guariglia et al. (2006) compared the multiple techniques for mapping shorelines using the different datasets [7, 8]. Li et al. (2003) used high-resolution satellite imagery for mapping shorelines by using the photogrammetry techniques [9]. Liu et al. (2009) and Choung et al. (2013) used the airborne topographic LiDAR (light detection and ranging) data for mapping shorelines by using geometric analysis [4, 10]. Lee (2012) utilized high-resolution satellite imagery for mapping shorelines by using the unsupervised segmentation method [6].

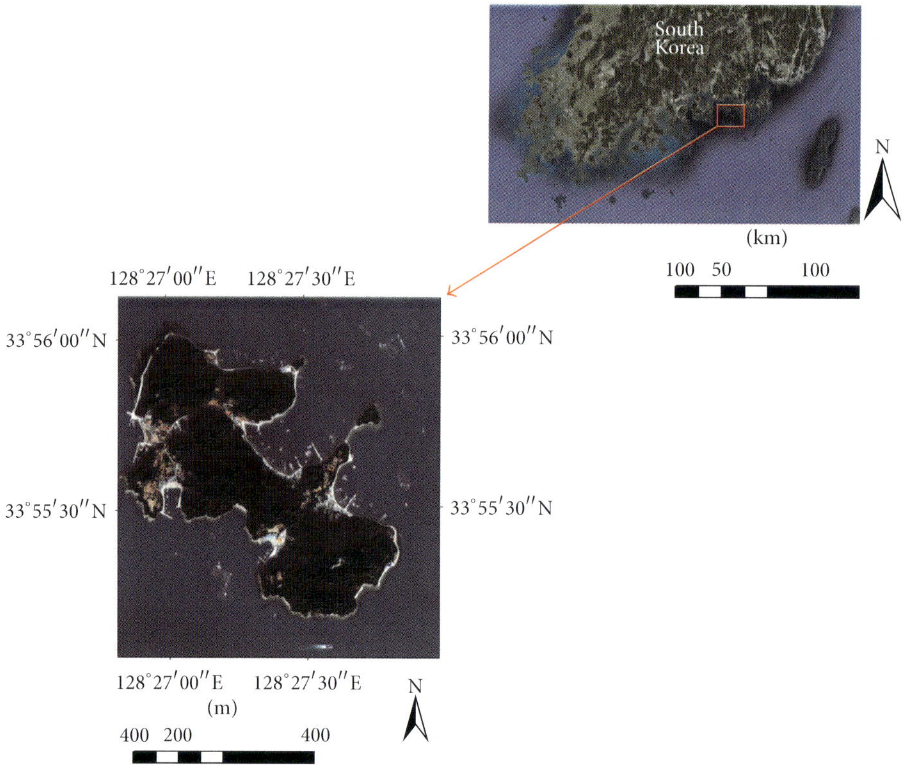

Figure 1: Hwado Island, South Korea, selected as the study area.

Hannv et al. (2013) and Masria et al. (2015) utilized Landsat imagery for mapping the shoreline by using the supervised classification methods [11, 12]. Sekovski et al. (2014) and Shenbagaraj et al. (2014) used high-resolution satellite and Landsat imagery, respectively, for mapping shorelines by using the unsupervised classification methods [13, 14]. Maglione et al. (2014) and Choung (2015) utilized a high-resolution satellite image for mapping shorelines using a water index [15, 16]. Bouchahma and Yan (2012) and Choung and Jo (2015) utilized Landsat imagery for mapping shorelines by using a water index [5, 17].

A recent research on mapping shorelines using various remote sensing data was carried out using two different approaches: (1) mapping shorelines by the supervised approach such as the machine learning techniques and (2) mapping shorelines by the unsupervised approach based on the water index derived from multispectral image sources. A comparison of these two approaches for mapping shorelines using the high-resolution image sources, however, has been limited. This research proposed a machine-learning-based method and compared the proposed method with the previous water-index-based method proposed by Choung and Jo (2015) for mapping accurate shorelines using a high-resolution satellite image.

2. Study Areas and Datasets

Hwado Island, South Korea, was selected as the study area in this research due to the data availability (see Figure 1). The coastal zones of Hwado Island have an approximately 7 km total shoreline length.

The orthorectified high-resolution satellite image was acquired by the WorldView-2 satellite on October 11, 2011. The given WorldView-2 image consists of the four available spectral bands (blue: 450–510 nm; green: 510–580 nm; red: 630–690 nm; and NIR (near infrared): 770–895 nm), and the ground resolution of the WorldView-2 image is 50 cm [18]. The horizontal datum of the given WorldView-2 image is WGS (World Geodetic System) 84, and the RMSE (root mean square error) of the image is 25 cm.

3. Methodology

This section illustrates the water-index-based method, the previous method, and the machine-learning-based method, the proposed method, for mapping shorelines using the given WorldView-2 image. Figure 2 presents a flowchart showing the procedure for mapping shorelines using the two different methods. In the water-index-based method, the NDWI (normalized-difference water index) image was generated from the WorldView-2 image, and then the first binary image separating the land and water features was generated from the NDWI image by an adaptive thresholding method. In addition, the first shoreline was extracted from the first binary image by selecting the boundary between the identified land and water features. In the machine-learning-based method, the coastal-surface classification map was generated from the given WorldView-2 image by the support vector machine

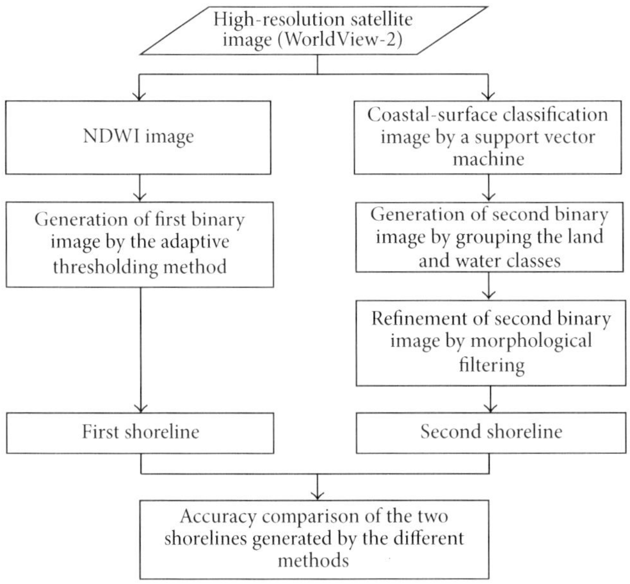

FIGURE 2: Flowchart showing the procedure for mapping shorelines using the two different methods.

(SVM) classifier, and then the second binary image was generated from the coastal-surface classification map by grouping the land (rock, vegetation) and water features. Morphological filtering was applied to refine the boundary between the land and water features in the second binary image, and then the second shoreline was extracted from the refined second binary image by selecting the boundary between the identified land and water features. Finally, the accuracy of both generated shorelines was measured using the checkpoints to select the most appropriate approach for the shoreline-mapping task using the given WorldView-2 image.

3.1. Water-Index-Based Method for Mapping the First Shoreline. NDWI is a remote-sensing-derived index for detecting water features such as oceans, rivers, lakes, and reservoirs from multispectral image sources by using their spectral bands [5, 16, 19, 20]. As the water features are enhanced while other features (e.g., land or vegetation features) are suppressed in NDWI, it is a widely used index for detecting water features from multispectral image sources [17–19]. In this study, the NDWI image was generated using the green and NIR bands of the WorldView-2 image given by (1) [19–22].

$$\text{NDWI image} = \frac{B_G - B_{\text{NIR}}}{B_G + B_{\text{NIR}}}, \quad (1)$$

where B_G and B_{NIR} are the reflectance of the green and NIR bands of the given WorldView-2 image, respectively. The generated NDWI image is shown in Figure 3. As the NDWI image is the ratio image in which one pixel of one spectral band is divided by the corresponding pixel of the other band, the minimum value in the NDWI image is −1, and the maximum value is 1. As can be seen in Figure 3, the pixels representing the water features (ocean) have relatively high values close to 1 while the pixels representing the land features (vegetation, roads, buildings, etc.) have relatively low values close to −1.

The next step was to convert the NDWI image into the first binary image separating the land and water features. In the water-index-based method, the adaptive threshold derived from the adaptive thresholding method is used to separate the land and water features in the NDWI image because it chooses an adaptive intensity threshold in the NDWI image for minimizing the intraclass variance of the white and black pixel groups in the converted binary image [5, 23]. Hence, the adaptive thresholding method was also used in this research to convert the NDWI image into the first binary image separating the land (the black pixels) and water (the white pixels) features (see Figure 4). As in the methodology proposed by Choung and Jo (2015), the first binary image was converted from the NDWI image using the adaptive thresholding method through the Matlab® program (Matlab 2013b, Mathworks, Inc., Natick, Massachusetts, USA). Figure 4 shows the first binary image converted from the NDWI image by the adaptive thresholding method.

Finally, the first shoreline was extracted from the first binary image by selecting the boundary between the land and water features. Figure 5 shows the first shoreline generated from the first binary image by selecting the boundary between the land and water features.

3.2. Machine-Learning-Based Method for Mapping the Second Shoreline. Machine learning is defined as "a branch of artificial intelligence in which a computer generates rules underlying or based on the raw data that have been fed into it," and the machine learning technique is defined as "the ability of a machine to improve its performance based on previous results" [24]. Machine learning has multiple advantages because it is useful for high-value prediction and real-time smart decision-making without human intervention [25]. The machine learning technique is generally used in many applications, such as fraud detection, face recognition, spam filtering, stock trading, and text categorization [26]. Recently, the machine learning techniques have been used in the remote sensing applications, for detecting important features and for classifying land covers from the remote sensing datasets [27].

Machine learning is classified into several methods according to the type of learning algorithm used, such as the supervised learning technique operated with training samples and the unsupervised learning technique not requiring training samples [28]. SVM, a machine learning technique, is a supervised learning algorithm for finding the appropriate hyperplane that maximizes the margins between the two classes in n-dimensional spaces [29]. The SVM classifier has been widely used for the land cover classification tasks using the remote sensing datasets because it produces a very accurate classifier and avoids classification noises by using a hyperplane [30]. Considering these advantages, the SVM classifier was utilized in this study for generating the

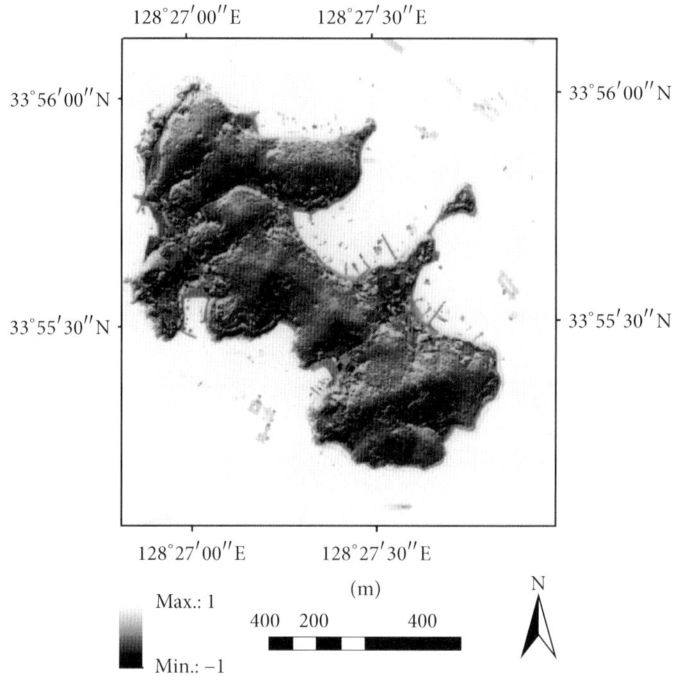

FIGURE 3: NDWI image generated from the given WorldView-2 image.

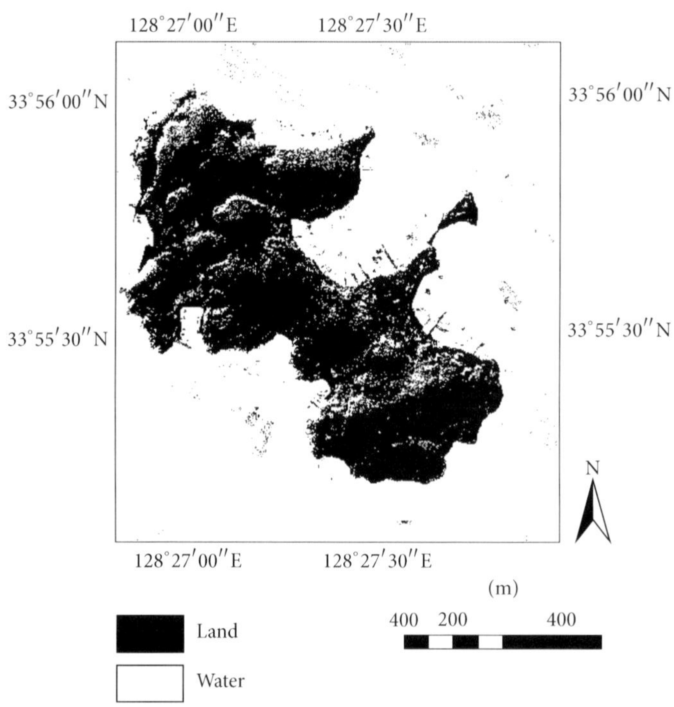

FIGURE 4: First binary image converted from the NDWI image through the adaptive thresholding method.

coastal-surface classification map from the given WorldView-2 image. As can be seen in Figure 1, the coastal zones in Hwado Island consist of multiple land covers, such as rocks, artificial structures (buildings, harbors, roads, seafarm facilities, ships, etc.), water, and vegetation. It was assumed that the main surfaces of the artificial structures were similar to the rock features. Hence, three training sample groups for rocks, water, and vegetation were generated, respectively, as the main objects in the coastal zones, and each training sample group was set to include the 15000 pixels for each land cover. Then the coastal-surface classification map was generated by the SVM classifier using ENVI 4.5 (Exelis

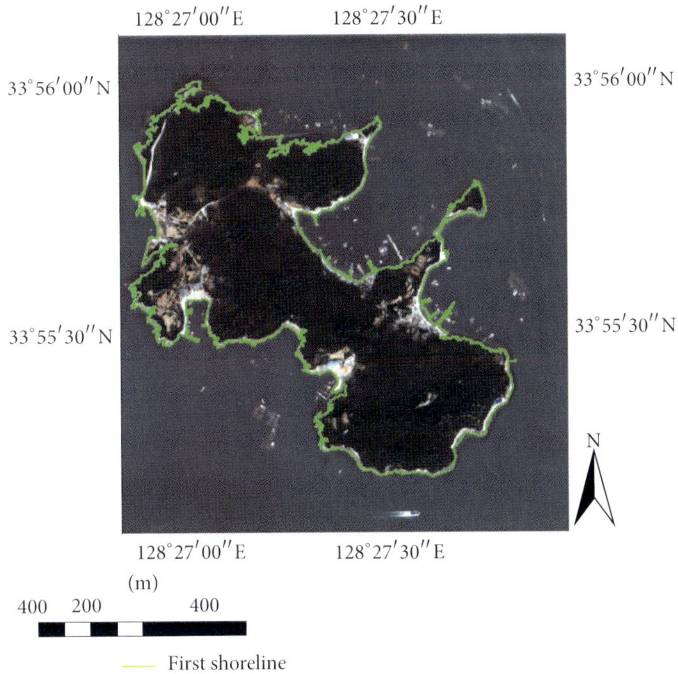

Figure 5: First shoreline extracted from the first binary image by selecting the boundary between the land and water features.

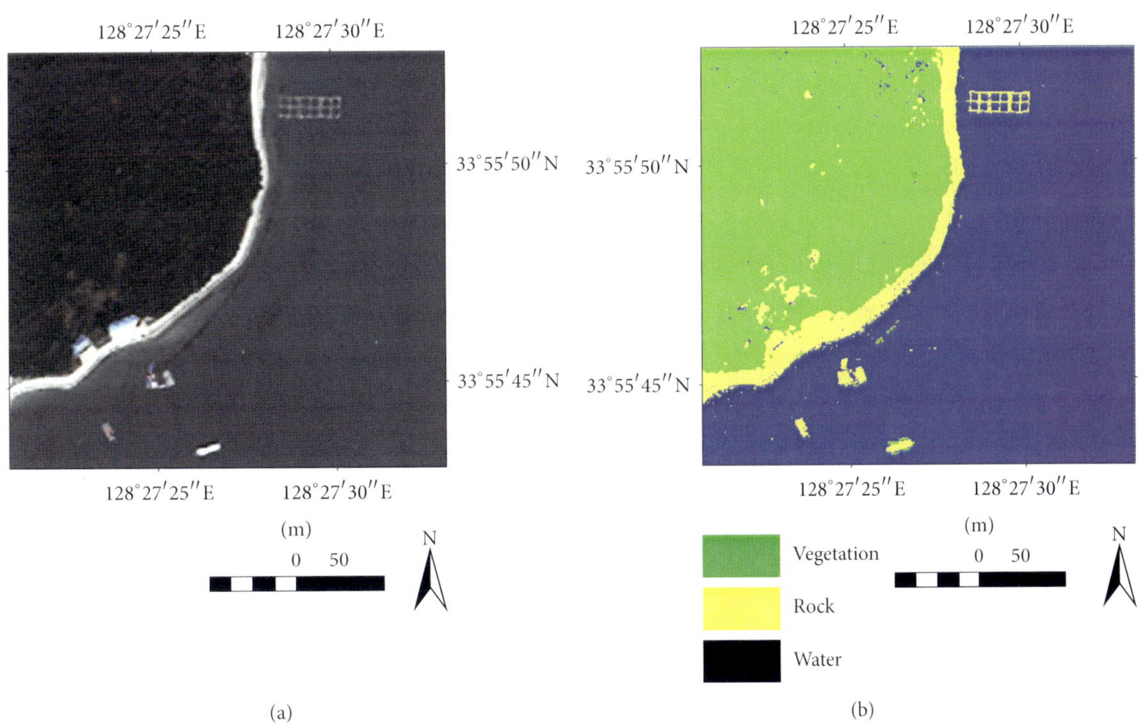

Figure 6: One section of the generated coastal-surface classification map: (a) one section of the given high-resolution satellite image; (b) one section of the generated coastal-surface classification map.

Visual Information Solution, Inc., Boulder, Colorado, USA). Figure 6 shows one section of the generated coastal-surface classification map.

The next step was to convert the coastal-surface classification map into the second binary image. As the rock and vegetation features represent the land features, these features were grouped into the land features in the converted binary image. Figure 7 shows the second binary image converted from the coastal-surface classification map. As seen in Figure 7, the land features were manually set to be

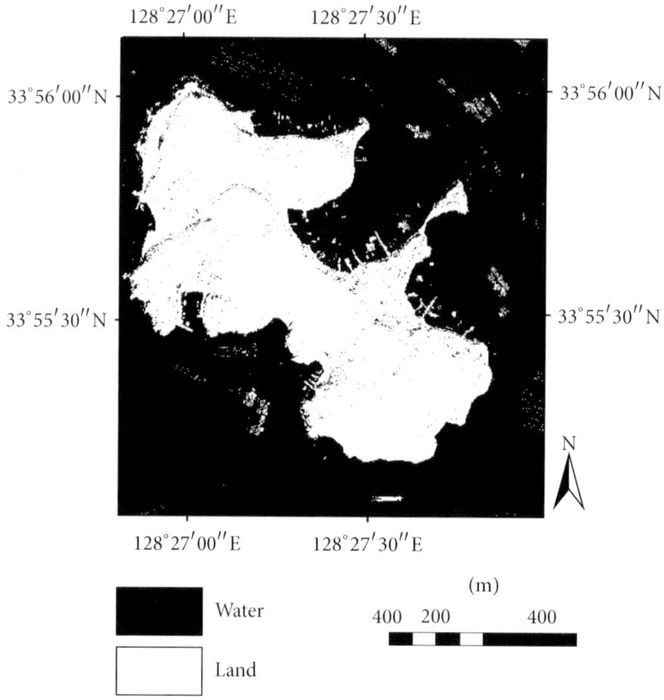

FIGURE 7: Second binary image converted from the coastal-surface classification map.

the white pixels while the water features were also manually set to be the black pixels, respectively, in the second binary image.

In the second binary image, the land features (the white pixels) often include small holes and gaps (the black pixels) due to the irregular shapes of the coastal zones, the continuous wave actions, or the misclassification errors of the SVM classifier, and they generally cause errors in mapping accurate shorelines. Hence, in this research, morphological filtering was applied on the second binary image to remove the small holes and gaps existing in the land features and to preserve the shapes of the land features. Morphological filtering is an image-processing technique for modifying the shapes of the input objects by running the structure elements with specific shapes over the input objects [31, 32]. It was recently used to create new shapes for the input objects for extracting the accurate boundaries of the important features from remote sensing datasets [32]. In this research, morphological filtering was done to remove the small holes in the land features and to preserve the land feature shapes through the two following steps (see Figure 8). In the first step, the image dilation filter was applied to the original binary image shown in Figure 8(a) to remove the small holes in the land features by expanding the land features using the structure element. As can be seen in Figure 8(b), the land features were expanded, and the small holes in the land features were removed by the image dilation filter. In the second step, the image erosion filter was applied to the dilated image for eroding the outside of the expanded land features to preserve their shapes using the same structure element. Finally, as can be seen in Figure 8(c), the outside of the expanded land features was eroded, and the shapes of the land features were preserved. The shape of the structure elements is also significant for reshaping the original objects through morphological filtering [32]. As the shorelines have a linear structure, the shape of the structure elements was set as a square. Also, the width of the structure element was set as 2 m based on empirical analysis. In this research, the entire process of the refinement of the second binary image by the morphological filtering was carried out using the Matlab program (Matlab 2013b, Mathworks, Inc., Natick, Massachusetts, USA).

After the eroded image was generated through the morphological filtering process, the second shoreline was extracted from the eroded image by selecting the boundary between the land and water features (see Figure 9).

4. Results

4.1. Accuracy Measurement of the Coastal-Surface Classification Map. In this section, the accuracy of the coastal-surface classification map is assessed using the 100 checkpoints, defined as the first checkpoint group, generated by manual digitization located around the second shoreline. Table 1 shows the accuracy of the identified coastal surfaces classified by the SVM classifier.

In the generated coastal-surface classification map, there were some misclassification errors owing to the following. First, some water features were misclassified into rock features due to the coastal materials located under the shallow water surfaces. Second, some rock features were misclassified into water features due to their similar reflectance characteristics caused by the shadows on the rock surfaces. Third, some

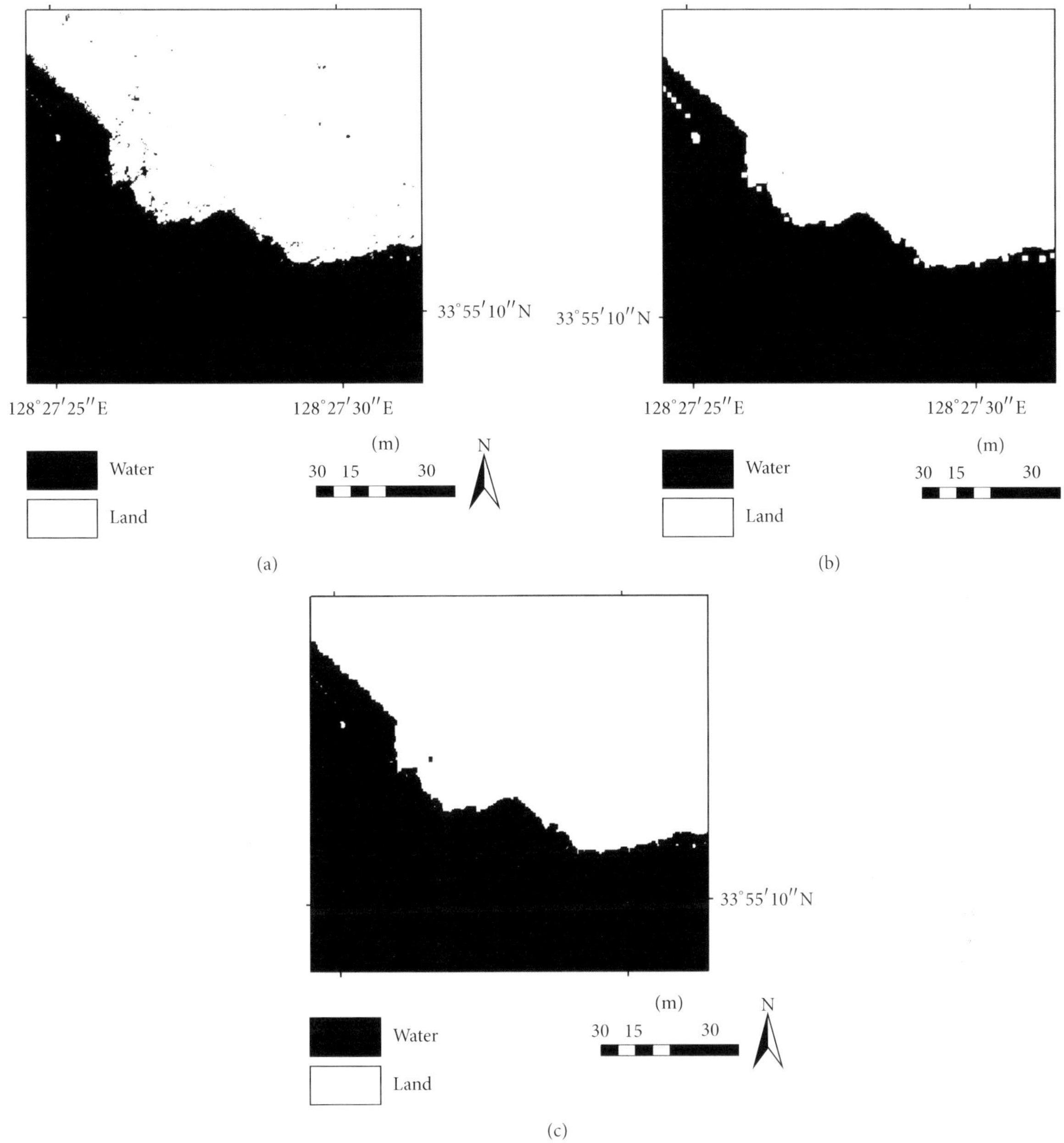

FIGURE 8: Process showing morphological filtering applied to the second binary image: (a) one section of the original binary image; (b) one section of the dilated image; and (c) one section of the eroded image.

TABLE 1: Accuracy of the coastal surfaces classified by the SVM classifier.

Overall accuracy		89%	
Producer's accuracy (Error of omission)		User's accuracy (Error of commission)	
Water	70%	Water	74%
Rock	96%	Rock	93%
Vegetation	75%	Vegetation	86%

vegetation features were misclassified into water features or vice versa, also due to their similar reflectance characteristics caused by the shades on their surfaces and so forth. Figure 10 shows examples of the misclassification errors that occurred in the coastal-surface classification map. Figure 10(a) shows an example of the misclassification in the region where the water features were misclassified into rock features, Figure 10(b) shows an example of the misclassification in the region where the rock features were misclassified into water features, and Figure 10(c) shows an example of the misclassification in the region where the vegetation features were misclassified into water features.

4.2. Accuracy Measurement of the First and Second Shorelines. In this section, the accuracy of the two shorelines generated by the water-index- and machine-learning-based methods,

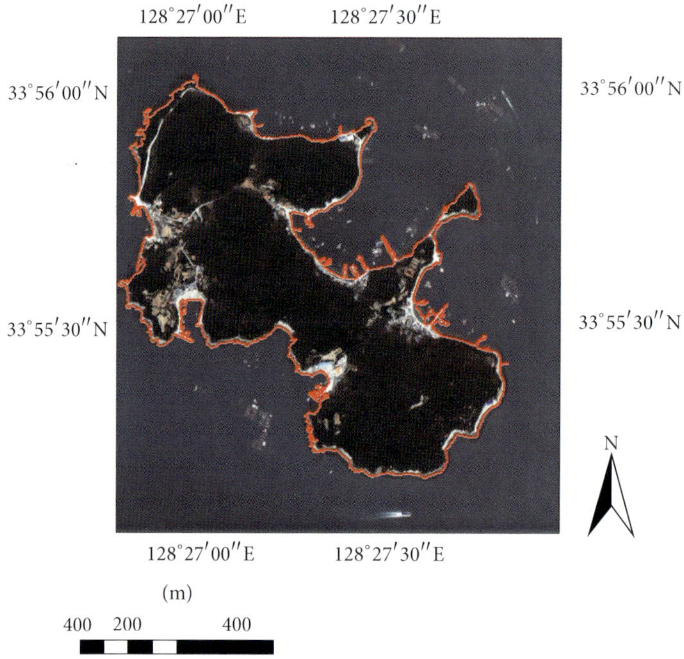

Figure 9: Second shoreline extracted from the eroded image by selecting the boundary between the land and water features.

Table 2: Accuracy of the first shoreline generated by the water-index-based method and the second shoreline generated by the machine-learning-based method.

Accuracies of both shorelines	First shoreline (m)	Second shoreline (m)
Mean	2.62	0.79
Standard deviation	5.27	2.99
Maximum	28.60	25.02

respectively, was measured through the following steps. First, 100 checkpoints, defined as the second checkpoint group, were generated by manual digitization, and the average distance of these checkpoints was 70 m (see Figure 11).

Then the accuracy of both shorelines was assessed by measuring the shortest distance from the checkpoints of the second checkpoint group to the first and second shorelines, respectively. Table 2 shows the accuracy of both shorelines generated by the water-index- and machine-learning-based methods, and Figure 12 shows the line graphs of the shortest distances from each checkpoint to the shorelines generated by the different methods.

5. Discussions

As can be seen in Table 2, the second shoreline generated by the machine-learning-based method had better overall accuracy than the first shoreline generated by the water-index-based method. For a detailed examination of the comparison results, checkpoint indices located in the regions where the first or second shoreline had significant errors were selected (see Figure 13(a)). In Figure 12, the first shoreline had serious errors in checkpoint indices 1, 18, 23, 24, 42, 43, 52, 55, 56, 70, 71, 76, 79, 85, 98, and 99 while the second shoreline had serious errors in checkpoint indices 24, 52, and 55. In general, both shorelines had high accuracy in the well-identified coastal zones where the boundary between land and water was easily recognized in the given high-resolution satellite image. The selected Region 1 shows an example of the coastal zones where both shorelines had high accuracy (see Figure 13(b)). The second shoreline generally had better accuracy than the first shoreline for the following reasons: (1) the irregular coastal shapes that were hardly recognized in the NDWI image due to the continuous wave actions and so forth (checkpoint indices 1, 18, 85, and 99) and (2) the shaded areas that were not identified in the NDWI image but were identified in the coastal-surface classification map generated by the SVM classifier (checkpoint indices 23, 42, 43, 56, 70, 71, 76, 79, and 98). The selected Region 2 (checkpoint index 85) shows an example of the coastal zones where the second shoreline had better accuracy than the first shoreline for the first reason (see Figure 13(c)), and the selected Region 3 (checkpoint indices 42 and 43) shows an example of the coastal zones where the second shoreline had better accuracy than the first shoreline for the second reason (see Figure 13(d)). Both shorelines, however, had low accuracy in the regions where the coastal zones were hardly identified due to the shaded areas that were identified neither in the NDWI image nor in the coastal-surface classification map (checkpoint indices 24, 52, and 55). The selected Region 4 (checkpoint index 24) shows an example of the coastal zones

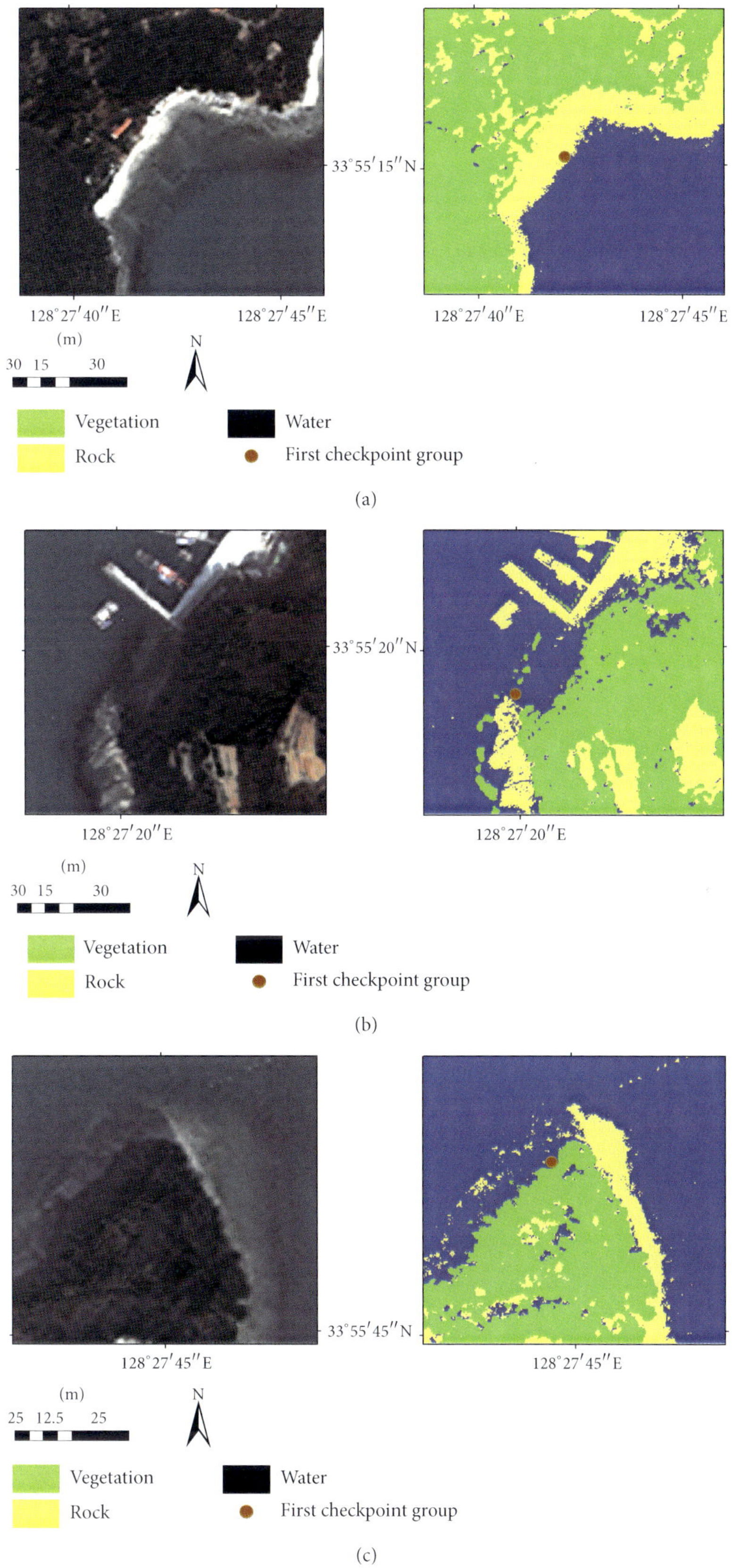

FIGURE 10: Examples of the misclassification errors that occurred in the coastal-surface classification map: (a) example of the misclassification in the region where the water features were misclassified into rock features; (b) example of the misclassification in the region where the rock features were misclassified into water features; and (c) example of the misclassification in the region where the vegetation features were misclassified into water features.

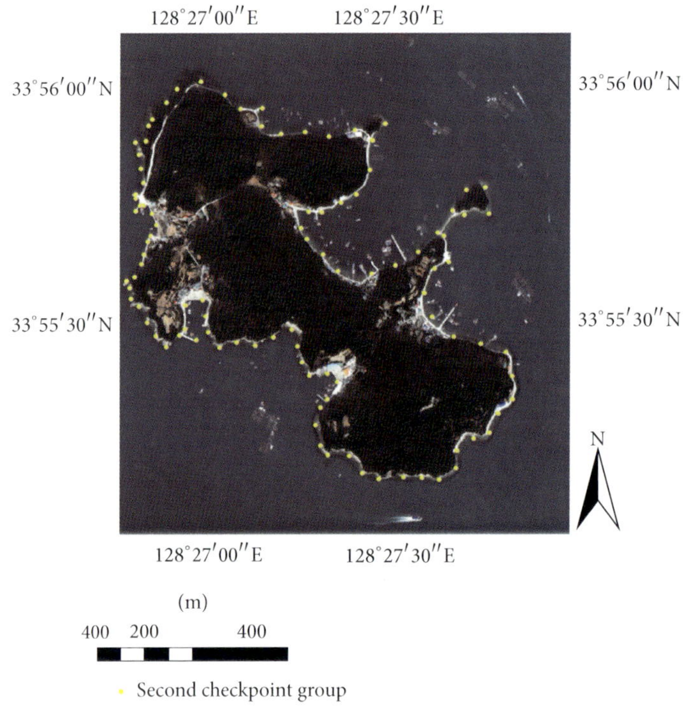

FIGURE 11: Locations of the second checkpoint group for the measurement of the accuracy of both shorelines.

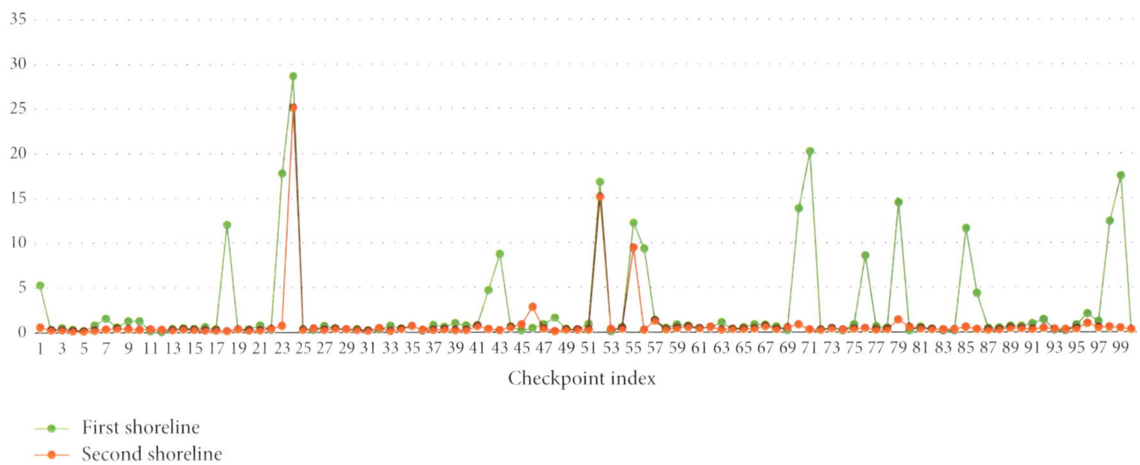

FIGURE 12: Line graphs of the shortest distance from each checkpoint to the shorelines generated by the different methods.

where both shorelines had low accuracy due to the shaded areas not identified in both the NDWI image and the coastal-surface classification map (see Figure 13(e)).

In conclusion, both shorelines had high accuracy in the well-identified coastal zones while the second shoreline generated by the machine-learning-based method had better accuracy than the first shoreline generated by the water-index-based method in the coastal zones with irregular shapes, light shades, and so forth. Both methods, however, showed inefficient performance for mapping the shorelines in the coastal zones with significant shades that were not identified in the NDWI image or the coastal-surface classification map. In general, the pixels representing the shaded areas have the intensity values lower than other pixels in all multispectral bands [33], and it causes the errors for identifying the features in the NDWI image or the coastal-surface classification map generated by using the multispectral bands of the image sources. Hence, the additional data not affected by the shadows would be needed for mapping the shorelines in the shaded areas.

6. Conclusions

The shoreline-mapping task using remotely sensed image sources is efficient for the estimation of the shoreline positions without human access. This research compared different

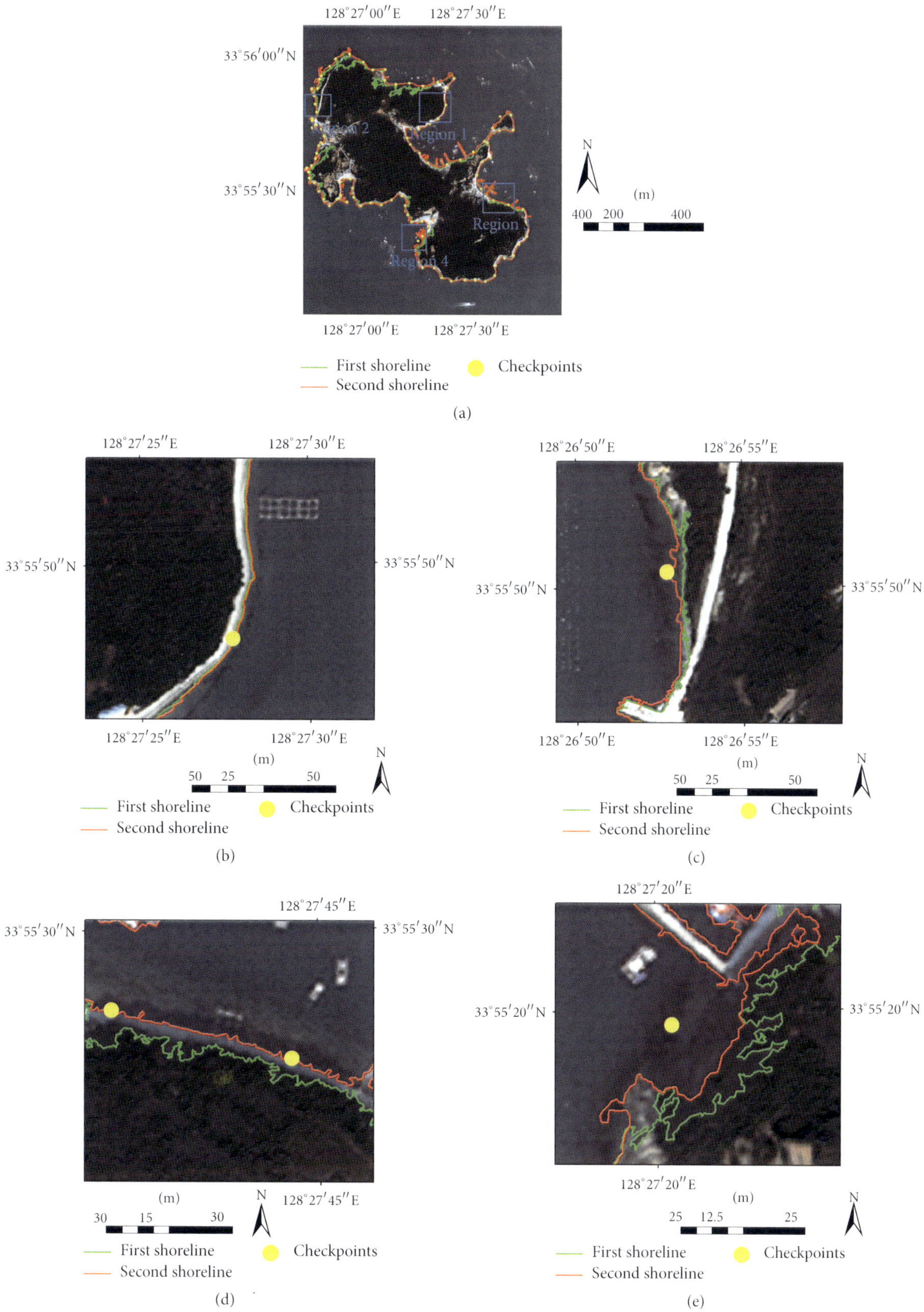

Figure 13: Detailed examination of comparison results: (a) locations of the selected Regions 1, 2, 3, and 4 in the entire study area; (b) Region 1, where both shorelines had high accuracy; (c) Region 2, where the second shoreline had better accuracy for the first reason; (d) Region 3, where the second shoreline had better accuracy for the second reason; and (e) Region 4, where both shorelines had low accuracy due to the shaded areas that were not identified in the NDWI image or the coastal-surface classification map.

methods (the water-index-based method and the machine-learning-based method) for mapping accurate shorelines using a high-resolution satellite image. The water-index-based method is useful for separating the land and water features from multispectral image sources, but it is limited for identifying the various land covers that constitute the coastal zones. The machine-learning-based method is useful for identifying these various coastal features with different spectral-reflectance characteristics, which means that using the machine-learning-based method is better than using the water-index-based method for mapping shorelines using multispectral image sources. There are significant improvements required, however, in future research for the development of an automatic shoreline-mapping process and the estimation of shoreline positions in various coastal zones. First, different machine learning algorithms or any other technique should be applied to generate a more accurate coastal-surface classification map for mapping accurate shorelines in various coastal zones. Second, additional datasets not influenced by the shadows should be integrated into the image sources for mapping accurate shorelines not only in the well-identified coastal zones but also in the shaded coastal zones. Third, the ground truths acquired by the ground-surveying method would be used for measuring accuracies of the generated shorelines and the coastal-surface classification map.

Conflicts of Interest

The authors declare that there are no conflicts of interest regarding the publication of this article.

Acknowledgments

This research was supported by Development of Space Core Technology through the National Research Foundation of Korea (NRF) funded by the Ministry of Science, Information and Communication Technology (ICT) and Future Planning (Grant no. NRF-2014M1A3A3A03067384).

References

[1] NOAA Shoreline Website, "Glossary," 2016, https://shoreline.noaa.gov/glossary.html.

[2] Y. Choung, *Extraction of blufflines from 2.5 dimensional Delaunay triangle mesh using LiDAR data [M.S. thesis]*, Columbus, Ohio, USA, 2012.

[3] Oxford Dictionaries, "Shoreline," 2014, https://en.oxforddictionaries.com/definition/shoreline.

[4] Y. Choung, R. Li, and M.-H. Jo, "Development of a vector-based method for coastal bluffline mapping using LiDAR data and a comparison study in the area of lake erie," *Marine Geodesy*, vol. 36, no. 3, pp. 285–302, 2013.

[5] Y.-J. Choung and M.-H. Jo, "Shoreline change assessment for various types of coasts using multi-temporal landsat imagery of the east coast of South Korea," *Remote Sensing Letters*, vol. 7, no. 1, pp. 91–100, 2016.

[6] I. Lee, *Instantaneous shoreline extraction utilizing integrated spectrum and shadow analysis from LiDAR data and high-resolution satellite imagery [Ph.D. thesis]*, The Ohio State University, Columbus, Ohio, USA, 2012.

[7] R. Li, K. Di, and R. Ma, "A comparative study of shoreline mapping techniques," in *Proceedings of the 4th International Symposium on Computer Mapping and GIS for Coastal Zone Management*, Halifax, Nova Scotia, Canada, June 2001.

[8] A. Guariglia, A. Buonamassa, A. Losurdo et al., "A multisource approach for coastline mapping and identification of shoreline changes," *Annals of Geophysics*, vol. 49, no. 1, pp. 295–304, 2006.

[9] R. Li, K. Di, and R. Ma, "3-D shoreline extraction from IKONOS satellite imagery," *Marine Geodesy*, vol. 26, no. 1-2, pp. 107–115, 2003.

[10] J. Liu, R. Li, S. Deshpande, X. Niu, and T. Shih, "Estimation of blufflines using topographic lidar data and orthoimages," *Photogrammetric Engineering and Remote Sensing*, vol. 75, no. 1, pp. 69–79, 2009.

[11] Z. Hannv, J. Qigang, and X. Jiang, "Coastline extraction using support vector machine from remote sensing image," *Journal of Multimedia*, vol. 8, no. 2, pp. 175–182, 2013.

[12] A. Masria, K. Nadaoka, A. Negm, and M. Iskander, "Detection of shoreline and land cover changes around rosetta promontory, Egypt, based on remote sensing analysis," *Land*, vol. 4, no. 1, pp. 216–230, 2015.

[13] I. Sekovski, F. Stecchi, F. Mancini, and L. Del Rio, "Image classification methods applied to shoreline extraction on very high-resolution multispectral imagery," *International Journal of Remote Sensing*, vol. 35, no. 10, pp. 3556–3578, 2014.

[14] N. Shenbagaraj, N. Mani, and M. Muthukumar, "Isodata classification technique to assess the shoreline changes of Kolachel to Kayalpattanam coast," *International Journal of Engineering Research & Technology*, vol. 3, no. 4, pp. 311–314, 2014.

[15] P. Maglione, C. Parente, and A. Vallario, "Coastline extraction using high resolution WorldView-2 satellite imagery," *European Journal of Remote Sensing*, vol. 47, no. 1, pp. 685–699, 2014.

[16] Y. J. Choung, "Mapping 3D shorelines using KOMPSAT-2 imagery and airborne LiDAR data," *Journal of the Korean Society of Surveying, Geodesy, Photogrammetry and Cartography*, vol. 33, no. 1, pp. 23–30, 2015.

[17] M. Bouchahma and W. Yan, "Automatic measurement of shoreline change on Djerba Island of Tunisia," *Computer and Information Science*, vol. 5, no. 5, pp. 17–24, 2012.

[18] LandInfo., "WorldView-2 50 cm global high-resolution satellite imagery," 2016, http://www.landinfo.com/WorldView2.htm.

[19] W. Li, Z. Du, F. Ling et al., "A comparison of land surface water mapping using the normalized difference water index from TM, ETM+ and ALI," *Remote Sensing*, vol. 5, no. 11, pp. 5530–5549, 2013.

[20] S. K. McFeeters, "The use of the Normalized Difference Water Index (NDWI) in the delineation of open water features," *International Journal of Remote Sensing*, vol. 17, no. 7, pp. 1425–1432, 1996.

[21] L. Ji, L. Zhang, and B. Wylie, "Analysis of dynamic thresholds for the normalized difference water index," *Photogrammetric Engineering & Remote Sensing*, vol. 75, no. 11, pp. 1307–1317, 2009.

[22] K. Rokni, A. Ahmad, A. Selamat, and S. Hazini, "Water feature extraction and change detection using multitemporal landsat imagery," *Remote Sensing*, vol. 6, no. 5, pp. 4173–4189, 2014.

[23] N. Otsu, "A threshold selection method from gray-level histograms," *IEEE Transactions on Systems, Man, and Cybernetics*, vol. 9, no. 1, pp. 62–66, 1979.

[24] DictionaryCom, "Machine learning," 2016, http://www.dictionary.com/browse/machine-learning.

[25] SAS, "Machine learning: what it is & why it matters," 2016, http://www.sas.com/it_it/insights/analytics/machine-learning.html.

[26] P. Domingos, "A few useful things to know about machine learning," *Communications of the ACM*, vol. 55, no. 10, pp. 78–87, 2012.

[27] D. J. Lary, A. H. Alavi, A. H. Gandomi, and A. L. Walker, "Machine learning in geosciences and remote sensing," *Geoscience Frontiers*, vol. 7, no. 1, pp. 3–10, 2016.

[28] Y. Zhang, "Types of machine learning algorithm. in new advances in machine learning," 2016, http://www.intechopen.com/books/new-advances-in-machine-learning.

[29] J. Talairach, "Tutorial on support vector machine (SVM)," 2016, http://www.ccs.neu.edu/course/cs5100f11/resources/jakkula.pdf.

[30] G. Mountrakis, J. Im, and C. Ogole, "Support vector machines in remote sensing: a review," *ISPRS Journal of Photogrammetry and Remote Sensing*, vol. 66, no. 3, pp. 247–259, 2011.

[31] R. Gonzalez and R. Woods, *Digital Image Processing*, Pearson Education, New Jersey, NJ, USA, 3rd edition, 2008.

[32] Y. Choung, "Mapping levees using LiDAR data and multispectral orthoimages in the Nakdong River Basins, South Korea," *Remote Sensing*, vol. 6, no. 9, pp. 8696–8717, 2014.

[33] C. Li, J. Yin, J. Zhao, and F. Ye, "Detection and compensation of shadows based on ICA algorithm in remote sensing image," *International Journal of Advancements in Computing Technology*, vol. 3, no. 7, pp. 46–54, 2011.

12

Wetland Change Detection using Cross-Fused-Based and Normalized Difference Index Analysis on Multitemporal Landsat 8 OLI

Yan Gao, Zeyu Liang, Biao Wang, Yanlan Wu, and Penghai Wu

School of Resources and Environmental Engineering, Anhui University, Hefei, Anhui 230601, China

Correspondence should be addressed to Biao Wang; wangbiao-rs@ahu.edu.cn

Academic Editor: Biswajeet Pradhan

Wetlands are one of the most important ecosystems on the Earth and play a critical role in regulating regional climate, preventing floods, and reducing flood severity. However, it is difficult to detect wetland changes in multitemporal Landsat 8 OLI satellite images due to the mixed composition of vegetation, soil, and water. The main objective of this study is to quantify change to wetland cover by an image-to-image comparison change detection method based on the image fusion of multitemporal images. Spectral distortion is regarded as candidate change information, which is generated by the spectral and spatial differences between multitemporal images during the process of image cross-fusion. Meanwhile, the normalized difference vegetation index (NDVI) and normalized difference water index (NDWI) were extracted from the cross-fused image as a normalized index image to enhance and increase the information about vegetation and water. Then, the modified iteratively reweighted multivariate alteration detection (IR-MAD) is applied to the generally fused images and normalized difference index images, providing a good evaluation of spectral distortion. The experimental results show that the proposed method performed better to reduce the detection errors due to the complicated areas under different ground types, especially in cultivated areas and forests. Moreover, the proposed method was tested and quantitatively assessed and achieved an overall accuracy of 96.67% and 93.06% for the interannual and seasonal datasets, respectively. Our method can be a tool to monitor changes in wetlands and provide effective technical support for wetland conservation.

1. Introduction

Wetlands are a unique ecosystem formed by the interaction between water and land, and they cover 6% of the Earth's surface [1]. Due to seasonal changes, the characteristics of wetlands vary among water, soil, and vegetation. This makes the wetland landscape more complex, and it becomes more difficult to extract information about changes in these regions. In addition, the combination of reflectance spectra of the underlying soil, the hydrologic regime, and atmospheric vapor makes optical classification more difficult, and these factors could introduce a reduction in spectral reflectance. Therefore, it is often difficult to achieve the expected results using a single method to extract information about wetland change [2].

Postclassification comparison (PCC), in which two multitemporal images are independently classified and then compared [3], is one of the methods used for wetland change detection. First, it is applied for detecting the trajectories of corresponding wetland cover types. More specifically, it includes many classification methods, such as the regression tree algorithm or the maximum likelihood classification [4, 5]. However, in some of these methods, high-accuracy classification and ground truth information are required [3–6].

The image-to-image (or direct) comparison change detection method, which is another method for wetland change detection, is used to obtain a difference image of spectral changes through the analysis and calculation of the spectral characteristics from multitemporal images, and then a binary image is generated in which the change areas are distinguished from unchanged areas [7]. The advantages of this method are first that it provides a faster comparison of images and second that it demands no ground truth information; however, it could not display the change trajectories of wetland cover types [8]. Change detection methods such as change vector analysis (CVA) [9], principal component analysis (PCA) [10], Erreur Relative Globale Adimensionnelle de Synthese (ERGAS) [11], and multivariate alteration detection (MAD) [12] directly calculate multitemporal images.

However, the results of the change detection method based on the difference image largely depend on the spectral characteristics and may include some false positives. For this reason, a change detection method based on cross-fusion and spectral distortion is proposed to improve the accuracy of change detection in flood zones [13]. Successful change detection results have been achieved in coal-mining subsidence areas; nevertheless, there are still some false detections in wetland areas [14].

To mitigate the false positive results, we employ an image-to-image method through NDVI and NDWI extraction based on a cross-fusion image to detect the change information of wetlands. In the case of wetlands, vegetation, soil, and water coexist, and the application of the cross-fusion method is beneficial for improving spatial resolution and enhancing information about wetland change. In addition, based on the cross-fused image, NDVI and NDWI are extracted in order to enhance the information about vegetation and water. We can then derive the change information using the modified IR-MAD algorithm, which is a well-established change detection method for multitemporal multispectral images [12]. Finally, the change area of wetlands is obtained using the automated threshold method.

2. Study Area and Dataset

In this study, we collected three Landsat 8 OLI multitemporal images covering the Shengjin Lake Nature Reserve area. These images represent the seasonal (25 July 2016 and 16 December 2016) and interannual (6 November 2013 and 16 December 2016) changes in the types of land cover, as shown in Figures 1(a)–1(c). In the preprocessing, the relevant bands were selected, including the 30-meter-resolution MS (multispectral) bands 2–7 and the 15-meter-resolution PAN (panchromatic) band 8. Then, the Shengjin Lake-protected area vector data were used to cut the images, with the specific parameters shown in Table 1.

The Shengjin Lake National Nature Reserve (30°15′~30° 30′N, 116°55′~117°15′E) is located in Chizhou City, Anhui Province. The protected region, with a total area of 333.40 km^2, consists of Shengjin Lake, cultivated areas, urban areas, forest, and bare land. The Shengjin Lake wetland ecological environment is well-preserved, with rich natural and cultural landscapes. It is one of the most complete areas in the wetland ecosystem of inland freshwater lakes in the lower reaches of the Yangtze River. It connects with the Yangtze River, and the water level of the lake is regulated by the Huangpen sluice. The location of the study area is shown in Figure 1(d). The water level of Shengjin Lake varies between 3.4 and 7.4 m due to the sluice. Water level changes lead to the largest lake area in the summer wet season and a smaller area in the winter dry season. During the dry season, two largest *Carex* meadows ("upper lake meadow" and "lower lake meadow") provide suitable living environments and food sources for Greater White-fronted Geese and Bean Geese, and it has become a critical winter habitat for rare birds [15, 16].

3. Methodology

In this section, we detail the process of extracting wetland change information from bitemporal images using a modified IR-MAD. We consider the two datasets $\mathrm{F}_{t_1}^{\mathrm{H}}$ and $\mathrm{F}_{t_2}^{\mathrm{H}}$ consisting of high-resolution PAN and low-resolution MS images acquired in the same geographical area at different times t_1 and t_2, respectively. The process flow is shown in Figure 2, and the details of each step are described below.

3.1. Cross-Fused Image Generation. Generally, the image fusion method fuses high-resolution PAN images and low-resolution MS images into a high-resolution MS image. In this paper, the PAN and MS images are generally fused using Gram-Schmidt adaptive (GSA) to produce high-resolution multispectral $F_{t_1}^{\mathrm{H}}$ and $F_{t_2}^{\mathrm{H}}$ images. The GSA algorithm is applied as a representative component substitution- (CS-) based fusion algorithm, and there is no limit to the number of bands of the fused image [17]. The major drawback of the CS-based fusion method is spectral distortion, also called color (or radiometric) distortion. This spectral distortion is caused by the mismatch between the spectral responses of the MS and PAN bands according to the different bandwidths [18]. In this study, the spectral distortion will be regarded as a candidate detection feature for wetland cover change [17]. To this end, we extract the high-resolution NIR image instead of high-resolution PAN image. The NIR band has a narrower bandwidth, and by using this band, the mismatch level of spectral response outside the NIR spectral range will increase, and the spectral distortion will become more pronounced [13]. At the same time, the NIR band is a very useful information source for detecting water or vegetation areas, because the water area appears dark since it has strong absorption characteristics, and the vegetation has the opposite characteristics and appears light. Thus, the NIR band is useful in extracting change information. Then, the cross-fused image $\mathrm{CF}_1^{\mathrm{H}}$ is generated where the $F_{t_1}^{\mathrm{H}}$ MS image is fused with the NIR band of $F_{t_2}^{\mathrm{H}}$ by using the GSA image fusion algorithm. $\mathrm{CF}_2^{\mathrm{H}}$ can be obtained in the same way as $\mathrm{CF}_1^{\mathrm{H}}$, and the formula is as follows:

$$\mathrm{CF}_1^{\mathrm{H}} = \mathrm{NIR}_{F_{t_2}}^{\mathrm{H}} + \mathrm{MS}_{t_1}^{\mathrm{L}}, \quad (1)$$

$$\mathrm{CF}_2^{\mathrm{H}} = \mathrm{NIR}_{F_{t_1}}^{\mathrm{H}} + \mathrm{MS}_{t_2}^{\mathrm{L}}. \quad (2)$$

3.2. NDVI and NDWI Extraction. The NDVI, which is a classic index used to monitor vegetation changes, is calculated from a normalized transform of the NIR and red reflectance ratio. Application of the NDVI strives to minimize the solar

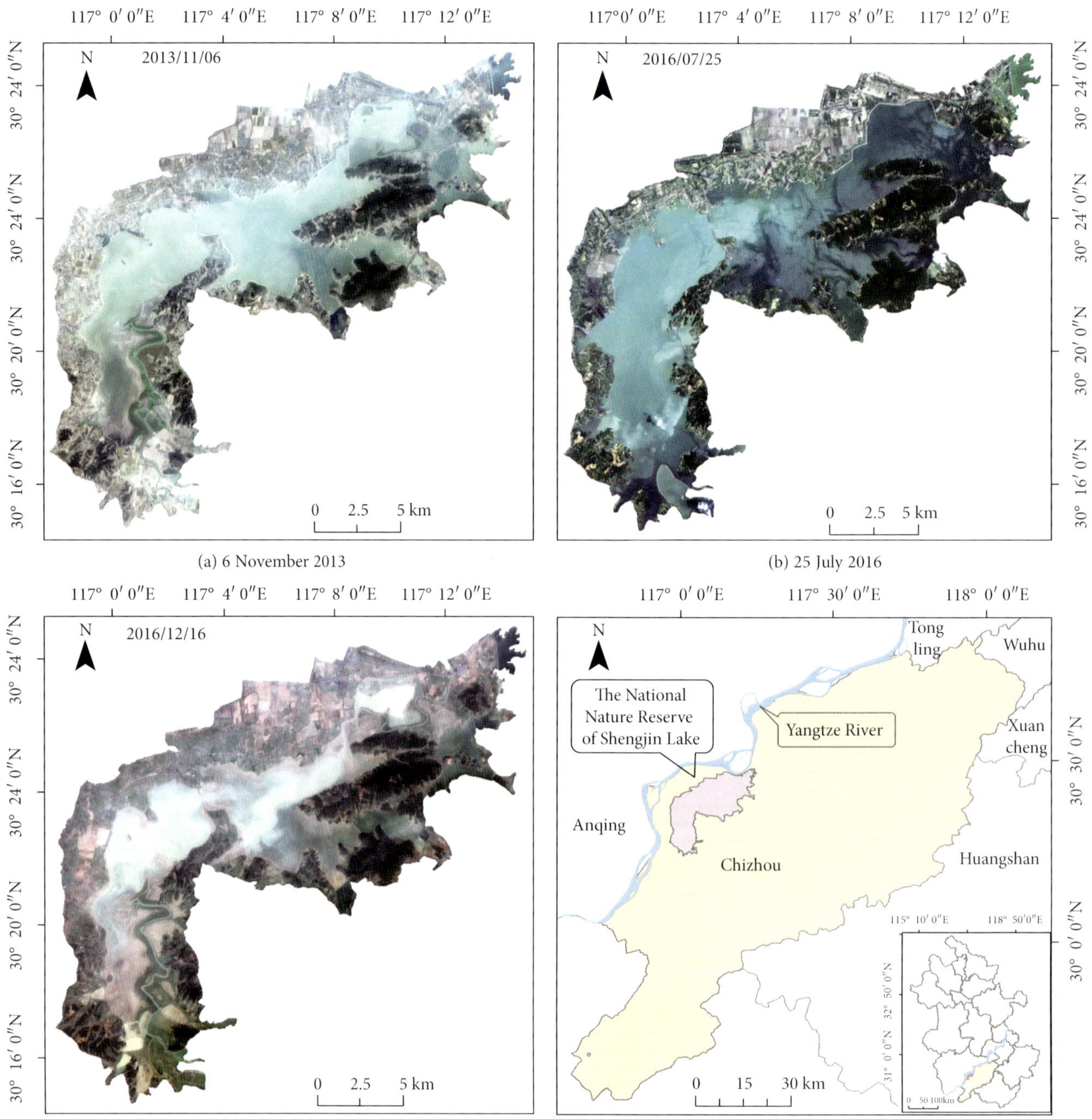

(a) 6 November 2013

(b) 25 July 2016

(c) 16 December 2016

(d) Location of the study area

Figure 1: Multitemporal images of Shengjin Lake used in the experiment.

Table 1: Data specifications.

Date	Spatial resolution	Image size (pixels)	Wavelength (μm)
06/11/2013 25/07/2016 16/12/2016	MS: 30 m	MS: 3650 × 3570	Band 2 blue: 0.450–0.515 Band 3 green: 0.525–0.600 Band 4 red: 0.630–0.680 Band 5 NIR: 0.845–0.885 Band 6 SWIR: 11.560–1.660 Band 7 SWIR: 12.100–2.300
	PAN: 15 m	PAN: 1825 × 1785	Band 8 pan: 0.500–0.680

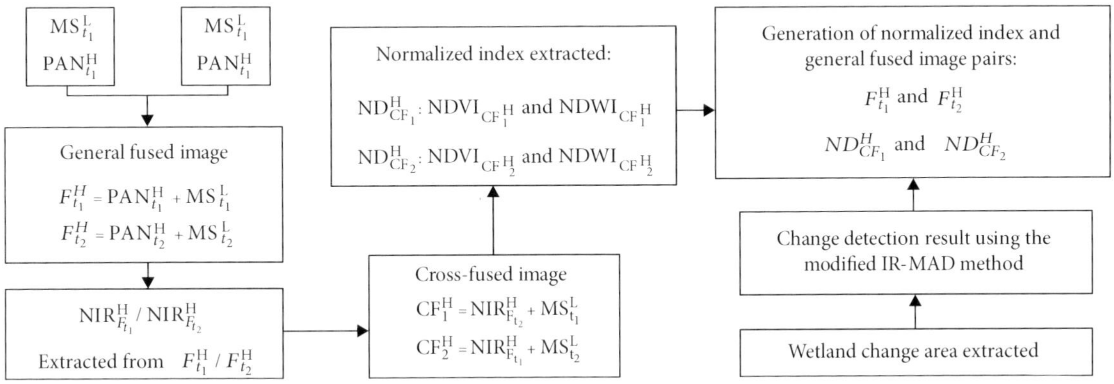

FIGURE 2: Workflow of the proposed methodology for wetland vegetation extraction using multitemporal satellite images. The superscript H and L represent high resolution and low resolution, respectively.

irradiance and soil background effects and improve the vegetation signal. Similarly, the NDWI is a commonly used remote sensing water monitoring index [19, 20].

The cross-fused image has high spatial and temporal resolution, and the vegetation and water details of the bitemporal images are preserved. By using the NDVI and the NDWI, the change information from water and vegetation is further optimized, and the distinction among water bodies, wet soil, and vegetation is improved. The spectral difference is enhanced twofold, and the detection sensitivity of the vegetation and water increases. The equation is as follows:

$$\mathrm{NDVI}_F = \frac{\mathrm{NIR}_F - \mathrm{R}_F}{\mathrm{NIR}_F + \mathrm{R}_F}, \tag{3}$$

$$\mathrm{NDWI}_F = \frac{\mathrm{green}_F - \mathrm{NIR}_F}{\mathrm{green}_F + \mathrm{NIR}_F}, \tag{4}$$

where NIR_F, R_F, and green_F are the NIR, red, and green bands of the cross-fused image, respectively. According to (3) and (4), the $\mathrm{ND}^{\mathrm{H}}_{\mathrm{CF}_1}$ and $\mathrm{ND}^{\mathrm{H}}_{\mathrm{CF}_2}$ images are generated from the cross-fused images $\mathrm{CF}^{\mathrm{H}}_1$ and $\mathrm{CF}^{\mathrm{H}}_2$ and include the NDVI and NDWI bands.

3.3. Wetland Change Area Extracted. The IR-MAD algorithm, which is based on canonical correlation analysis, considers two K band multispectral images R and T of the same area, but acquired at two different times, and it has an important application in the field of multitemporal multispectral image change detection. The random variables U and V, generated by any linear combination of the intensities of the spectral bands using coefficient matrices **a** and **b**, are defined. The equation is as follows [21].

$$\begin{aligned} U &= \mathbf{a}^T R, \\ V &= \mathbf{b}^T T, \end{aligned} \tag{5}$$

where the superscript T is the transpose of each matrix. The task is to find suitable vectors **a** and **b** by maximizing the variance of $U - V$. This leads to solving two generalized eigenvalue problems for **a** and **b** from a canonical matrix analysis.

The MAD variate M_k, which is generated by taking the paired difference between U and V, represents the changed information [22], where the equation is as follows (6).

$$\begin{aligned} M_k &= U_{K-k+1} - V_{K-k+1}, \\ k &= \mathbf{a}_{K-k+1}{}^T R - \mathbf{b}_{K-k+1}{}^T T, \\ K &= 1, 2, \ldots, K. \end{aligned} \tag{6}$$

In this study, corresponding to the two multitemporal normalized difference index images ($\mathrm{ND}^{\mathrm{H}}_{\mathrm{CF}_1}$ and $\mathrm{ND}^{\mathrm{H}}_{\mathrm{CF}_2}$) and two general fused images ($F^{\mathrm{H}}_{t_1}$ and$F^{\mathrm{H}}_{t_2}$), MAD variate M_{p} is generated by using optimal coefficients **a** and **b** through (7) and (8):

$$M_1 = \mathbf{a}^T_{(1)} F^{\mathrm{H}}_{t_1} - \mathbf{b}^T_{(1)} F^{\mathrm{H}}_{t_2}, \tag{7}$$

$$M_2 = \mathbf{a}^T_{(2)} \mathrm{ND}_{\mathrm{CF}_1} - \mathbf{b}^T_{(2)} \mathrm{ND}_{\mathrm{CF}_2}, \tag{8}$$

where M_1 is the MAD variate of the generally fused image with the same temporal data. Meanwhile, two normalized index images ($\mathrm{ND}^{\mathrm{H}}_{\mathrm{CF}_1}$ and $\mathrm{ND}^{\mathrm{H}}_{\mathrm{CF}_2}$) in M_2 are extracted from the cross-fusion image.

The probability of the changed information for pixel j, which is calculated by the sum of the squares of the standardized MAD variate, is defined in (9).

$$Z_j = \sum_{k=1}^{K} \left(\frac{M_{kj}}{\sigma_{M_k}} \right)^2, \tag{9}$$

where variable Z_j represents a weight for the probability of changes in each pixel to identify a greater chi-square value, M_{kj} is the MAD variate of the kth band for pixel j, and σ_{M_k} is the variance of the no-change distribution. The above results can be regarded as the weights of the observations. The iteration process would continue for a number of fixed iterations or until there is no significant change in the canonical correlations. The latter is used in this study [22]. Then, the optimal matrices **a** and **b** are recalculated by the weight factor.

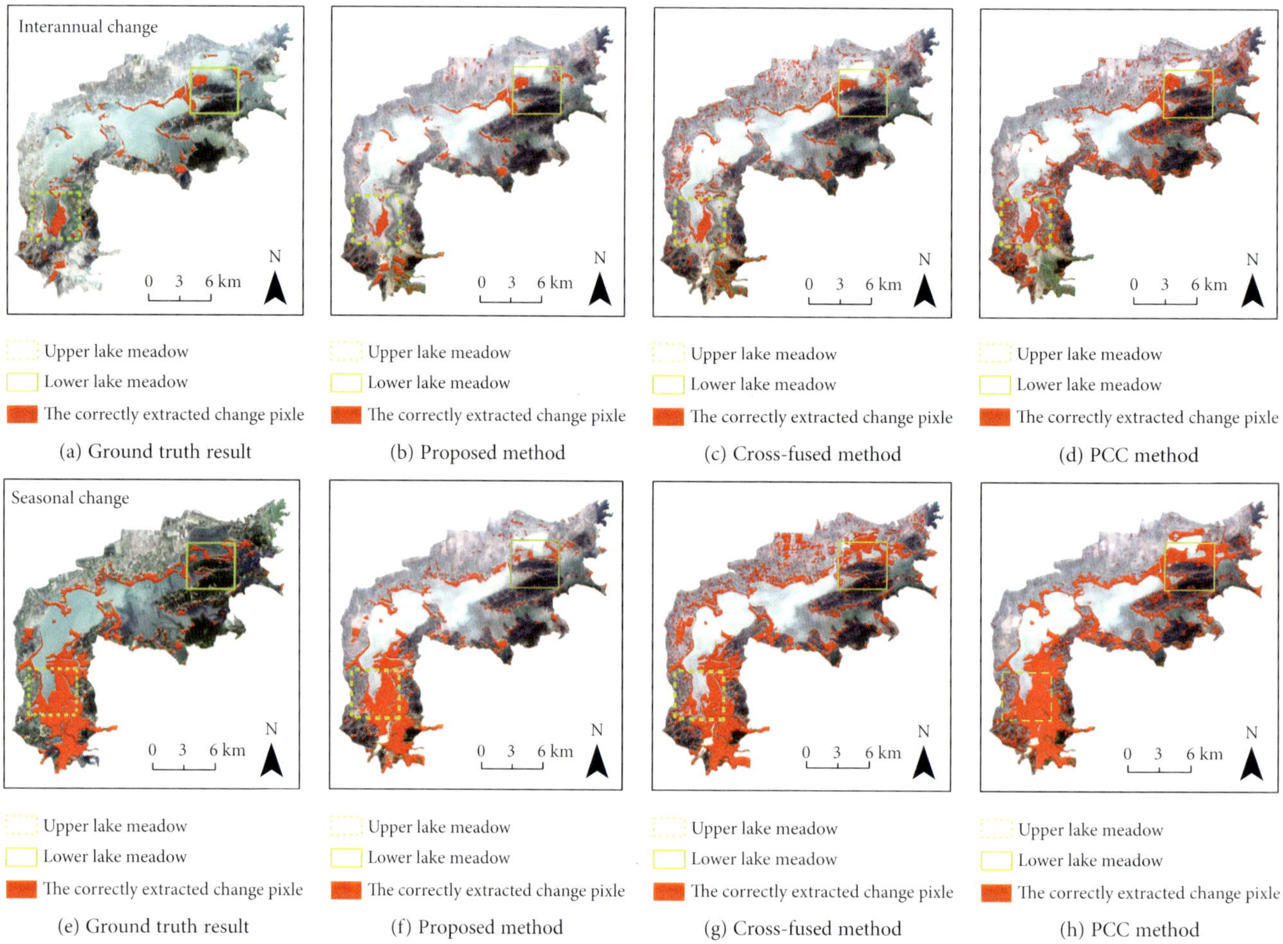

FIGURE 3: Results of wetland change area by using the tested methods: (a)–(d) interannual change detection result; (e)–(h) seasonal change detection result.

In this study, based on the combination of M_1 and M_2, the final change detection index Z'_j is calculated.

$$Z'_j = \sum_{p=1}^{2} \sum_{k=1}^{K} \left(\frac{M_{(p)kj}}{\sigma_{(p)M_k}} \right)^2. \qquad (10)$$

$M_{(p)kj}$ is the kth band MAD variate of the pth pair of fused images for pixel j. This method can effectively reduce the falsely detected changes by considering Z'_j values twice in (10).

This modified IR-MAD algorithm can not only alleviate the problem of spectral distortion that caused massive false change alarms in the process of using bitemporal images to generate the cross-fused images but also reduce the interaction between bands of multispectral images [14, 21]. Therefore, this algorithm can yield better change detection results in multitemporal images. Finally, the Otsu thresholding algorithm, which is based on histogram image segmentation [23], has effective performance and easy application, and it was applied to the modified IR-MAD image to obtain binary data of the changed and the unchanged area [23].

4. Experimental Result and Discussion

To evaluate the effectiveness of this method, we analyze and discuss the seasonal and interannual variations in the study area. We use the cross-fused and PCC change detection methods to compare with our result. In the cross-fused change detection method, original IR-MAD is applied between CF_1^H and CF_2^H images to extract the changed area. In the process of PCC, three general fused images are classified into 7 classes (water, bare land, meadow, cultivated area, city, forest, and mudflats) through the maximum likelihood classification.

To quantitatively compare the performance of these methods, ground truths of seasonal and interannual variation images were generated from GSA-fused images ($F_{t_1}^H$ and $F_{t_2}^H$) by manually digitizing the changed areas of Shengjin Lake Nature Reserve as shown in red in Figures 3(a) and 3(e). The results are overlain on the multispectral images of 6 November 2013 and 25 July 2016, respectively. In the quantitative analysis process, the confusion matrix method was applied to evaluate the statistical accuracy of the tested methodologies, and some indices such as overall accuracy (OA), kappa coefficient (KC), commission error (CE), omission error (OE), and false alarm rate (FAR) were calculated

TABLE 2: Change detection accuracy results: overall accuracy (OA), kappa coefficient (KC), commission error (CE), omission error (OE), and false alarm rate (FAR).

		OA (%)	KC	CE (%)	OE (%)	FAR
Interannual	Proposed method	96.67	0.72	27.60	25.02	0.02
	Cross-fused method	92.07	0.52	58.97	15.02	0.07
	PCC	86.26	0.39	70.22	11.13	0.14
Seasonal	Proposed method	93.06	0.76	14.88	24.00	0.03
	Cross-fused method	85.28	0.59	43.06	13.93	0.15
	PCC	84.45	0.57	44.56	16.36	0.15

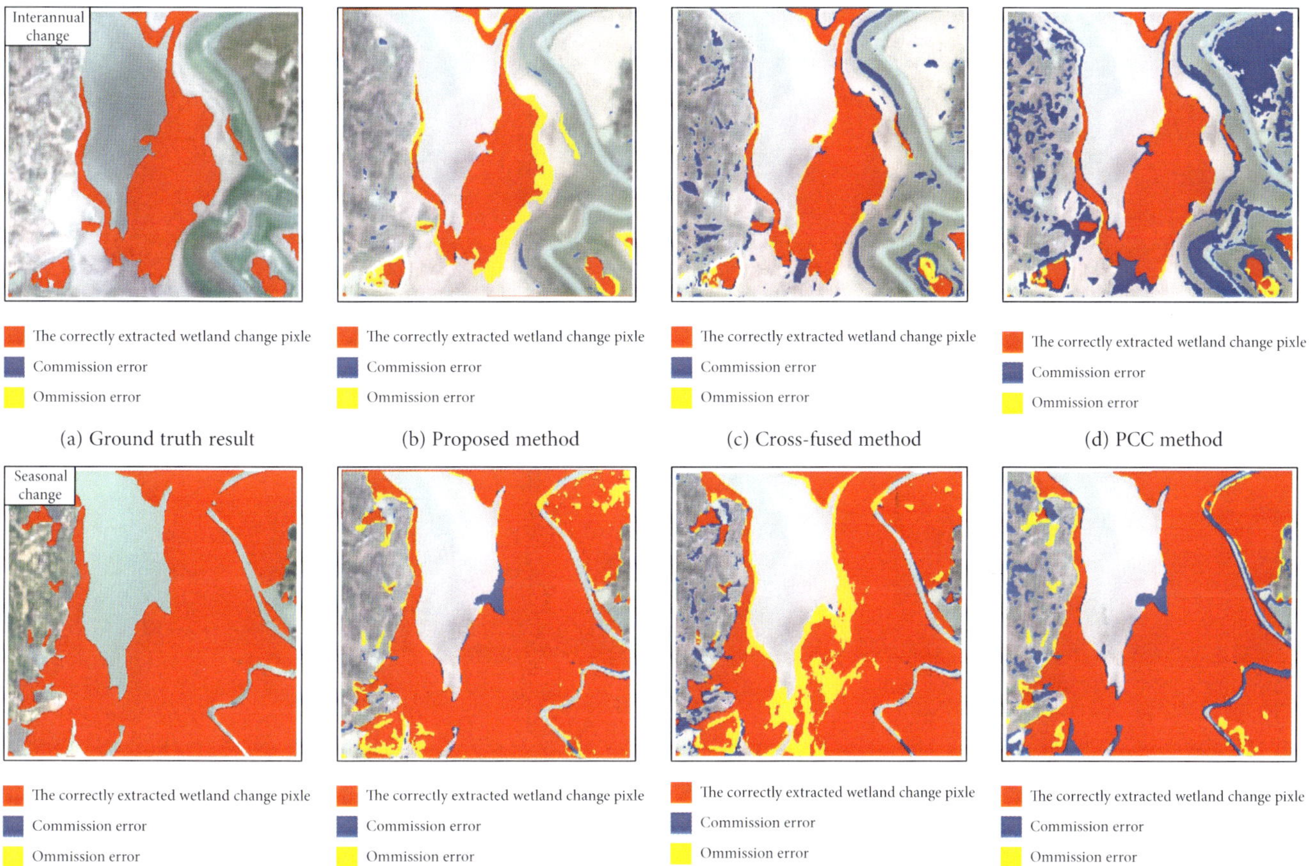

(a) Ground truth result (b) Proposed method (c) Cross-fused method (d) PCC method

(e) Ground truth result (f) Proposed method (g) Cross-fused method (h) PCC method

FIGURE 4: Detailed images from upper lake meadow regions of Figure 3: (a)–(d) interannual change detection result; (e)–(h) seasonal change detection result.

[24]. The detailed quantitative change detection accuracy assessment results for each method are shown in Figure 3 and Table 2. The red color indicates the change pixels extracted from the change detection results of the different methods, and the results are overlain on the multispectral image of 25 December 2016.

Through the observation and analysis of the results in Figure 3, in areas where different ground types coexist (water, bare land, meadow, cultivated area, city, forest, and mud-flats), the proposed method, compared to PCC based on the general fused image and cross-fused method, can more accurately detect the wetland change information and effectively reduce the change detection errors for interannual or seasonal wetland cover change. In the results of the PCC and cross-fused methods, some parts of the unchanged area are considered to be changed areas. As shown in Table 2, the CE value of PCC reaches 70%. In addition, the results of our study are more accurate than the PCC; the OA value reaches 90% and the FAR value reaches 0.02.

Figure 3 includes the whole study area, allowing for an initial visual assessment of the results of wetland change extent extraction. Figures 4 and 5 show the subimages extracted from "upper lake meadow" and "lower lake meadow" regions of Figure 3.

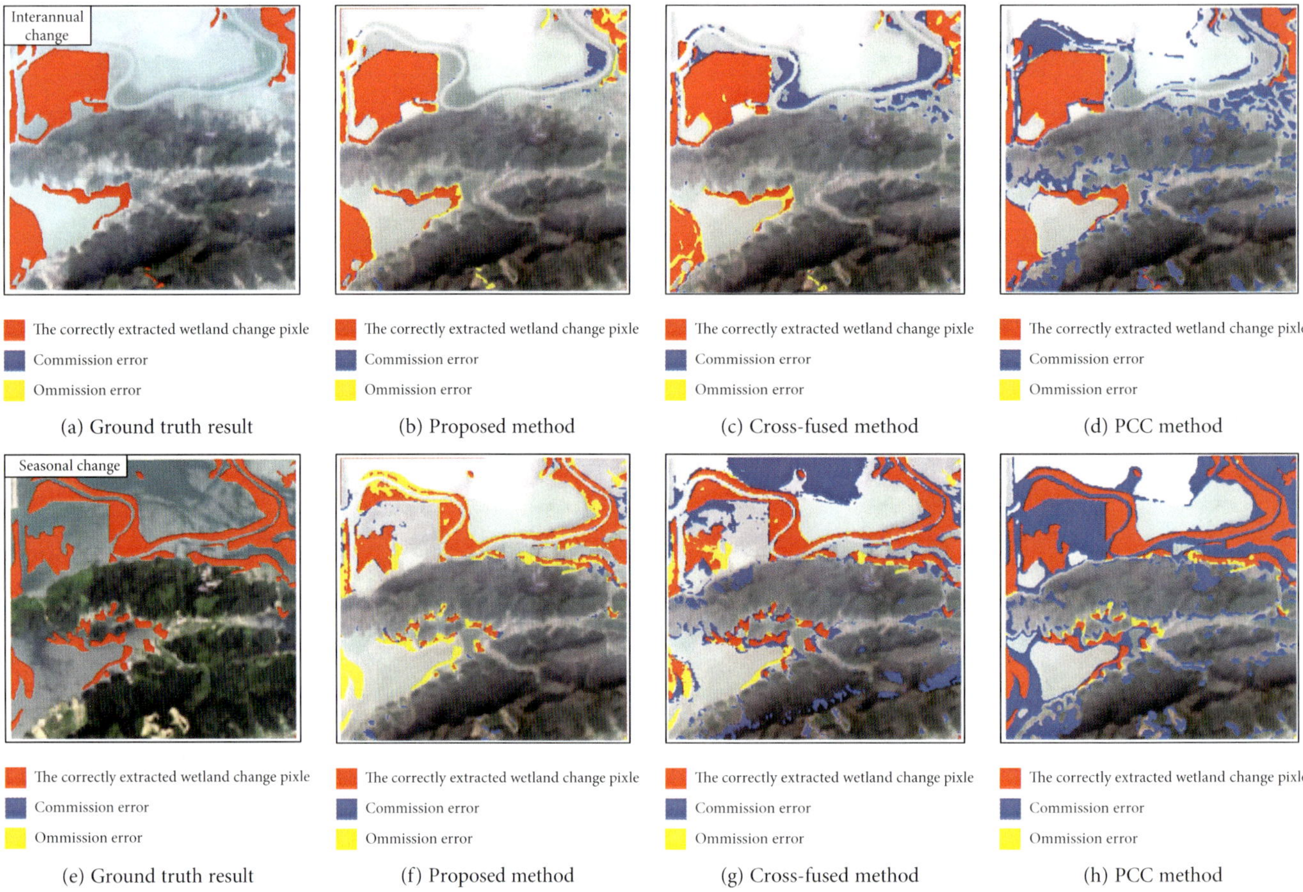

FIGURE 5: Detailed images from lower lake meadow regions of Figure 3: (a)–(d) interannual change detection result; (e)–(h) seasonal change detection result.

Figure 4 mainly consists of a complex area composed of water, meadow, cultivated area, city, and mudflats. In addition, the area shown in Figure 5 mainly consists of complex areas composed of water, meadow, forest, and mudflats. The blue and yellow colors indicate the commission and omission error, respectively, and the results are overlain on the multispectral image of 25 December 2016.

As shown in Figures 4 and 5, the PCC and cross-fused method are not sensitive to artificially cultivated areas and forest areas affected by seasonal changes, and they detect too many false positives caused by similar spectral characteristics.

The proposed method efficiently detects the changed area that corresponds to the complex area with similar spectral characteristics, improves the accuracy of wetland change detection, and minimizes the impacts of seasonality and artificiality. The proposed method also has good performance in meadow areas. However, it produces false positives in some mudflat edge regions, such as yellow areas in Figures 4(b) and 5(f). On the one hand, this is because spatial inconsistency occurred due to the different look angles of bitemporal imagery. On the other hand, the proposed method is based only on the NDWI, which is sensitive to water. As a result, some omission errors can occur in areas where the water level falls and the mudflats are exposed because the mudflats still contain a certain amount of water.

5. Conclusions

In this paper, we proposed an image-to-image change detection method using multitemporal images to quantify wetland cover changes; the method is based on a combination of a cross-fusion image and normalized difference index image. For multitemporal Landsat 8 OLI images, the GSA fusion method is used to generate cross-fusion images, and then NDVI and NDWI are extracted. The optimal change information was calculated through the modified IR-MAD, which used pairs of normalized difference index images and general fused images. The experimental results showed that the proposed method increases the accuracy of change detection and minimizes the error detection in the complex areas under different ground types. Especially in the cultivated area affected by manmade alterations, change information can be identified more accurately, and a lower FAR can be achieved. This allows us to help wetland managers implement effective management plans. Further, our method is of guiding importance in the monitoring of wetland health and wetland conservation.

Conflicts of Interest

The authors declare no conflict of interest.

Acknowledgments

This research was supported in part by the Natural Science Foundation of Anhui Province under Grant no. 1608085MD83, the Science and Technology project of the Department of Land and Resources of Anhui Province: Study on the Data Integration Technology for the Real Estate Registration (no. 2016-K-12), the Educational Commission of Anhui Province of China (no. KJ2018A0007), and the Department of Human Resources and Social Security of Anhui: Innovation Project Foundation for Selected Overseas Chinese Scholar.

References

[1] G. Liu, L. Zhang, Q. Zhang, Z. Musyimi, and Q. Jiang, "Spatio-temporal dynamics of wetland landscape patterns based on remote sensing in Yellow River Delta, China," *Wetlands*, vol. 34, no. 4, pp. 787–801, 2014.

[2] E. Adam, O. Mutanga, and D. Rugege, "Multispectral and hyperspectral remote sensing for identification and mapping of wetland vegetation: a review," *Wetlands Ecology and Management*, vol. 18, no. 3, pp. 281–296, 2010.

[3] A. Singh, "Review article digital change detection techniques using remotely-sensed data," *International Journal of Remote Sensing*, vol. 10, no. 6, pp. 989–1003, 1989.

[4] L. Yang, C. Homer, J. Brock, and J. Fry, "An efficient method for change detection of soil, vegetation and water in the Northern Gulf of Mexico wetland ecosystem," *International Journal of Remote Sensing*, vol. 34, no. 18, pp. 6321–6336, 2013.

[5] C. Munyati, "Wetland change detection on the Kafue Flats, Zambia, by classification of a multitemporal remote sensing image dataset," *International Journal of Remote Sensing*, vol. 21, no. 9, pp. 1787–1806, 2000.

[6] F. Bovolo, S. Marchesi, and L. Bruzzone, "A framework for automatic and unsupervised detection of multiple changes in multitemporal images," *IEEE Transactions on Geoscience and Remote Sensing*, vol. 50, no. 6, pp. 2196–2212, 2012.

[7] D. Lu, P. Mausel, E. Brondízio, and E. Moran, "Change detection techniques," *International Journal of Remote Sensing*, vol. 25, no. 12, pp. 2365–2401, 2004.

[8] D. Renza, E. Martinez, I. Molina, and D. M. Ballesteros L., "Unsupervised change detection in a particular vegetation land cover type using spectral angle mapper," *Advances in Space Research*, vol. 59, no. 8, pp. 2019–2031, 2017.

[9] J. Chen, X. Chen, X. Cui, and J. Chen, "Change vector analysis in posterior probability space: a new method for land cover change detection," *IEEE Geoscience and Remote Sensing Letters*, vol. 8, no. 2, pp. 317–321, 2011.

[10] M. Hussain, D. Chen, A. Cheng, H. Wei, and D. Stanley, "Change detection from remotely sensed images: from pixel-based to object-based approaches," *ISPRS Journal of Photogrammetry and Remote Sensing*, vol. 80, pp. 91–106, 2013.

[11] D. Renza, E. Martinez, and A. Arquero, "A new approach to change detection in multispectral images by means of ERGAS index," *IEEE Geoscience and Remote Sensing Letters*, vol. 10, no. 1, pp. 76–80, 2013.

[12] A. A. Nielsen, "The regularized iteratively reweighted MAD method for change detection in multi- and hyperspectral data," *IEEE Transactions on Image Processing*, vol. 16, no. 2, pp. 463–478, 2007.

[13] Y. Byun, Y. Han, and T. Chae, "Image fusion-based change detection for flood extent extraction using bi-temporal very high-resolution satellite images," *Remote Sensing*, vol. 7, no. 8, pp. 10347–10363, 2015.

[14] B. Wang, J. Choi, S. Choi, S. Lee, P. Wu, and Y. Gao, "Image fusion-based land cover change detection using multi-temporal high-resolution satellite images," *Remote Sensing*, vol. 9, no. 8, p. 804, 2017.

[15] C. Li, G. Beauchamp, Z. Wang, and P. Cui, "Collective vigilance in the wintering hooded crane: the role of flock size and anthropogenic disturbances in a human-dominated landscape," *Ethology*, vol. 122, no. 12, pp. 999–1008, 2016.

[16] M. Barter, L. Cao, L. Chen, and G. Lei, "Results of a survey for waterbirds in the lower Yangtze floodplain, China, in January-February 2004," *Forktail*, vol. 21, pp. 1–7, 2005.

[17] B. Aiazzi, S. Baronti, and M. Selva, "Improving component substitution pansharpening through multivariate regression of MS +Pan data," *IEEE Transactions on Geoscience and Remote Sensing*, vol. 45, no. 10, pp. 3230–3239, 2007.

[18] C. Thomas, T. Ranchin, L. Wald, and J. Chanussot, "Synthesis of multispectral images to high spatial resolution: a critical review of fusion methods based on remote sensing physics," *IEEE Transactions on Geoscience and Remote Sensing*, vol. 46, no. 5, pp. 1301–1312, 2008.

[19] R. B. Myneni, F. G. Hall, P. J. Sellers, and A. L. Marshak, "The interpretation of spectral vegetation indexes," *IEEE Transactions on Geoscience and Remote Sensing*, vol. 33, no. 2, pp. 481–486, 1995.

[20] A. K. Bhandari, A. Kumar, and G. K. Singh, "Improved feature extraction scheme for satellite images using NDVI and NDWI technique based on DWT and SVD," *Arabian Journal of Geosciences*, vol. 8, no. 9, pp. 6949–6966, 2015.

[21] B. Wang, S.-K. Choi, Y.-K. Han, S.-K. Lee, and J.-W. Choi, "Application of IR-MAD using synthetically fused images for change detection in hyperspectral data," *Remote Sensing Letters*, vol. 6, no. 8, pp. 578–586, 2015.

[22] P. R. Marpu, P. Gamba, and M. J. Canty, "Improving change detection results of IR-MAD by eliminating strong changes," *IEEE Geoscience and Remote Sensing Letters*, vol. 8, no. 4, pp. 799–803, 2011.

[23] A. S. Abutaleb, "Automatic thresholding of gray-level pictures using two-dimensional entropy," *Computer Vision Graphics and Image Processing*, vol. 47, no. 1, pp. 22–32, 1989.

[24] R. G. Congalton, "A review of assessing the accuracy of classification of remotely sensed data," *Remote Sensing of Environment*, vol. 37, no. 1, pp. 35–46, 1991.

13

Spark Sensing: A Cloud Computing Framework to Unfold Processing Efficiencies for Large and Multiscale Remotely Sensed Data, with Examples on Landsat 8 and MODIS Data

Hai Lan,[1] **Xinshi Zheng**,[2] **and Paul M. Torrens**[1,2]

[1]*Department of Computer Science and Engineering, Tandon School of Engineering, New York University, Brooklyn, NY 11201, USA*
[2]*Center for Urban Science + Progress, New York University, Brooklyn, NY 11201, USA*

Correspondence should be addressed to Hai Lan; hai.lan@nyu.edu

Academic Editor: Victor Mesev

Inquiry using data from remote Earth-observing platforms often confronts a straightforward but particularly thorny problem: huge amounts of data, in ever-replenishing supplies, are available to support inquiry, but scientists' agility in converting data into actionable information often struggles to keep pace with rapidly incoming streams of data that amass in expanding archival silos. Abstraction of those data is a convenient response, and many studies informed purely by remotely sensed data are by necessity limited to a small study area with a relatively few scenes of imagery, or they rely on larger mosaics of images at low resolution. As a result, it is often challenging to thread explanations across scales from the local to the global, even though doing so is often critical to the science under pursuit. Here, a solution is proposed, by exploiting Apache Spark, to implement parallel, in-memory image processing with ability to rapidly classify large volumes of multiscale remotely sensed images and to perform necessary analysis to detect changes on the time series. It shows that processing on three different scales of Landsat 8 data (up to ~107.4 GB, five-scene, time series image sets) can be accomplished in 1018 seconds on local cloud environment. Applying the same framework with slight parameter adjustments, it processed same coverage MODIS data in 54 seconds on commercial cloud platform. Theoretically, the proposed scheme can handle all forms of remote sensing imagery commonly used in the Earth and environmental sciences, requiring only minor adjustments in parameterization of the computing jobs to adjust to the data. The authors suggest that the "Spark sensing" approach could provide the flexibility, extensibility, and accessibility necessary to keep inquiry in the Earth and environmental sciences at pace with developments in data provision.

1. Introduction

Data provided by remote sensing have long presented as a critical resource in monitoring, measuring, and explaining natural and physical phenomena. Indeed, remote sensing might justly be characterized as one of the first "big data" sciences [1]. Steadfastly, for the advances in the sensing capabilities of remote, Earth-observing platforms have continued to produce more and more data, with increasing observational breadth and finesse of detail. These developments carry a dual benefit and problem: analysis and inquiry in the environmental and Earth sciences not only are routinely awash with data but also often struggle to match pace in building empirical knowledge from those data because the data are incoming with such haste and heft. Strategies to manage big remotely sensed data are required to fully exploit the benefits those data hold for applied scientific inquiry, and the topic of how computing might be leveraged to ease pathways between science and sensing holds significant currency across many fields, with particularly rapid adoption of high-performance computing [2] and cloud computing in the geographical sciences [3].

Remote sensing imagery is a commonly used source to support those studies of sustainable ecosystems, such as ecosystem dynamics, grassland degradation, and urban ecosystem restoration, especially in large areas [4]. Traditionally, studies with pure remotely sensed data involved only a few scenes of data in a limited study area, or they rely on low-resolution remotely sensed images in large-area experiments [5]. Those traditions are changing as new data have dramatically altered the underlying substrate for analysis. For example, in the past few decades, the space-borne and air-borne Earth observation sensors are continually providing large-volume datasets. For example, Landsat 8, the latest Landsat mission launched in 2013, can collect more than 700 images per day, corresponding to approximately 86 terabytes of data per year [6], which is 14 times as much as that in the 1980s [4]. Processing the massive volume of remotely sensed data is now not the only problem: the intrinsic complexity of those data is also an important issue that must be considered.

The sensors that are actuated in remote sensing are usually designed to serve specific requirements of analysis for different fields of study. To fulfill those different needs, sensors are usually tasked to capture images at different resolutions. For example, high-resolution satellite sensors such as *WorldView-4* can produce imagery with a spatial resolution of 0.31 m in panchromatic vistas and 1.24 m spatial resolution in multispectral vistas [7], while the *QuickBird* platform can image the Earth with 0.61 m spatial resolution in panchromatic form and 2.44 m spatial resolution in multispectral form [8]. These resolutions, on the order of fractions of a meter to a few meters in spatial resolution, presented significant opportunities to monitor the Earth and represent the state of the art in Earth-observing imaging detail.

Concurrently, other remote sensing platforms are tasked with refreshing observations of the whole Earth's surface, aiming for coverage of large areas with temporal consistency, rather than small-area detail. For example, relatively medium-resolution sensors, such as *Landsat*, and relatively low-resolution sensors, such as *MODIS*, are deployed as long-term Earth observatories. Landsat provides 15 m panchromatic and 30 m multispectral imagery, which is very widely used in studies of large-area grassland degradation and urban land cover dynamics [9]. MODIS offers 250 m multispectral imagery and can build a mosaic view of the entire Earth once every few days. MODIS data has been widely adopted in global-scale research studies, particularly those trained on studying vegetation canopies for investigation of worldwide forest cover dynamics [10].

Many sensors support multispectral imaging. For example, WorldView-4 data includes four spectral bands, and Landsat 8 OLI/TIRS provides 11 spectral bands. For some spectrally sensitive studies, higher spectral resolution imagery is required. In those studies, hyperspectral sensors (such as *Hyperion*, which generates 220 bands between 0.4–2.5 μm [11]) can produce detailed spectral data over a very small wavelength range. Furthermore, different sensors offer different temporal resolutions in their rate of imaging as well as the timing of their coverage of subjects under their purview. For example, the WorldView-4 satellite is capable of revisiting views every 4.5 days (sometimes sooner), while Landsat can deliver repeat views every 16 days [12]. Higher temporal resolution in return views facilitates the study of dynamics on the Earth surface, so that the time series and the time interval between visits become significant attributes of the observation, alongside the spatial resolution and spectral range. Therefore, methods to streamline a feasible, effective, and efficient approach to processing archived and continually incoming multispatial, multispectral, and multitemporal remote sensing data are an ongoing requirement across many potential applications of remote sensing to applied scientific inquiry.

In this paper, possible scalable solutions are introduced to address issues of processing multiscale large-volume remote sensing datasets in multispatial, multispectral, and multitemporal cases. The aim is to implement a tool that can process different proposed remote-sensed tasks with only minor adjustments rather than fully rebuild new toolkits. Furthermore, this solution should be fully capable of exploiting benefits from cutting-edge cloud computing technologies, resources, and platforms to help researchers process and analyze large remotely sensed datasets that are difficult to process on local machines in an effective and efficient manner.

1.1. Cloud Computing as a Resource for Big Data Processing. Many researchers have made significant progress in advancing feasible, effective, and efficient processing for multiprong attributes of remotely sensed data, using developments in computer engineering. In particular, research into how graphical processing units (GPUs) and cluster-based high-performance computing (HPC) might be leveraged to advance image processing for remote sensing has been particularly fruitful [13]. More recently, cloud computing is increasingly being considered as a resource in processing remotely sensed imagery, largely because of cloud computing's native abilities to scale computing in kind as the data being processed also the scale. Furthermore, significant cloud computing resources are now available commercially, on a "pay as you go" model, from providers such as *Amazon Web Services* (AWS) [14], *Microsoft Azure* [15], and *Google's Compute Engine* [16]. These resources can be brought to bear on image processing tasks as IaaS (infrastructure as a service), PaaS (platform as a service), or SaaS (software as a service).

Cloud computing is useful in providing some of the flexibility required to match pace between incoming data, large existing data silos, and evolving analytical needs in image processing that authors alluded to in the introduction. Cloud computing affords this flexibility by allowing users to allocate and share software and hardware resources on the Internet in a distributed fashion, by splitting large computational tasks into many small parallel computing tasks, then assigning them to as many computing instances as are required to achieve computing goals based on data size, data fusion, resource use, or computing time. After all the distributed nodes of the cloud service have completed their assigned tasks, the results are bundled and returned to the users' local database. In this way, virtual instances, applications, and software are provided on an as-requested basis, and users may pay for those services as demanded. This affords a user access to a theoretically limitless size computing capacity (although very strong limits of available financial budgets to pay for the services quickly dock theoretical capacities to tangible practical realities in many real instances).

A promising community of computing frameworks has codeveloped alongside cloud computing hardware, and several of these frameworks hold significant promise for processing remotely sensed imagery of the Earth's surface. For example, MapReduce was introduced by Dean and Ghemawat [17], ten years ago. In the decade since, a number of open-source implementations of the MapReduce model have emerged as promising frameworks for mediating the computing between image processing for remotely sensed data and cloud resources that are available to distribute and/or accelerate that computing on commercial (or user-run) clouds. Chief among these open-source implementations of MapReduce is *Apache Hadoop*. While Hadoop MapReduce relies on reading and writing data to a disk, another variant, Apache Spark [18] maintains data partitions in memory (a so-called in-memory computing framework). Spark also provides a network buffer for each reducing task, rather than merging outputs into a single partition, with the result that Spark can be one hundred times faster than Hadoop MapReduce on some big data tasks [19]. Nevertheless, one advantage that Hadoop might hold over Spark is that Hadoop allows parallel processing of large amounts of data that are bigger in physical storage size than the available memory. In fact, many remote sensing datasets are of a size that is so massive that they exceed the memory available in local machines or small clusters. Furthermore, physical disk resources are usually much less expensive in financial cost (of owning or accessing) than memory resources are. So, in cases for which limited memory may become a constraining factor, Hadoop presents as a better option in some cases for processing large amounts of remote sensing data.

However, cloud computing frameworks are agile relative to resource constraints. And that novel advances in cloud computing technologies and cloud platforms allow Spark to leverage resources from and across different cloud computing platforms, with the possibility that memory limitations may no longer loom as large a constraint for big remote sensing data processing scenarios. Consider memory as a resource that can be drawn upon on an as-needed basis, researchers can access theoretically unlimited memory on the cloud. Furthermore, Spark can run on a single workstation, as well as local computing clusters and cloud platforms. And Spark could access diverse data sources, such as *Amazon S3*, *Hadoop Distributed File System* (HDFS), *Cassandra*, and *HBase*. In other words, Spark not only can access local private data warehouses but also can reach cloud-stored big remote sensing datasets and do so via the cloud platform directly, with the ability to process those data on the cloud and then stream back the required results.

1.2. Cloud Computing for Processing Remotely Sensed Images. One of the common computing solutions to the burden of processing, analyzing, and managing large-scale remote sensing data is to parallelize the remote sensing processing tasks: to spread the burden over multiple computing units to reduce the overall processing time [6]. For example, Huang and her colleagues used the message passing interface (MPI) as a computing framework for their work on dust storm simulation and forecasting on the *Amazon EC2* commercial cloud service [13]. They deployed MPI on the Amazon cloud and applied loosely coupled nested models to process a high-resolution dust storm dataset. Their performance tests showed efficient and economical results. Cavallaro et al. [20] used GPUs to implement a support vector machine (SVM) classifier with MPI and *openMPI* frameworks. As an alternative to using HPC computing frameworks, other researchers have developed their own bespoke parallel large-scale remote sensing data processing platforms. For example, Wang et al. [21] developed *pipsCloud*, which is a cloud-based HPC approach to process remote sensing on-demand and in real time. To further enhance performance, Wang et al. [21] used Hilbert R^+-tree indexing.

The turn toward development of tools by remote sensing scientists, for remote sensing scientists, leveraging computing techniques but departing in ways that are special to remote sensing applications, is a wonderful development for remote sensing science. Nevertheless, bespoke solutions (particularly in academic settings) cannot feasibly contribute to the massive levels of computing available now commercially, with the result that absolute performance will always lag behind that which might otherwise be available on the marketplace. Also, applying MPI or self-developed systems often requires significant research and development effort into programming, debugging, and tuning the computing system and environment, and one might perhaps make an argument that the time devoted to these tasks could be used on the applied science instead. In some cases, building these systems on a bespoke basis is very challenging. For example, consider MPI, programming tasks designed for serial computing and converting them to parallel form can be significantly burdensome and particularly so for some complex image processing algorithms. Moreover, the networking security, near ubiquitous availability, and fast-moving hardware compatibility (e.g., in shared memory clusters) of commercial platforms offer significant practical advantages. Some existing work points to the potential advantages that are obtainable in cloud processing of big geospatial data. For example, Chen and Zhou [22] demonstrated that Apache Hadoop can be leveraged for partitioning using a mean shift algorithm. With a local mode test, they successfully increased the processing speed by ~2 times [22]. Also, Giachetta [23] introduced a Hadoop-based geospatial data management and processing toolkit, *AEGIS*, which he compared against many existing MapReduce-based frameworks, such as *SpatialHadoop*, *Hadoop-GIS*, *HIPI*, and *MrGeo* with spatial join, query, and aggregation operations [23].

Compared to MapReduce-based approaches, solutions based on Apache Spark can usually generate results at higher efficiency, as mentioned in "Introduction." For example, Sun et al. [24] used *MLlib* in Spark to test the multi-iteration singular value decomposition (SVD) algorithm on high-resolution hyperspectral remote sensing images. Compared with the *Apache Mahout* (MapReduce) approach, they found that the Spark approach can essentially trounce MapReduce in their tests, once Spark is able to access enough hardware resources. Another study using Spark to process massive remote sensing data, by Huang and his colleagues [25], demonstrated a series of comprehensive performance tests, using Spark to implement different types of algorithms

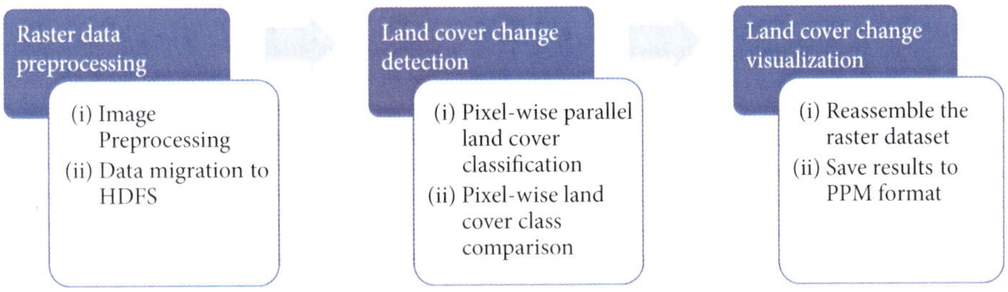

FIGURE 1: Spark-based remote sensing image processing workflow.

in remote sensing. Huang et al. [25] discussed the performance of each of the tests on different running environments, including local, standalone, and *yet another resource negotiator* (YARN). They also proposed a self-defined, strip-based partitioning approach to replace the default hash partitioning method [25].

2. Methodology and Data Sources

In this paper, authors propose to extend upon recent developments in cloud computing as a resource for processing remotely sensed imagery. Specifically, a Spark-based large-scale remote sensing image processing tool that can be deployed on cloud platforms will be introduced. In the following sections, authors will discuss how to design experiments by applying two commonly used classifiers—normalized difference vegetation index (NDVI) and normalized difference water index (NDWI)—as a testbed to assess the feasibility and performance of our tool. A scheme for designing a change detection scheme based on classification results generated by the proposed tool on a time series will be also introduced. One of the key advantages of this approach is in its ability to easily and straightforwardly support users' access to processed land cover changes within a large set of spatial-temporal images. Moreover, the paper will demonstrate a visualization approach to save text-based output from the analysis in common picture format, allowing users to examine results easily and quickly.

2.1. Methodology. The approach presented in this paper is based on the YARN cloud environment. The goal, in using YARN, is to facilitate the availability of the processing environment in ways that are widely applicable to real-world scenarios. YARN has been widely used in many current cloud environments [26]. Unlike traditional versions of Hadoop MapReduce, YARN allows for allocation of system resources as containers to the various applications. In other words, different computing frameworks can be deployed on a single physical cluster in a noninterfering manner. In YARN mode, a Spark framework is composed of a master instance and many workers. The master instance is responsible for negotiating with the YARN *Resource Manager* to request enough computing resources, as needed, by analyzing the Spark applications submitted by clients. Once the master instance is loaded, it will schedule the tasks to executors with allocated containers. Until all tasks have been finished, Resource Manager will revoke all allocated resources for further possible tasks.

Resilient distributed datasets (RDD) form the basic abstraction of a dataset in Spark. RDDs can be created from an external dataset, or from existing RDDs. For cases in this research, RDDs will be created from the original remote sensing images stored in HDFS. For each image, three RDDs will be created for green, red, and near-infrared bands for NDVI and NDWI land cover classification at the initial stage. RDDs contain the data partitions and the metadata records. According to the Spark mechanism, RDDs will go through a series of transformations and action operations to process the data partitions in a distributive manner. The transformation operations of RDDs will be executed on individual partitions of an RDD and those operations will return a new RDD. Action operations will summarize information from an RDD by user-defined functions and return a result. For example, the joining operation, as a commonly used transformation operation in Spark, involves joining partitions belonging to two RDDs and creating a new one. In this case, the data blocks will not be moved at the current stage because of the lazy mechanism. However, action operations, such as count, will process the data partitions in RDD and perform real computing.

Based on the features of Spark, a workflow has been designed as illustrated in Figure 1. The first step in that workflow is to extract each band of a raster dataset by using the *Geospatial Data Abstraction Library* (GDAL), which is an open-source translator library for raster and vector geospatial data formats [27]. Each line of the extracted data contains the geographical coordinates and the digital number (DN) value of the raster cell in the original raster dataset. These files are then put into the HDFS as data input sources for distributive processing.

The second step is to perform parallel processing of image classification and land cover change detection. As introduced before, for each image that must be processed by this tool, three RDDs will be created for its green, red, and near-infrared bands. The basic units of the RDD are key-value pairs. In this case, the geographical coordinates are set as the key, and the DN value of a raster cell will be the value. In this manner, parallel land cover classifications can be conducted across many computer units in a cluster. The extent of parallel processing depends on the Spark application configuration and the cluster hardware specifications. Because the basic parallel processing unit is pixel-wise, this method can be highly flexible and scalable, allowing raster datasets to be partitioned in any manner, regardless of the spatial structures of the raster. However, partitioning

```
Input= TIF_1, TIF_2, ... , TIF_k; Dimension = X*Y
For k = 1 to K-1 do:
    InputTIF_1 = TIF_k
    InputTIF_2 = TIF_k+1
    For m = 1 to 2 do:
        For x in 1 to X
            For y in 1 to Y:
                ndvi_xy = (InputTIF_xy.NIR – InputTIF_xy.R)/(InputTIF_xy.NIR + InputTIF_xy.R)
                ndwi_xy = (InputTIF_xy.G – InputTIF_xy.NIR)/(InputTIF_xy.G + InputTIF_xy.NIR)
                landcover_xy = CLASSIFIED(ndvi_xy, ndwi_xy)
                landcover_m[x, y] = landcover_xy
            endFor
        endFor
    endFor
    landcover_change_k = CHANGE(landcover_1, landcover_2)
    Output = landcover_cover_k
endFor
```

Algorithm 1: NDVI/NDWI classification and changing detection.

strategies still must be seriously considered, because an appropriate partitioning strategy will help to optimize the performance with better usage of the computing resources of the cluster. Spark supports hash partitioning and range partitioning by default; these applications are appropriate for many cases in the real world. However, according to the results of the study of Huang et al. [25], the partition scale cannot be too small or too large. A very small partition scale will result in low-performance computing and even increase the fault recovery cost. A very large partition scale may lead to the out-of-memory error. Inspired by their strip-based partitioning method, splitting data into chunks with a configured HDFS block size is needed in following experiments.

In Algorithm 1, the detailed algorithm of using NDVI and NDWI as classifiers on remote sensing images has been illustrated. Several transformation operations are performed on each partition in RDDs. For k input *GeoTiFF* remote sensing images, the algorithm will first parse the dimension of them. Moreover, those created RDDs will be marked by a time sequence in the time series. For each data chunk, in each pair of image RDDs (e.g., a pair of images in 2013 and images in 2014), the NDVI and NDWI classifier will be applied to each pixel to calculate the indicated values. The NDVI [28] can be calculated according to its definition as

$$\text{NDVI} = \frac{\text{NIR} - \text{red}}{\text{NIR} + \text{red}}. \qquad (1)$$

The range of NDVI is from negative one to one, which can use different ranges to denote sparse vegetation, dense vegetation, barren rock and sand, and water. However, NDVI may not always correctly distinguish the water body, especially when there is a noisy signal in the water area (e.g., mud). To further improve the accuracy of classification, NDWI is applied according to McFeeters [29] as follows:

$$\text{NDWI} = \frac{\text{green} - \text{NIR}}{\text{green} + \text{NIR}}. \qquad (2)$$

NDWI can distinguish the water feature with the positive value indicator. From the perspective of algorithm implementation, RDDs of three bands will be joined as a single RDD for NDVI and NDWI calculation of each raster cell with mapping operations. Land cover classification can, therefore, be conducted in parallel by using calculated NDVI and NDWI results. Once the land cover features are classified by NDVI and NDWI indicators, each feature can be labelled with a class ID and pass those IDs when changing the detection function. The changing detection function will compare the feature IDs from each image pair and summarize how those land features change from the former year to the current year.

The final step is to visualize land cover change, as illustrated in pseudo code in Algorithm 2. With all k numbers of land cover change results generated by Algorithm 1 for each image pair, visualization images are created in PPM format from the obtained output RDD, which is a lowest common denominator color image file format [30]. Although redundant, PPM is an easy format to write and manage text-based outputs into human-readable figures. The two RDDs of land cover classifications for two images from Algorithm 1 will be joined as a single RDD, and the values of key-value pairs will be converted to colors according to a user-defined RGB color scheme. This joined RDD will then be sorted to the original order and reduced in a manner such that each key-value pair represents a row of the original raster image. Finally, to produce a PPM-format image, this RDD is appended to the PPM file header to create a complete PPM image.

Input = $landcover\_change_1$, $landcover\_change_2$, … , $landcover\_change_K$; Dimension = X^*Y
For k = 1 to K **do:**
    $InputKV_k = landcover\_change_k$
    $rgbKV_k = RGB(inputKV_k)$
    Sort $rgbKV_k$ by two dimensions
    $rgbV_k$ = values of $rgbKV_k$
    Reshape $rgbV_k$ to dimensions X^*Y
    Output = $rgbV_k$
    Save $rgbV_k$ to files
endFor

ALGORITHM 2: Visualization.

An overview of how the whole process works is illustrated in Figure 2. Four steps are applied in this design: (1) read and parse each pair of input images and create RDDs for green, red, and near-infrared bands; (2) split data into chunks and process them in parallel; (3) gather results and assign labels to each pixel; and (4) reconstruct whole images by sorting output RDDs and visualize them in PPM format.

2.2. Data Sources. To prove the applied utility of the proposed scheme, three experiments are performed, and two different remote-sensed imagery datasets are involved. The first experiment is using the Landsat 8 operational land imager (OLI) dataset in three different scales to test the workflow of classification and change detection and visualization algorithm with Spark on local cloud environment. To further study the tuning performance of this scheme, the second experiment is designed to run this tool under different execution configurations with the same Landsat 8 dataset. The last experiment is set to prove this tool can consume different source remote-sensed imagery datasets with only minor parameter adjustments. By using the same algorithm presented above, a MODIS dataset on Amazon EC2 which is a real commercial cloud platform is processed.

Landsat 8 scans the entire planet surface in a 16-day period [31]. Equipped with the latest OLI sensor, Landsat 8 provides unprecedented spectral information with two additional spectral bands in the whole Landsat instrument family. In addition to offering a 15 m panchromatic band and a 30 m multispectral band, as in many previous products, Landsat 8 also includes a quality assessment (QA) band to support pixel-based cloud, shadow, and terrain occlusion filtering. A relatively short revisit period, medium spatial resolution, and seven spectral bands make Landsat 8 OLI products a commonly used freely accessible remote sensing imagery datasets for science that relies upon spatial, spectral, and temporal Earth attributes for environmental research. Furthermore, the Landsat 8 dataset is currently open to the public on cloud platforms, such as Amazon S3 [32] and *Google Earth Engine* (GEE), making it a great data source to serve cloud-based large-scale remote sensing processing in the cloud.

Like Landsat 8 OLI, MODIS satellite datasets are landed on as Amazon S3 [33] and GEE since 2017. MODIS provides a variety of planet observation products with daily temporal resolution. In this paper, MODIS/Terra Surface Reflectance Daily L2G Global 1 km and 500 m SIN Grid V006 (MOD09GA) will be used as second data input, which can provide 7 band surface spectral reflectance with 500-meter spatial resolution. Besides, a bunch of 1-kilometer resolution observation and geolocation statistic bands are offered in this product.

The coverage of each of the datasets that are used is illustrated in Figure 3. Three scales of Landsat 8 imagery datasets of small, medium, and large scale were used in experiments, as the input data source. This multiscale approach allowed us to assess whether our tool can perform multiscale remote sensing image processing with only minor parameter adjustments, to assess whether a wide array of remote sensing imagery datasets can be processed by our proposed framework in real scenarios. Because of remote sensing images, which are collected by different sensors at different spatial resolutions, with various bands of spectral information, and at diverse temporal scales are essentially formed by pixels, pixel-based algorithms implemented via Spark in the cloud should be capable of performing a very wide array of pixel-based processing and analysis. However, to further prove the presented tool can handle multisource remote sensing images in real cases, a MODIS dataset that fits for the exact same coverage of large-size Landsat 8 dataset is imported.

To generate a time series for change detection, one image per year has been chosen from a dataset spanning 2013 to 2017. The small-size images and medium-size images are approximately in one single scene of a Landsat image. Hence, those images that were acquired on a similar date for each year are preferably selected to reduce the land cover changes caused by seasonal variation factors. However, a large dataset is formed by over 15 scenes of original images. In this case, it cannot guarantee that each tile of this dataset can be filled by same-date images for each year. Hence, by broadening the filter criteria to the month level, obtaining enough tiles can be ensured to form the whole area. Another important factor is that the chosen images are preferably high quality to reduce the cloud and haze problems. For MODIS dataset, with relatively coarse spatial resolution and large coverage per scene, it takes about half of single scene to fit the large-size Landsat dataset coverage. Also, the acquisition time of it follows the date of the small- and medium-size

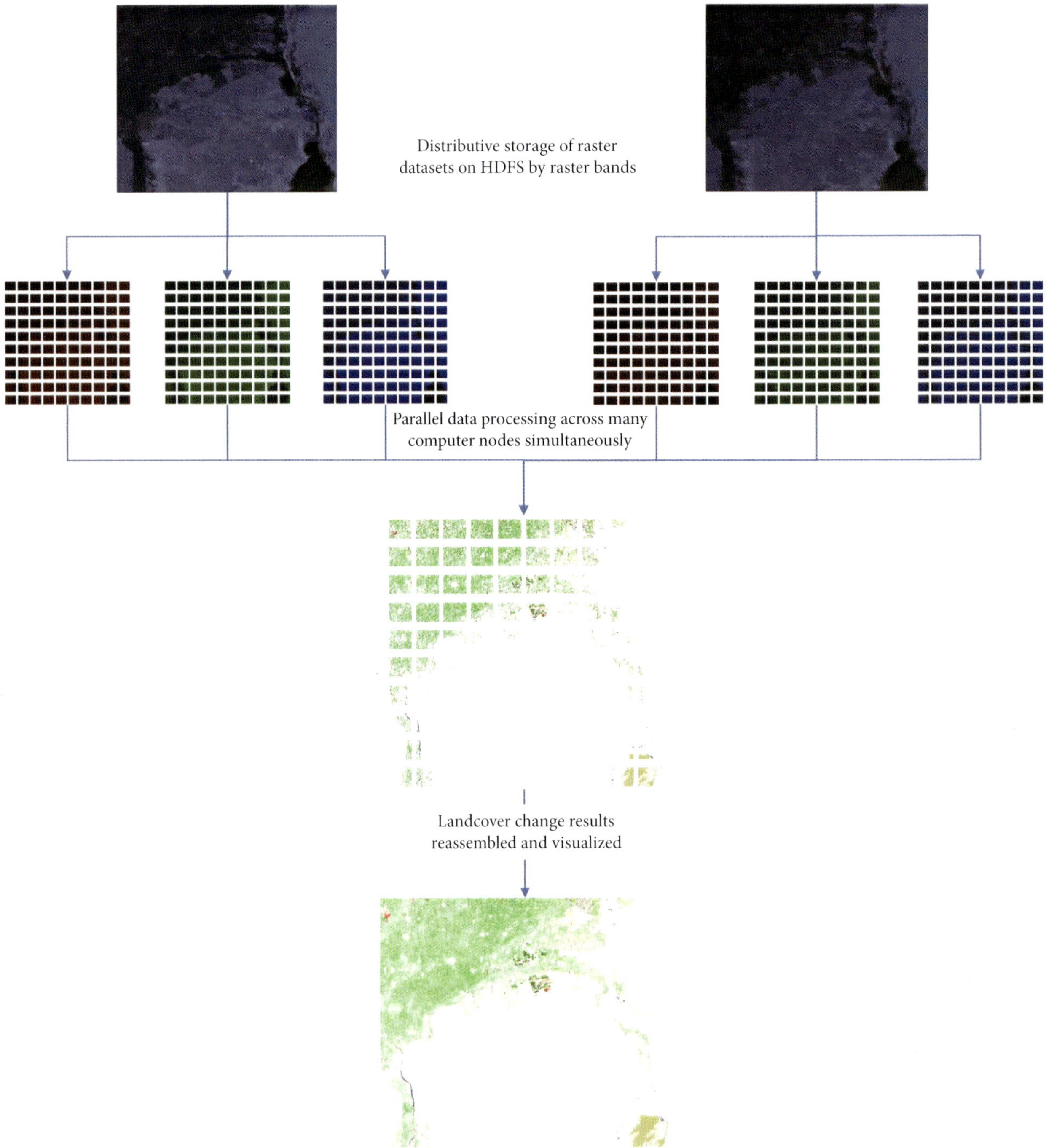

Figure 2: Detailed workflow for image classification and changing detection.

Landsat dataset because MODIS can provide daily products. The detailed acquisition time of each dataset is listed in Table 1.

The reason for the choice of the study area shown in Figure 3 is that, in the Suez Canal area, the land cover mainly includes sparse and dense vegetation, a natural water body, the Suez Canal (with water), and bare sands and rocks. Thus, the study area is a great test area for our NDVI/NDWI classifier to detect detailed land cover and generate relatively accurate change detection results. Another reason for the choice of the study area is that Suez Canal was expanded since 2013 to build another branch [34]. This project can be clearly monitored by the presented change detection process.

In those experiments, all remote sensing image datasets were exported from GEE, which is a cloud-based platform that can serve remote sensing data source with customized criteria [35], for example, setting (1) the region boundaries to acquire data belonging to our study area and (2) the cloud mask to filter the cloud pixels on images before exporting the data to our cloud drive. Landsat 8 and MODIS (MOD09GA) both provide surface reflectance produced on GEE. If other datasets are used, a strictly image-based atmospheric

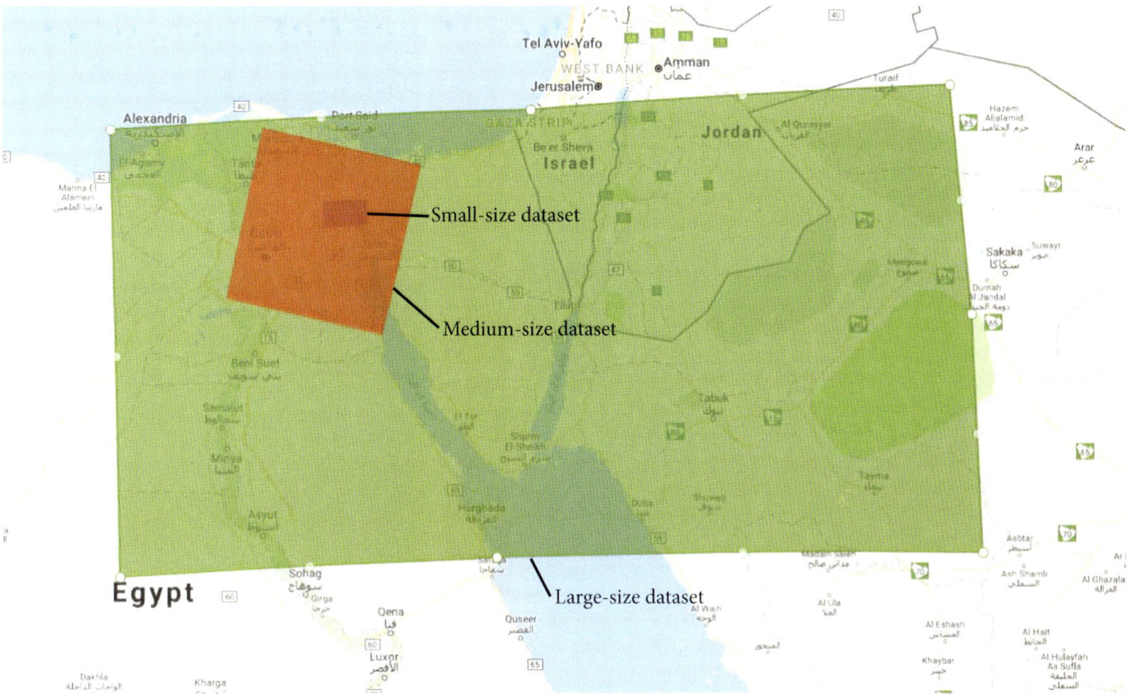

Figure 3: Landsat 8 image dataset in three different scales.

Table 1: Detailed acquisition time of each dataset.

	2013	2014	2015	2016	2017
Landsat_small	4/29	3/31	4/19	3/4	4/5
Landsat_medium	4/29	3/31	4/19	3/4	4/5
Landsat_large	Apr	Mar	Apr	Apr	Apr
MODIS	4/29	3/31	4/19	3/4	4/5

Table 2: Landsat data size and processing time.

	Scenes for each year	Size	Processing time
Small	~0.15	~1.2 G	67 s
Medium	~1	~10 G	276 s
Large	~15	~107.4 G	1018 s

correction should be followed to remove the haze impact on those images [36].

2.3. Experiment Environment. Two experiment environments are involved in this research for local cloud environment test and real commercial cloud platform test, respectively. The local computing cluster contains a total of 20 nodes. Two of them are master nodes, which equipped with 2 Intel Xeon E5-2680v4 2.4 GHz CPUs and 256 G memory each. 18 computing nodes are equipped with 2 Intel Xeon E5-2690v4 2.6 GHz CPUs (28 cores) and 256 G memory each. 10 Gbit network is assigned. In total, there are 1008 computing vcores and 5.12 Tb memory available. 2 Pb HDFS is configured and the block size is 128 M as default. Linux (Centos 6.9) is running on this cluster. Java 1.8.0_152 64-Bit Server VM is installed. Spark version is 2.2.0 with Cloudera release 1 and CDH 5.12.0.

The Amazon EC2-based computing cluster is another computing environment. Based on the computing needs of the third experiment in this paper, a 3-node cluster is built with 1 t2.xlarge instance with 4 vCPU Intel Broadwell E5-2686v4 2.3 GHz as master nodes and 2 t2.large instances with 2 vCPU Intel Broadwell E5-2686v4 2.3 GHz as slaves. All nodes are involved in computing. The total number of vcores is 8 and the overall memory is 32 G. 20 G storage per node is attached with default 128 M block size. The default HDFS replica is set as 3. Ubuntu Server 14.04 LTS (HVM) is the operation system. Similar to local computing cluster, Java 1.8.0_152 64-Bit Server VM, Spark 2.2.0 with Cloudera release 1, and CDH 5.12.0 are configured.

In the following parts of this paper, how authors applied the aforementioned workflow on real multiscale and different source remote sensing datasets to test the performance and feasibility of the Spark-sensing scheme will be discussed. The performance of the tool and the visualization of the experimental results will also be shown. Moreover, the flexibility, extensibility, and accessibility gleaned by using a Spark-based solution in remote sensing image dataset processing will be further discussed.

3. Results and Discussion

The first experiment is performed on a 20-node YARN computing cluster. In this experiment, a multiscale remote sensing image dataset with the Spark-based classification and change detection algorithm has been successfully processed.

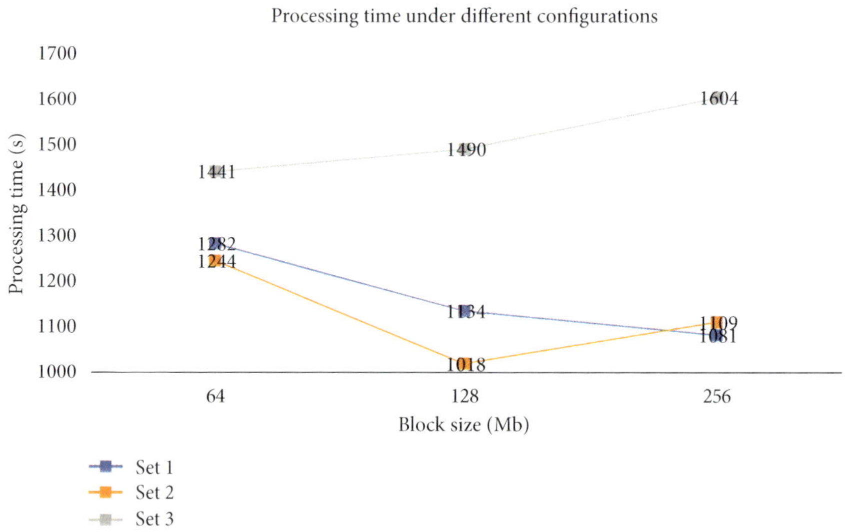

FIGURE 4: Processing time of large-size Landsat dataset under different configurations.

The detailed data size and processing time are shown in Table 2. The processing time is recorded under configurations with default block size, 50 executors with 20 cores each, and 20 G executor memory. Note that Spark can offer highly efficient processing for remote sensing images. Especially for a relatively large cluster, a sufficient computing resource can support the entire distributed computing process, ensuring that the processing time does not significantly increase as the data size increases. From the perspective of algorithm design, by anatomizing the overall processing time in each operation time-consuming segment, the joining operations are found to be very time-consuming, especially for a dataset with large pixels. In contrast, mapping operations for classification and change detection are much faster in comparison to joining operations. Hence, designing an image processing algorithm with fewer join operations and appropriate partitioning to reduce the data block moving may enhance the performance. Besides, repartitioning operations should also be avoided because it will result in a very time-consuming shuffle process.

Except enhancing the overall performance from algorithm design, the second experiment is developed to further study if the execution configurations may effect on the processing performance. Only the large-size Landsat 8 dataset is applied in this experiment to assess the processing time with different configurations. According to Spark performance tuning official documents [37], the executor numbers, executor, cores and executor memory are three main factors that may effect on performance of Spark-based applications. Budgeting available computing resources in advance is usually needed for users to gain satisfying processing performance. Here, three different configuration sets are assigned as follows: (1) 10 executors with 100 G memory and 100 cores each as set 1, (2) 50 executors with 20 G memory and 20 cores each as set 2, and (3) 500 executors with 2 G memory and 2 cores as set 3. Those three configuration sets are designed with the same total computing vcores and memory. It is also worth to point out that block size may sometimes effect on performance as well. As discussed above, repartition operations (especially for increase partitions) usually should be avoided to eliminate unnecessary shuffle process. Under this circumstance, partitions will be mainly decided by block size during I/O process when RDDs are created. If the block size were too small, massive number of partitions will be created especially with very large input dataset, which will lead to increasing the overhead of task management, though coalesce operation may be applied to decrease the partitions sometimes without shuffling in some cases. If the block size were too large, only a few partitions will be created so that not all the cores in the available computing resource can be sufficiently utilized. In other words, the feature of parallelism is not fully exploited to enhance the performance. Here, three different block sizes are set during the experiments. In Figure 4, the performance of processing the dataset under different execution configurations and different block sizes is represented. The shortest run is offered by 50 executors with 20 G memory and 20 cores each under default block size. With the same executor configuration set, the performance shows a bit lower with 64 M block size, which may result from the total task number which is increased with a smaller block size. However, the execution time for each task is not significantly reduced with corresponding settings. The performance of the same configuration set with 256 M block size is also beaten by it with a default block size. The reason is the partitions generated under this block size are too few so that parts of the executors are not active during processing. This problem shows more obviously when applying configuration set 3, because the number of executors is far more than the tasks under this case and too many computing resources are in idle, which leads to the lower performance. The execution configuration set 1 shows very similar performance with set 2 under each block size setting. However, by monitoring the core utilization of set 1, it is lower than it is with set 2. This may present that the increasing number of executors and decreasing the cores per executor may lift the utilization of cores and may

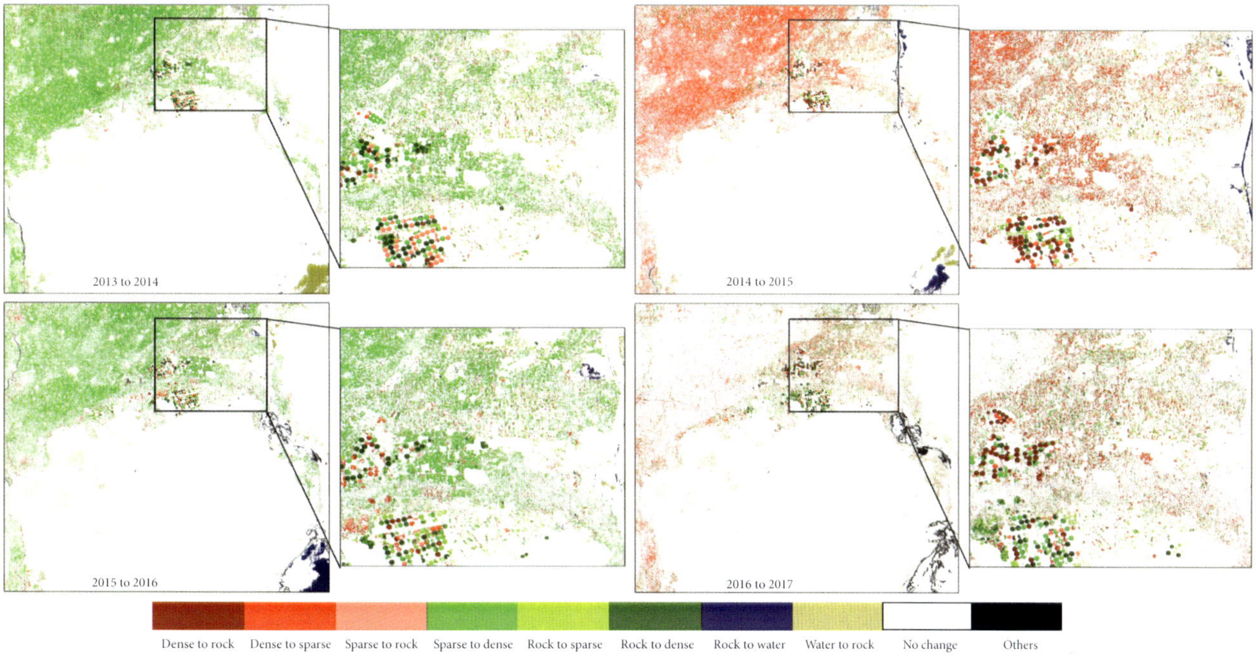

FIGURE 5: Changing detection visualization results of 2013 to 2014, 2014 to 2015, 2015 to 2016, and 2016 to 2017.

also help enhance the overall performance in some cases especially for very large dataset with relatively limited computing resources.

Here, the land cover feature is defined in four classes: dense vegetation, sparse vegetation, rock/sand, and water. In the legend of Figure 5, colors are set for the changes of each pixel in the study area from one feature to another. For example, "rock to water" indicates that the land cover feature was observed to have changed from sand or rock to water body in the past year. Conditions of no change indicate there are no detected changes in the past year. Others indicate those parts in the images with no data or with erroneous data. In Figure 5, the changes for each pair of images can be clearly seen. Moreover, users can generate a "final" result directly from 2013 to 2017 with a single parameter change. Taking the changing map of 2014 to 2015 in Figure 5 as an example, users can monitor the new branch of the Suez Canal being built and filled with water. It also shows that some new vegetation is growing along the Canal; such vegetation may be a new farmland because most of the vegetation regions are in artificial-like shapes.

As discussed before, the presented tool can be deployed on commercial cloud platform with no change (if the hardware configuration is highly different, strategies of balancing workload may need to be reconsidered). The third experiment is designed to use the proposed framework on Amazon EC2 to process MODIS dataset. Different from Landsat 8 products with band 4 as red, band 5 as NIR and band 3 as green, MODIS (MOD09GA) product sets red band as band 1, NIR as band 2, and green as band 4. Except changing the band index for input, the processing tool is ready to launch with no additional adjustment needed in coding. This experiment run on the 3-node Amazon EC2 cloud computing cluster. By applying 2 executors with 2 cores and 4 G memory each, the experiment is successfully completed in 54 seconds. It is worth to point out that 54-second processing time is not fast with such a small input dataset and that ~0.5 scene MODIS image is only about 50 M. Because in this case, lots of time are occupied by job submission, task management, resource allocation, and so on, the performance of Spark applications can only show significantly with relatively large dataset. However, this experiment still proves the idea that the proposed framework can be deployed on real commercial platforms to process multisource remote sensing images with only minor parameter adjustments.

Those experiments represent a robust solution for constructing remote sensing image processing tools for multiple purposes that are flexible (the ability of the tool to fit multisource datasets in different scales), extensible (the ability of the tool to grow in size and accommodate the volume of computing to resolve), and accessible (the ability of the tool to access data from multiple storage platforms and locations on local resources, data centers, and the cloud). All tested image datasets in those experiments are Landsat 8 with 30-meter spatial resolution and MODIS with 500-meter spatial resolution. However, the reader should note that the input dataset can be any raster-based images in different spatial, temporal, and spectral resolutions, because the algorithm implemented in our experiment is a pixel-based processing algorithm. Following a similar mechanism, all pixel-based algorithms can be implemented by processing the DN values of each pixel to generate required results under this framework. Hence, regardless of applying it to an existing large amount of data or filling it with a next-horizon dataset that will be collected in the future with current sensors or new sensors, this solution is capable of handling the tasks with only minor change in coding, thereby saving the high cost usually invoked in reprogramming different tools for different research goals.

With the Spark-based implementation, as demonstrated in the experiment, the main structures of this tool do not require modification as the volume of dataset changes. For example, it is becoming possible that, even if a Landsat 8 dataset covering a whole year (47.33 Tb) [6] is involved in processing at the same time, by the support of cloud platforms, users can always gain sufficient computing resources (memory especially, for Spark) in theory. Consider that Amazon EC2 now provides the "x1e.32xlarge" instance, which contains 128 virtual CPUs, ~4 Tb memory, and ~4 Tb storage. Users can apply fewer than 20 instances to implement in-memory computing with Spark-based approaches for this dataset in many different processing purposes. Nevertheless, to maintain the high performance of the tool, the partitioning strategies should be tailored based on real execution environments, especially for clusters with unevenly distributed computing resources and networking performance.

As discussed in the introduction, Spark-based approaches can easily access HDFS, Amazon S3, Cassandra, and HBase. Benefitting from cloud storage and management development, an increasing amount of data has been stored in the cloud and is open to the public. The approach as demonstrated provides cloud-based data resources to support users in performing a "pure" cloud analysis and in creating new products from it, without transferring unnecessary original and intermediate data to local storage before the final results are generated. This solution can also support data from multiple sources at the same time. For example, users can access cloud-stored public data as part of their data source and can also access and load their private data stored on local HDFS in a single Spark application.

4. Conclusions

In this paper, authors argue that the current state of the art for big data remote sensing, involving massive datasets being generated from existing and next-generation satellites and observation platforms, is, in many cases, proceeding at paces that outstrip our analytical capabilities to keep up with information products atop those data. While data and analysis are out of alignment, researchers perhaps miss opportunities to build the necessary science that might otherwise be attainable if data and analysis could be better connected. In this paper, authors discussed the current widely used approaches that have been developed by existing studies as a means to cater to the call of the community for the development of an effective and efficient large-scale processing framework to process large-volume remote sensing datasets. Based on the reviews of a comparison with other possible approaches, using Spark to build a robust scalable tool on a cloud environment is possibly an important and practical option to match data with analysis at pace. To this end, a Spark-based multiscale large remote sensing classification and change detection tool has been introduced, and its successful deployment and experimental testing in a cloud environment have been shown.

The approach in this paper suggests several promising advantages. First, the scheme for Spark-sensing offers considerable flexibility for processing big remote sensing datasets in multispatial, multispectral, and multitemporal cases. Indeed, shifting between resolutions and spectrums is possible with slight adjustment, thereby significantly saving the time cost of reprogramming brand new toolkits for different purposes. Second, this scheme makes it possible to exploit the benefits of cloud platforms to gain (theoretically) unlimited computing resources, with highly efficient performance. Third, the presented approach is natively highly accessible to multisource data storage, even in the cloud, which is useful in reducing data transformation costs.

The tool discussed in this paper is obviously just a prototypical framework. Thus, significant improvements could be made. The work here serves to prove the general principle and mechanisms necessary to get going with experiments in this area, and hopefully it can encourage others in the community to build on this foundation. An obvious extension of our approach could include the implementation of more complex remote sensing image processing algorithms, especially in classification, to better match the real cases in different research areas. Another improvement could be explored in designing better partitioning strategies to further enhance the computing performance. Moreover, using a dataset from the cloud directly may reduce the unnecessary data transferring from the data source to the cloud environment, which may better fit the usage habits in real-world problem solving.

Conflicts of Interest

The authors declare that they have no conflicts of interest.

Acknowledgments

The authors are grateful to Professor Huy T. Vo in the Department of Computer Science and Engineering at New York University for his advice in building the partitioning components of the work.

References

[1] M. F. Goodchild, H. Guo, A. Annoni et al., "Next-generation digital earth," *Proceedings of the National Academy of Sciences*, vol. 109, no. 28, pp. 11088–11094, 2012.

[2] C. Yang, M. Yu, F. Hu, Y. Jiang, and Y. Li, "Utilizing cloud computing to address big geospatial data challenges," *Computers, Environment and Urban Systems*, vol. 61, pp. 120–128, 2017.

[3] C. A. Lee, S. D. Gasster, A. Plaza, C. I. Chang, and B. Huang, "Recent developments in high performance computing for remote sensing: a review," *IEEE Journal of Selected Topics in Applied Earth Observations and Remote Sensing*, vol. 4, no. 3, pp. 508–527, 2011.

[4] M. A. Wulder and N. C. Coops, "Make earth observations open access: freely available satellite imagery will improve science and environmental-monitoring products," *Nature*, vol. 513, no. 7516, p. 30, 2014.

[5] Z. Wang, J. Zhong, H. Lan, Z. Wang, and Z. Sha, "Association analysis between spatiotemporal variation of net primary productivity and its driving factors in Inner Mongolia, China during 1994–2013," *Ecological Indicators*, 2017.

[6] Y. Ma, H. Wu, L. Wang et al., "Remote sensing big data computing: challenges and opportunities," *Future Generation Computer Systems*, vol. 51, pp. 47–60, 2015.

[7] Satellite Imaging Corporation, "WorldView-4 satellite image gallery," March 2018, https://www.satimagingcorp.com/gallery/worldview-4/.

[8] Satellite Imaging Corporation, "Quick Bird satellite sensor," March 2017, https://www.satimagingcorp.com/satellite-sensors/quickbird/.

[9] H. Lan and Y. Xie, "A semi-ellipsoid-model based fuzzy classifier to map grassland in Inner Mongolia, China," *ISPRS Journal of Photogrammetry and Remote Sensing*, vol. 85, pp. 21–31, 2013.

[10] M. A. Friedl, D. K. McIver, J. C. F. Hodges et al., "Global land cover mapping from MODIS: algorithms and early results," *Remote Sensing of Environment*, vol. 83, no. 1-2, pp. 287–302, 2002.

[11] US Geological Survey, "Sensors-Hyperion," March 2018, https://eo1.usgs.gov/sensors/hyperion.

[12] W. Turner, S. Spector, N. Gardiner, M. Fladeland, E. Sterling, and M. Steininger, "Remote sensing for biodiversity science and conservation," *Trends in Ecology & Evolution*, vol. 18, no. 6, pp. 306–314, 2003.

[13] Q. Huang, C. Yang, K. Benedict, S. Chen, A. Rezgui, and J. Xie, "Utilize cloud computing to support dust storm forecasting," *International Journal of Digital Earth*, vol. 6, no. 4, pp. 338–355, 2013.

[14] E. Amazon, "Amazon web services," 2015, November 2012, http://aws.amazon.com/es/ec2/.

[15] B. Wilder, *Cloud Architecture Patterns: Using Microsoft Azure*, O'Reilly Media, Inc., 2012.

[16] D. Sanderson, *Programming Google App Engine: Build and Run Scalable Web Apps on Google's Infrastructure*, O'Reilly Media, Inc., 2009.

[17] J. Dean and S. Ghemawat, "MapReduce: simplified data processing on large clusters," *Communications of the ACM*, vol. 51, no. 1, pp. 107–113, 2008.

[18] Apache Software Foundation, "Apache Spark," April 2018, https://spark.apache.org/.

[19] M. Guller, *Big Data Analytics with Spark: a practitioner's Guide to Using Spark for Large Scale Data Analysis*, Springer, 2015.

[20] G. Cavallaro, M. Riedel, C. Bodenstein et al., "Scalable developments for big data analytics in remote sensing," in *2015 IEEE International Geoscience and Remote Sensing Symposium (IGARSS)*, pp. 1366–1369, Milan, Italy, July 2015.

[21] L. Wang, Y. Ma, J. Yan, V. Chang, and A. Y. Zomaya, "pipsCloud: high performance cloud computing for remote sensing big data management and processing," *Future Generation Computer Systems*, vol. 78, pp. 353–368, 2018.

[22] X. Chen and L. Zhou, "The remote sensing image segmentation mean shift algorithm parallel processing based on MapReduce," in *International Conference on Intelligent Earth Observing and Applications 2015, 98083T*, Guilin, China, December 2015.

[23] R. Giachetta, "A framework for processing large scale geospatial and remote sensing data in MapReduce environment," *Computers & Graphics*, vol. 49, pp. 37–46, 2015.

[24] Z. Sun, F. Chen, M. Chi, and Y. Zhu, "A spark-based big data platform for massive remote sensing data processing," in *ICDS 2015 Proceedings of the Second International Conference on Data Science*, pp. 120–126, Sydney, Australia, August 2015.

[25] W. Huang, L. Meng, D. Zhang, and W. Zhang, "In-memory parallel processing of massive remotely sensed data using an apache spark on Hadoop YARN model," *IEEE Journal of Selected Topics in Applied Earth Observations and Remote Sensing*, vol. 10, no. 1, pp. 3–19, 2017.

[26] X. Fan, B. Lang, Y. Zhou, and T. Zang, "Adding network bandwidth resource management to Hadoop YARN," in *2017 Seventh International Conference on Information Science and Technology (ICIST)*, pp. 444–449, Da Nang, Vietnam, April 2017.

[27] C.-Z. Qin, L.-J. Zhan, and A.-X. Zhu, "How to apply the geospatial data abstraction library (GDAL) properly to parallel geospatial raster I/O?," *Transactions in GIS*, vol. 18, no. 6, pp. 950–957, 2014.

[28] J. W. Rouse Jr., R. H. Haas, J. A. Schell, and D. W. Deering, "Monitoring vegetation systems in the Great Plains with ERTS," in *NASA. Goddard Space Flight Center 3d ERTS-1 Symp*, College Station, TX, USA, January 1974.

[29] S. K. McFeeters, "The use of the normalized difference water index (NDWI) in the delineation of open water features," *International Journal of Remote Sensing*, vol. 17, no. 7, pp. 1425–1432, 1996.

[30] A. C. Frery and T. Perciano, "Image data formats and color representation," in *Introduction to Image Processing Using R. SpringerBriefs in Computer Science*, pp. 21–29, Springer, London, 2013.

[31] US Department of the Interior, and US Geological Survey, "Landsat 8," April 2018, https://landsat.usgs.gov/landsat-8.

[32] Amazon Web Services Inc, "Landsat on AWS," March 2018, https://aws.amazon.com/public-datasets/landsat/.

[33] Amazon Web Services Inc, "MODIS on AWS," June 2018, https://docs.opendata.aws/modis-pds/readme.html.

[34] Suez Canal Authority, "New Suez Canal," June 2017; https://www.suezcanal.gov.eg/English/About/SuezCanal/Pages/NewSuezCanal.aspx.

[35] N. Gorelick, M. Hancher, M. Dixon, S. Ilyushchenko, D. Thau, and R. Moore, "Google Earth Engine: planetary-scale geospatial analysis for everyone," *Remote Sensing of Environment*, vol. 202, pp. 18–27, 2017.

[36] P. S. Chavez, "Image-based atmospheric corrections-revisited and improved," *Photogrammetric Engineering and Remote Sensing*, vol. 62, no. 9, pp. 1025–1035, 1996.

[37] Apache Software Foundation, "Tuning Spark," June 2018, http://spark.apache.org/docs/latest/tuning.html#tuning-spark.

14

Estimating Aboveground Biomass on Private Forest using Sentinel-2 Imagery

Askar,[1] Narissara Nuthammachot,[1] Worradorn Phairuang,[1] Pramaditya Wicaksono,[2] and Tri Sayektiningsih[3]

[1]*Faculty of Environmental Management, Prince of Songkla University (PSU), 90110 Songkhla, Thailand*
[2]*Cartography and Remote Sensing, Faculty of Geography, Gadjah Mada University, 55281 Yogyakarta, Indonesia*
[3]*Research and Development Institute for Natural Resource Conservation Technology, Ministry of Environment and Forestry of Indonesia, 76112 East Kalimantan, Indonesia*

Correspondence should be addressed to Narissara Nuthammachot; narissara.n@psu.ac.th

Academic Editor: Franz L. Dickert

Private forests have a crucial role in maintaining the functioning of the Indonesian forest ecosystem especially because of the continuous degradation of natural forests. Private forests are a part of social forestry which becomes a tool for the Indonesian government to reduce carbon dioxide (CO_2) emission by 26% by 2030. The United Nations Programme on Reducing Emissions from Deforestation and Forest Degradation has encouraged the Indonesian government to establish a forest monitoring system by estimating forest carbon stock using a combination of forest inventory and remote sensing. This study is aimed at assessing the potential of vegetation indices derived from Sentinel-2 for estimating aboveground biomass (AGB) of private forests. We used 45 sample plots and 7 vegetation indices to evaluate the ability of Sentinel-2 in estimating AGB on private forests. Normalised difference index (NDI) 45 exhibited a strong correlation with AGB compared to other indices ($r = 0.89$; $R^2 = 0.79$). Stepwise linear regression fitted for establishing the model between field AGB and vegetation indices ($R^2 = 0.81$). We also found that AGB in the study area based on spatial analysis was 72.54 Mg/ha. A root mean square error (RMSE) value from predicted and observed AGB was 27 Mg/ha. The AGB value in the study area is higher than the AGB value from some of forest types, and it indicates that private forests are good for biomass storage. Overall, vegetation indices from Sentinel-2 multispectral imagery can provide a good result in terms of reporting the AGB on private forests.

1. Introduction

The Indonesian tropical forest is home to myriad flora and fauna, including charismatic species such as orangutans (*Pongo pygmaeus*), Sumatran tigers (*Panthera tigris sumatrae*), and rhinos (*Rhinoceros sondaicus*) [1–4]. Unfortunately, the Indonesian tropical forest is under threats due to degradation and deforestation. In a time span of a decade from 2000 to 2010, 14.7 Mha of Indonesian tropical forests has disappeared [5]. Sumatra, during 1990–2010, lost 7.54 Mha of its primary tropical forests [6], and, likewise, Kalimantan between 1996 and 2002 lost nearly 3 Mha of forest cover [7]. Many factors are blamed for triggering deforestation such as inappropriate transmigration policy, mining, and palm oil expansion and the increase in agricultural areas [8].

Deforestation in Indonesia is eventually becoming a global concern. This phenomenon has not only driven Indonesia to lose forest areas but also increased the greenhouse gas (GHG) emission, which in turn can lead to the accumulation of GHG in the atmosphere. Majority of CO_2 emission in Indonesia is from land use, land use change and forestry (LULUCF) sectors [9]. In response to that, the Indonesian government through the 21st of Conference of the Parties (COP) in Paris has committed to reducing CO_2 emission at 29% by 2030 [10]. One of the strategies applied to achieve the goal is through social forestry development [10].

Based on the Ministry of Environment and Forestry regulation, social forestry in Indonesia can be divided into six parts including hutan tanaman rakyat (people's planting

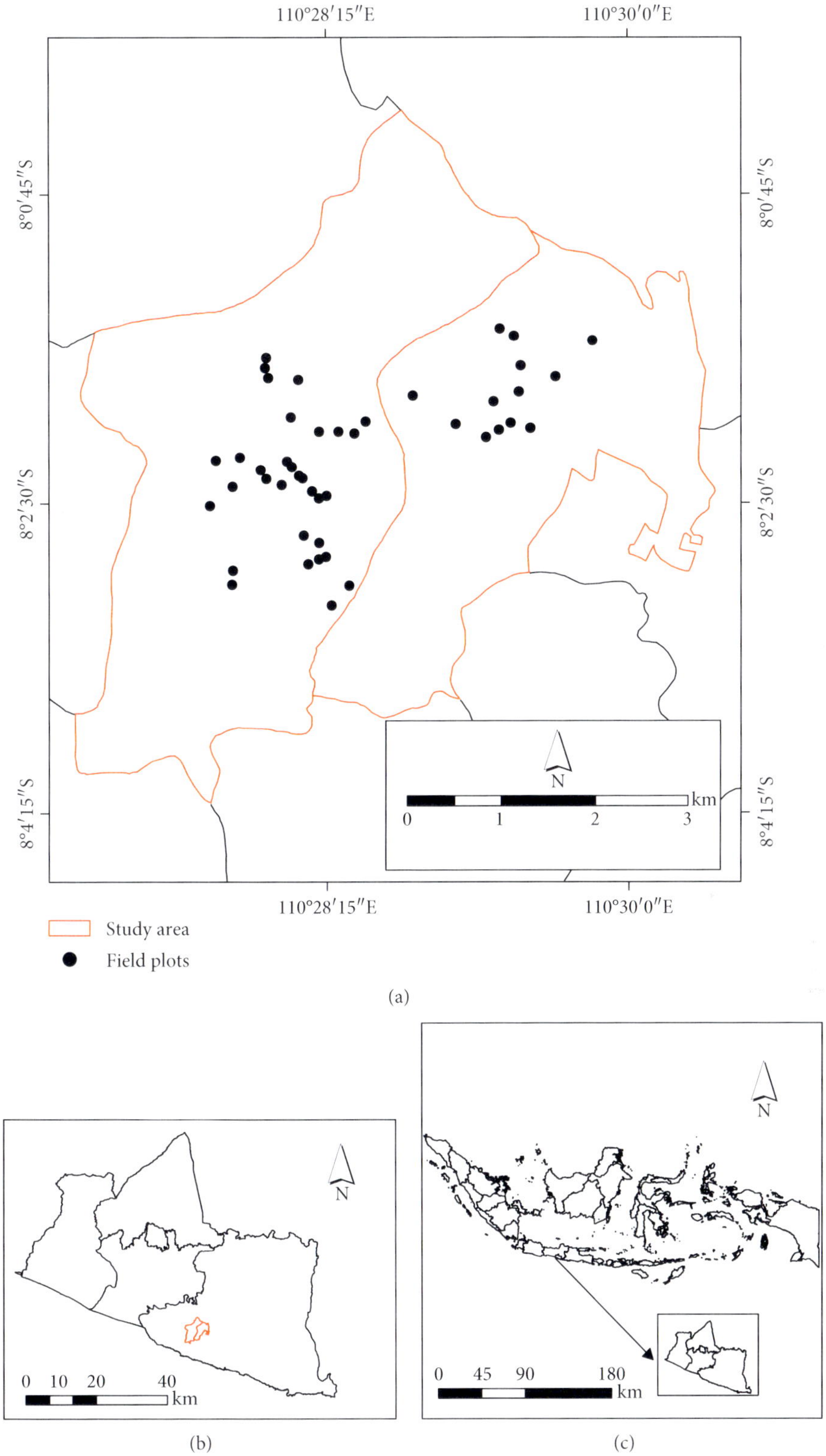

FIGURE 1: The study area: (a) map of Girisekar and Jetis; (b) the location of study area in Yogyakarta Province; (c) the location of Yogyakarta Province in the Indonesian map.

forest), hutan kemasyarakatan (community forest), hutan desa (village forest), kemitraan (partnership), hutan adat (customary forest), and hutan rakyat (private forest) [11]. Private forests are grown by farmers on their own land; others are facilitated by the government on natural forests. Private forests have a crucial role in the forest ecosystem especially since degradation of natural forests is continuing in Indonesia. Recognition of efforts to reduce greenhouse gas emission outside the natural forest including the private land (e.g., private forest) originated from the 16th COP held in Mexico in 2010 resulted in a concept of Reducing Emissions from Deforestation and Forest Degradation (REDD) + [12]. A REDD scheme has encouraged United Nations Framework Convention on Climate Change (UNFCCC) members to establish a forest monitoring system by estimating forest carbon stock using a combination of forest inventory and remote sensing [13]. In addition, REDD+ in the private forest community provides a chance to gain international funding as long as they can increase carbon sink.

Considering the important role of the forest monitoring system, understanding the spatial distribution of aboveground biomass (AGB) is crucial [14]. AGB represents a majority of biomass values on the terrestrial ecosystem and is a useful parameter to measure the velocity of forest succession [15–17]. Furthermore, AGB also provides valuable information for forestry strategic planning [18]. AGB accumulation in the earth can be increased through expanding plantation areas such as private forests. Private forests have a potential to store AGB reaching a capacity of up to 300 tons/ha [19]. The number of private forests in Indonesia is likely to increase annually in accordance with the ambition of the government to expand social forestry areas at 12.7 million ha by 2019 [20]. The capabilities of remote sensing for assessing AGB have been tried in some types of forest in Indonesia like tropical forest [21] and mangrove [22]. However, information and methods for estimating AGB in private forests through remote sensing have not been treated in much detail.

Remote sensing based on vegetation indices has been widely used for estimating AGB. The vegetation index is enhanced by strong reflectance of near infrared (NIR) due to leaf internal scattering and high chlorophyll absorption by the red region of wavelength. One of the vegetation indices used to estimate biomass is the normalised difference vegetation index (NDVI) [23–26]. However, there are certain problems associated with the use of NDVI. One of these is that NDVI has a saturation problem particularly for dense vegetation which tends to have a high level of biomass [27, 28]. Utilisation of vegetation indices based on wavelengths located in the red edge is then a method that is proposed to overcome that problem [28].

Sentinel-2 is a new generation of multispectral satellite imagery that has been launched on 23 June 2015 by the European Space Agency (ESA). Sentinel-2 is a continuing image data from Landsat and SPOT, offering 13 spectral bands with 3 spatial resolutions (10 m, 20 m, and 60 m), a wide swath of 290 km, a radiometric resolution of 12 bits, and 5 days of revisit times by two satellites [29]. Utilisation of Sentinel-2 to assess AGB on private forests is interesting since the availability of red edge bands. Sentinel-2 can be applied for mapping and monitoring forest areas and measuring biophysical structures of vegetation like AGB and leaf area index (LAI) [30, 31]. However, there is lack of evidence about the utilisation of Sentinel-2 for predicting AGB on private forests.

Generally, the purpose of this study was to assess the potential of vegetation indices derived from Sentinel-2 imagery for estimating AGB of private forests. Specifically, it is aimed at (1) modelling the relationship between vegetation indices resulted from Sentinel-2 and AGB derived from field measurement from private forests and (2) determining the AGB value on the Girisekar and Jetis private forest management unit. AGB from this study area then was compared with AGB from other forest types to evaluate the ability of private forests on the global carbon cycle and to support the Indonesian government mission for reducing CO_2 emission.

TABLE 1: Allometric equations for calculation AGB in private forest.

No.	Species	Allometric equation
1	Teak (*Tectona grandis*)	$AGB = 0.0149(D^2H)^{1.0835}$
2	Acacia (*Acacia auriculiformis*)	$AGB = 0.0775(D^2H)^{0.9018}$
3	Mahogany (*Swietenia mahagoni*)	$AGB = 0.9029(D^2H)^{0.684}$
4	Other trees	$AGB = 0.0240(D^2H)^{0.7817}$

Remarks: *D*: stem diameter at breast height; *H*: total tree height.

2. Materials and Methods

2.1. Study Area. The study area lies in 8°01′15″N and 110°27′30″ E (Figure 1). The area is located in the Jetis and Girisekar private forest management unit in Gunung Kidul Region, Yogyakarta Province, Indonesia. In terms of topography, Gunung Kidul is dominated by small limestone hills. Soils are shallow and prone to erosion. Annual precipitation is between 1700 and 2500 mm with around 122 rainy days [32].

In the past, massive forest loss, poverty, and soil erosion were common problems experienced by community living in Gunung Kidul [33]. This condition encouraged groups of farmers to start rehabilitation of the barren land through a planting tree program by the central government in 1963 [32]. Therefore, small-scale teak plantations of local farmers were widespread by the mid-1960s and became more attractive to the farmers in 1980 [34]. Nowadays, Gunung Kidul has become the main area of private forests in Indonesia with an annual production around 80,000 to 100,000 m^3 [32]. There are several trees commonly planted by farmers such as teak (*Tectona grandis*), mahogany (*Swietenia mahagoni*), and earleaf acacia (*Acacia auriculiformis*).

2.2. Field Data Collection. Field data was taken in September 2017 and the end of November 2017. A total of 45 plots were set up. We used 30 plots for establishing the model and another 15 plots for validation. A stratified random sampling method was applied to select the plots based on accessibility, size, and type of trees. This sampling method was used to ascertain that areas with low and high AGB in community forests would be sampled. Each plot had a dimension of 20×20 m. In each plot, the height and diameter at breast height (DBH) of trees with diameter ≥ 8 cm were measured. Based on the tree DBH and height, field AGB was calculated by applying specific allometric equations developed by Indonesian researchers [35] (Table 1).

TABLE 2: Vegetation indices used to establish the AGB model.

No.	Vegetation indices	Formula	References
1	NDVI	$B7 - B4/B7 + B4$	[40]
2	EVI	$G * B7 - B4/B7 + C1 * B4 - C2 * B2 + L$ Note: $C1 = 6$; $C2 = 7.5$; $L = 1$; $G = 2.5$	[41]
3	MSR	$(B7/B4 - 1)/((B7 + B4)^{1/2} + 1)$	[42]
4	SR	$B7/B4$	[43]
5	NDI45	$B5 - B4/B5 + B4$	[44]
6	S2REP	$705 + 35 * ((((B7 + B4)/2) - B4)/(B6 - B5))$	[30]
7	GNDVI	$B7 - B3/B7 + B3$	[45]

2.3. Sentinel-2 Data Acquisition and Preprocessing. Sentinel-2 level 1-C data that covered the study area was acquired on 19 May 2017. The image was freely downloaded through the Copernicus Scientific Data Hub website. It had been scaled to top of atmosphere (TOA) level including orthorectification and spatial registration on a global reference system [29]. Sentinel-2 level 1-C was processed to level 2-A to gain a bottom of atmosphere- (BOA-) corrected reflectance image using the ATCOR algorithm through Sen2Cor plugin in Sentinel Application Platform (SNAP) software [36]. The image was resampled to 20 m spatial resolution using the nearest neighbour method to adjust the size of the sample plots. Finally, subsetting was done for the image to obtain the study area.

2.4. Private Forest Map. All land use patterns in the study area must be selected to classify the image data into land use and land cover categories. The land use pattern was classified into two groups such as private forest and nonprivate forest. Private forests are close to mixed plantation forests which are established by some trees and used for industrial and nonindustrial purposes [37]. These are planted by the farmers, and each farmer has a minimum of 0.25 ha of private forest land [38]. Nonprivate forests include state forest, settlement, agricultural land, water bodies, bare soil, road, and others. Contrary to private forests, state forests in our study area are established by the government and determined as production forests.

The private forest map of the study area was created from the Sentinel-2 image through the classification process. Supervised classification using maximum likelihood classification was employed to distinguish private forest and nonprivate forest. True colour combination from Sentinel-2 in particular bands 4, 3, and 2 in 10 m resolution was used to classify the image. Accuracy of the classified private forest map was assessed through the confusion matrix.

2.5. Modelling Relationship between Field AGB and Vegetation Indices. In this study, 30 field AGB and 7 vegetation indices were applied to assess the correlation between field AGB and vegetation indices shown in Table 2. Vegetation indices used in this study were divided into two. They were Sentinel-2 vegetation indices and traditional vegetation indices. Sentinel-2 vegetation indices are comprised of normalised difference index (NDI) 45 and Sentinel-2 red edge position (S2REP) whereas traditional indices involved NDVI, simple ratio (SR), modified simple ratio (MSR), green normalised difference vegetation index (GNDVI), and enhanced vegetation index (EVI). Traditional indices were selected based on simplicity and robustness. NDVI, SR, MSR, and GNDVI work through a simple algorithm, whereas GNDVI uses a green band instead of a red band. GNDVI has sensitivity on variation chlorophyll content. EVI is the robust index and has sensitivity to high biomass regions because it uses a correction factor to eliminate influence of aerosol and canopy background.

Coordinates of each plot were used to extract pixel values of vegetation indices using ArcGis 10.5. Linear regression was employed for exploring the relationship between AGB and vegetation indices. All vegetation indices were evaluated based on Pearson correlation coefficient (r) and coefficient of determination (R^2). The indices with high r and R^2 were indicated to fit into AGB. Furthermore, AGB equation modelling was conducted by applying a stepwise linear regression method using SPSS version 17 by plotting AGB as a dependent variable and vegetation indices as independent variables. R^2, root mean square error (RMSE), and multicolinearity of variables, tolerance, and variance inflation factor (VIF) were calculated to confirm the reliability of the model [39].

2.6. Mapping the AGB over the Study Area. AGB stored in the study site could be calculated through an AGB map prediction. The model derived from stepwise linear regression was used to generate the map in ArcGis 10.5. A total of 15 sample plots observing AGB were plotted against 15 predicted AGB to validate the AGB map. R^2 and RMSE were calculated during this process (Figure 2).

The main purpose of community forest development is to meet the necessity of farmers; therefore, harvesting the intensity on the private forest is high. It may lead to unfavorable environmental benefits including a decrease in carbon stocks. Accordingly, AGB stored in our study area was compared to AGB in other types of forest.

3. Results

3.1. Private Forest Composition and Field AGB. The number of trees recorded in 45 (1.8 ha) sample plots was 1451. A total of 8 species of trees were found, namely, *Tectona*

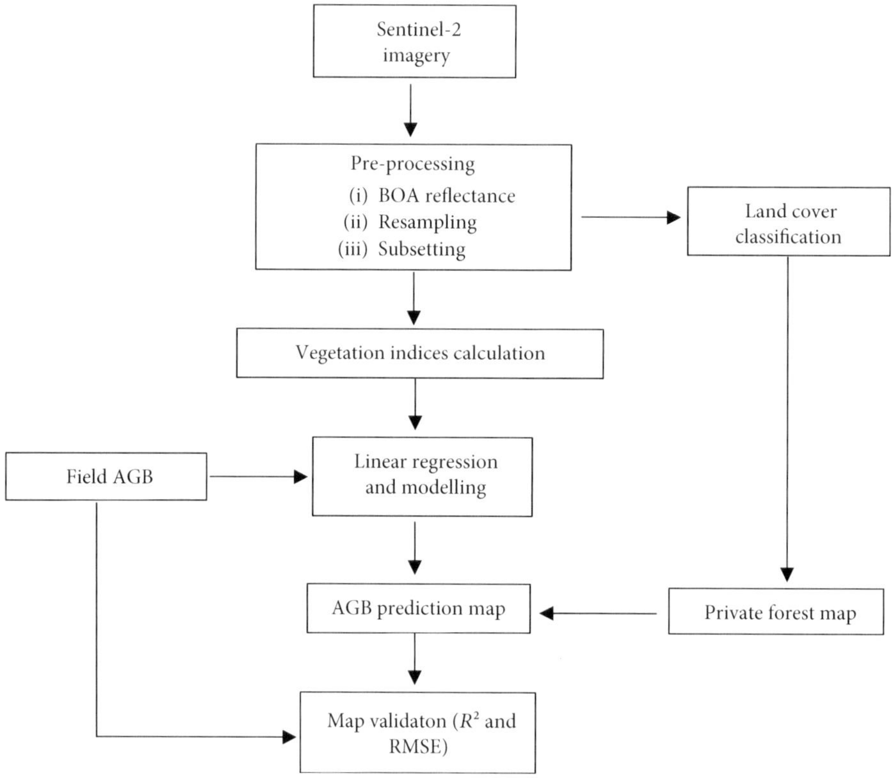

Figure 2: Flow chart of this study.

Table 3: Important value index for each species found in private forest.

No.	Species	Count	Density	RD (%)	Frequency	RF (%)	RD1 (%)	IVI
1	*Tectona grandis*	891	19.80	63.82	0.93	42.43	87.00	193.25
2	*Swietenia mahagoni*	302	6.71	20.04	0.58	26.26	8.29	54.59
3	*Acacia auriculiformis*	232	5.16	14.45	0.40	18.18	4.59	37.22
4	Other trees	26	0.58	1.69	0.29	13.13	0.12	14.94
	Total	**1451**	**32.24**	**100**	**2.20**	**100**	**100**	**300**

RD: relative density; RF: relative frequency; RD1: relative dominance. IVI is calculated from RD + RF + RD1 [46].

grandis, *Swietenia mahagoni*, *Acacia auriculiformis*, and other trees such as *Samanea saman*, *Gnetum gnemon*, *Alstonia scholaris*, *Parkia speciosa*, and *Tamarindus indica*. The dominant species in the Girisekar and Jetis forest management unit was *Tectona grandis* with a total of trees and important value index (IVI) of 891 and 193.25, respectively (Table 3). The IVI for *Swietenia mahagoni* and *Acacia auriculiformis* being the second and third dominant species were 54.59 and 37.22, respectively. Other trees group constituted the lowest in terms of IVI in private forest.

The mean of field AGB was 80 Mg/ha, with minimum and maximum values of 21 Mg/ha and 226 Mg/ha, respectively. Majority of AGB plots were spread evenly, ranging from 50 to 100 Mg/ha. We found 8 sample plots with field AGB < 50 Mg/ha and few sample plots with field AGB > 150 Mg/ha (Figure 3).

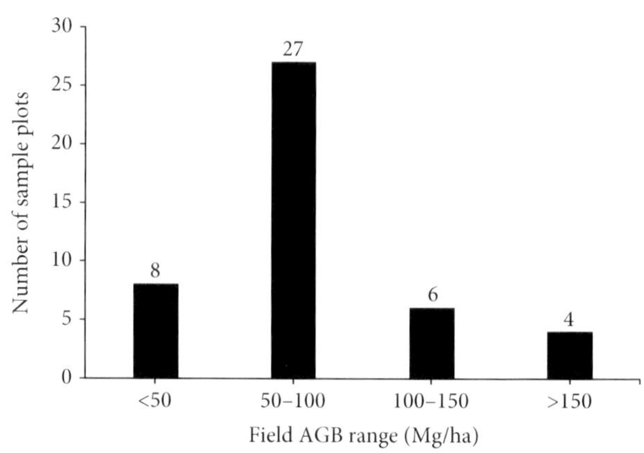

Figure 3: Distribution of field AGB within the sample plot.

3.2. Land Cover Classification Result. As many as 160 ground control sample points were selected for accuracy

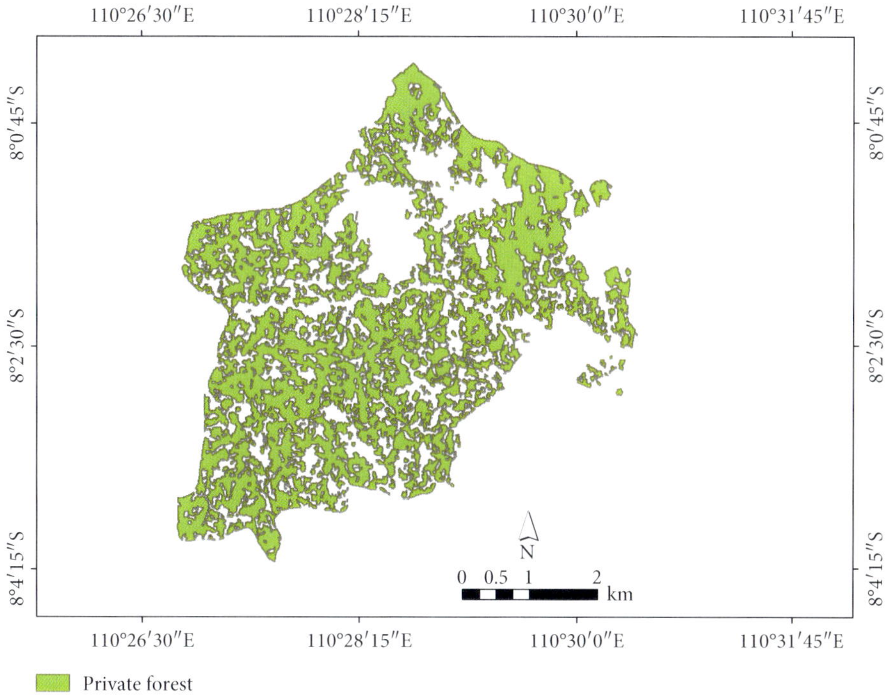

Figure 4: Girisekar and Jetis forest management unit map derived from supervised classification.

Table 4: Result of linear regression between vegetation indices and AGB.

No.	Vegetation indices	r	R^2
1	NDI45	0.89**	0.79
2	MSR	0.86**	0.75
3	SR	0.85**	0.73
4	NDVI	0.80**	0.65
5	GNDVI	0.70**	0.49
6	EVI	0.48**	0.23
7	S2REP	0.44**	0.19

** refers to a significant correlation at 0.01 level.

Table 5: Statistics of the AGB model.

Model variable	R^2	RMSE	p level	Tolerance	VIF
($n = 30$)	0.81	19.44	0.001		
NDI45				0.876	1.14
EVI				0.876	1.14

assessment of the Girisekar and Jetis private forest management unit map (Figure 4). User's accuracy of the private forest map was 95%, while producer's and overall accuracies were 95% and 94%, respectively. Based on the classification result, the large part of the Girisekar and Jetis private forest management unit was 1427 ha.

3.3. Modelling the Relationship between Vegetation Indices and Field AGB. The results of linear regression analysis between AGB and vegetation indices derived from Sentinel-2 are shown in Table 4. The r value of vegetation indices ranges from 0.44 to 0.89, and R^2 varied between 0.19 and 0.79. All vegetation indices showed a significant and positive correlation with AGB. NDI45 was the best vegetation index which corresponded to AGB ($r = 0.89$ and $R^2 = 0.79$) followed by MSR, SR, NDVI, GNDVI, EVI, and S2REP. All of the data for modelling is available in Supplementary Materials Section 1.

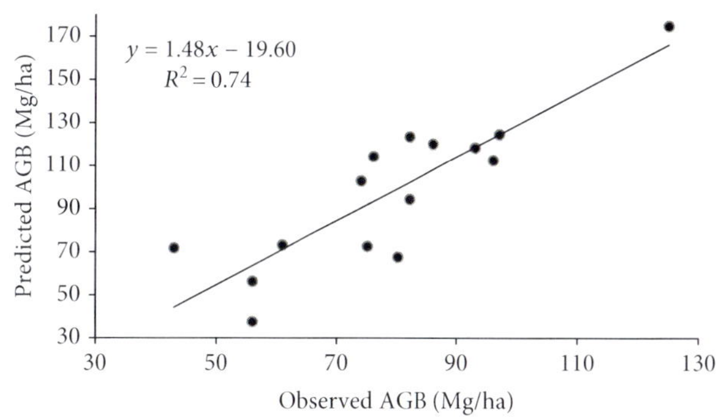

Figure 5: Scatter plot between predictive versus observed AGB.

Based on stepwise linear regression, a model for estimating AGB in this study is expressed as

$$\text{AGB} = (537 * \text{NDI45}) + (158.42 * \text{EVI}) - 353.66. \quad (1)$$

A developed model from NDI45 and EVI fitted for estimating AGB ($R^2 = 0.81$, $p < 0.05$) (Table 5). R^2 81% meant that as much as 81% of AGB variability could be explained by the model. RMSE model was 19.44 Mg/ha,

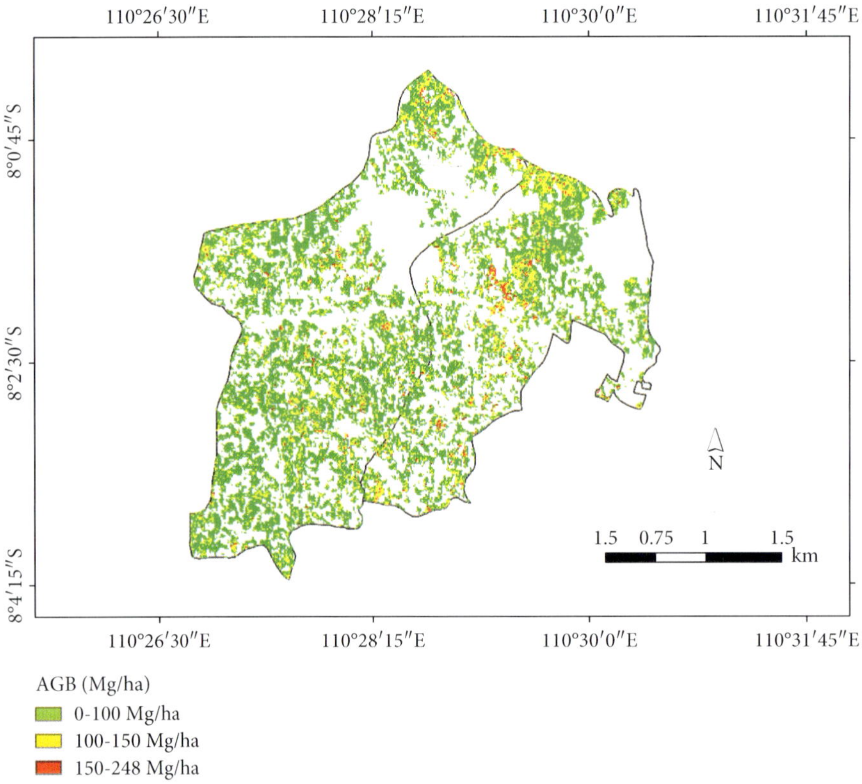

FIGURE 6: The AGB map in Girisekar and Jetis private forest management unit.

and it was not a multicolinearity problem as the tolerance value was more than 0.1 and VIF was less than 10.

3.4. AGB Map Prediction. Simple linear regression was developed to validate the AGB map from 15 plots. The correlation between the predicted and observed AGB gave a strong coefficient of determination, $R^2 = 0.74$. It indicated that approximately 74% of the observed AGB was explained by the predicted AGB according to this model (Figure 5). RMSE of the predicted and observed AGB values was 27 Mg/ha. Data used to validate the AGB map is available in Supplementary Materials Section 2.

Figure 6 illustrates AGB map prediction resulted from the stepwise linear regression model between AGB field and vegetation indices (NDI45 and EVI). The number of AGB predicted for the Girisekar and Jetis private forest management unit from spatial analysis was 72.54 Mg/ha. The AGB values varied from 0 to 248 Mg/ha. Using the 0.5 conversion factor from biomass to carbon [47], the aboveground carbon biomass estimated from the study area was 36.27 Mg/ha.

4. Discussion

Pearson correlation was employed for accessing the relationship between AGB and vegetation indices derived from the Sentinel-2 image. NDI45, MSR, SR, and NDVI had a strong correlation with AGB. NDVI is most widely used to measure biophysical properties of vegetation. Therefore, we compared it with other indices on this research. Once NDI45 and NDVI were compared, NDI45 was more powerful than NDVI since NDVI had a saturation problem at a higher value of biomass (Figure 7). Saturation resulted from a slight change of the NDVI ratio due to high reflectance of NIR and decreasing red reflectance in an area with close canopy cover, and therefore, it would lead to a weak correlation with biomass [48]. Substitution of NIR to the red edge 1 on NDI45 at the Sentinel-2 image is able to improve the relationship between satellite data and biophysical properties of the vegetation. This is consistent with the result of Frampton et al. [30] where they found that the correlation of NDI45 was higher than NDVI on measuring canopy chlorophyll content (CCC). NDI45 created from Sentinel-2 B4 (665 nm) and red edge 1 B5 (705 nm) is more robust in measuring biophysical parameters of vegetation than other band combination in Sentinel-2 [44].

MSR and SR outperformed NDVI in this study. It might be because the relationship of MSR and SR with biophysical properties of the vegetation was more linear than NDVI [30, 42]. NDVI is much affected by leaf optical and geometry effect from a sun view angle; hence, linearity to parameters of vegetation is lower than MSR [42]. EVI is more reliable than NDVI to measure AGB on dense vegetation because of its ability to reduce the effect of atmosphere and canopy background. However, EVI showed poor correlation to AGB in this research. A possible explanation is that the slope of the plots in the study area varies from flat to slightly inclined (slope range of the sample plots between 0° and 19°). EVI is highly influenced by various terrain conditions [49, 50]. The soil adjustment factor becomes the

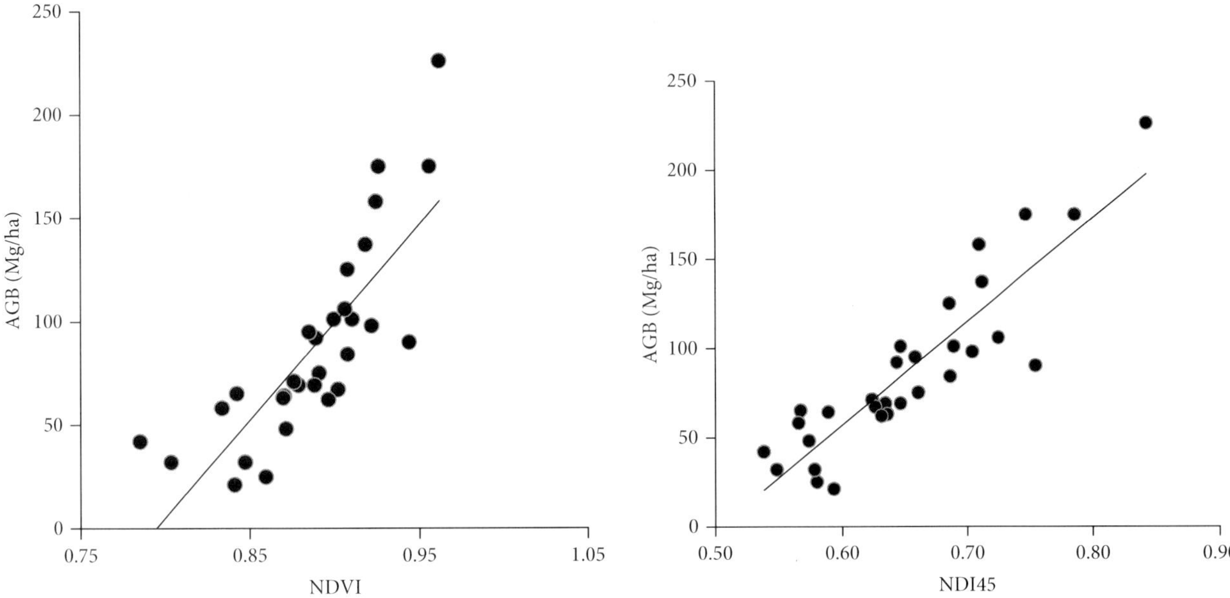

FIGURE 7: Scatter plot comparison between NDVI and NDI45 in estimating AGB.

TABLE 6: Comparison between AGB from private forest and other forest types; 1 Mg/ha = 1 ton/ha.

No.	Type of forest	AGB	Sensor	References
1	Tropical	382 Mg/ha	Lidar	[51]
2	Private	72.54 Mg/ha	Sentinel-2	This study
3	Mangrove	250.53 Mg/ha	GeoEye-1	[52]
4	Mediterranean	38 Ton/ha	QuickBird	[53]
5	Deciduous	58 Ton/ha	ALOS PALSAR	[54]
6	Boreal	39.5 Mg/ha	Landsat ETM+	[55]
7	Conifer and broad leaf	106.45 Mg/ha	Landsat 8 and Radarsat-2	[56]

limitation of EVI because it is very sensitive to topography than indices which are based on the simple ratio algorithm such as SR and NDVI [49].

Private forests are a potential source of AGB, having an important role in climate change mitigation. Therefore, comparing AGB from private forests and other forest types is imperative. AGB predicted in this study area was 72.54 Mg/ha. The estimates of AGB given in the current study are higher than the estimates calculated from deciduous forests in India (58 Mg/ha) [54] (Table 6). It also almost doubles compared to the value from Mediterranean forests in Italy (38 Mg/ha) [53] and the AGB value from boreal forests in Alaska (39.5 Mg/ha) [55]. However, the current AGB is lower than the biomass of mangrove forests in Thailand (250.53 Mg/ha) [52], tropical forests in Borneo (382 Mg/ha) [51], and conifer broadleaf forests in China (106.45 Mg/ha) [54]. The AGB of tropical forests is higher than that of private forests because natural forests store a large amount of biomass on terrestrial ecosystems which accumulates over a long period of time. On the other hand, the sustainability of private forests depends on the success of the silviculture system and harvesting time by the farmers. Yet, AGB on private forests is still higher than AGB from some types of forest, and this is unexpected and suggests that private forests can be an alternative as biomass and carbon reservoir.

This research showed the potential of vegetation indices derived from Sentinel-2 to predict AGB on private forests. A number of researchers have reported AGB in plantation forests with different sensors [56–58]. For instance, Baig et al. [59] utilised ALOS-2 PALSAR to model and map *Dalbergia sissoo* forest plantation in Pakistan using nonlinear regression between field data and SAR backscatter. Dube and Mutanga [57] demonstrated the ability of WorldView-2 to retrieve AGB on *Eucalyptus* plantation in South Africa. However, the researchers have used mainly commercial satellite data instead of a cheap data source. This issue is problematic specially for developing countries or forest researchers who have limitation of budget. Therefore, making this methodology used in this research is very vital as it can address the cost issue as it uses free satellite image and is supported by open-source software, SNAP.

Furthermore, the sample plots of this study are limited as a result of accessibility and cost. Nevertheless, the sample plots were successful in establishing a model ($R^2 = 0.81$) and predicted AGB map on the study area. The reason why the model on this research shows good performance can be attributed to the average of field AGB, which is 80 Mg/ha where only one sample plot had field AGB more than 200 Mg/ha and the others are below 200 Mg/ha. It is too difficult to find areas with high AGB value on private forests because the farmers harvest mature trees. Therefore, it is plausible if majority of the sample plots would be only in low to medium AGB areas. Although NDVI was more saturated than NDI45, saturation data did not significantly influence our work due to limited sample plots in high biomass areas. In the case of Landsat TM, Lu et al. [60] stated that data saturation occurred when biomass reaches 100–150 Mg/ha depending on the complexity of the vegetation structure. There are abundant rooms for further research in evaluating the ability of Sentinel-2 imagery in tropical humid forests which tend to have a high value of biomass.

5. Conclusion

Sentinel-2 multispectral imagery can be utilised to estimate AGB on the Girisekar and Jetis private forest management unit. Vegetation indices were obtained from the Sentinel-2 data image. Normalised difference index (NDI) 45, which is established from red and red-edge 1 bands, had a strong correlation with AGB in comparison with other indices. The study demonstrated that the AGB model derived from stepwise linear regression was robust ($R^2 = 0.81$). R^2 between observed and predicted AGB from the Sentinel-2 private forest model was 0.74 indicating that AGB can be predicted with high accuracy using remote sensing data. The AGB predicted in this research is higher than the AGB from some types of forest like deciduous and Mediterranean. The result suggested that private forests are a reliable source to reducing CO_2 emission.

Conflicts of Interest

The authors declare that they have no conflicts of interest.

Acknowledgments

Prince of Songkla University (PSU) has accorded technical support and funding in the accomplishment of this study through Thailand's Education Hub for ASEAN Countries (TEH-AC) program. We thank the Girisekar and Jetis private forest owner for allowing data collection.

Supplementary Materials

The data aforementioned is used to correlate between AGB and the vegetation indices via linear regression. The data is also used to establish the AGB prediction model through multilinear regression. AGB is a field data which is collected through plot measurement in the private forest. Vegetation indices are derived from a Sentinel-2 satellite image through mathematical function. The UTM coordinates are coordinates of each sample plot in the field. The coordinates are used to connect between field data and vegetation index values. The results of this data can be seen in Modelling the Relationship between Vegetation Indices and Field AGB in this manuscript. Section 2: In this section, we provide data for AGB map validation. Output of this research is an AGB map which represents real AGB in the study area. The AGB map was produced from mapping software using the AGB prediction model in Section 1. To ensure the reliability of our map, it has to be validated. For this analysis, the data can be divided into two groups: observed and predicted AGB. Each group has fifteen data. Observed AGB was real AGB and were taken from plot measurement in the private forest while predicted AGB was AGB from the predicted map which had the same coordinate, having observed AGB. The coefficient of determination (R^2) and root mean square error (RMSE) were used to validate the performance of the model. The results of AGB map validation can be seen in AGB Map Prediction in this manuscript. *(Supplementary Materials)*

References

[1] A. E. Russon, A. Erman, and R. Dennis, "The population and distribution of orangutans (*Pongo pygmaeus pygmaeus*) in and around the Danau Sentarum Wildlife Reserve, West Kalimantan, Indonesia," *Biological Conservation*, vol. 97, no. 1, pp. 21–28, 2001.

[2] T. G. O'Brien and M. F. Kinnaird, "Caffeine and conservation," *Science*, vol. 300, no. 5619, pp. 587–587, 2003.

[3] H. T. Wibisono and W. Pusparini, "Sumatran tiger (*Panthera tigris sumatrae*): a review of conservation status," *Integrative Zoology*, vol. 5, no. 4, pp. 313–323, 2010.

[4] B. Loken, S. Spehar, and Y. Rayadin, "Terrestriality in the bornean orangutan (Pongo pygmaeus morio) and implications for their ecology and conservation," *American Journal of Primatology*, vol. 75, no. 11, pp. 1129–1138, 2013.

[5] S. A. Abood, J. S. H. Lee, Z. Burivalova, J. Garcia-Ulloa, and L. P. Koh, "Relative contributions of the logging, fiber, oil palm, and mining industries to forest loss in Indonesia," *Conservation Letters*, vol. 8, no. 1, pp. 58–67, 2015.

[6] B. A. Margono, S. Turubanova, I. Zhuravleva et al., "Mapping and monitoring deforestation and forest degradation in Sumatra (Indonesia) using Landsat time series data sets from 1990 to 2010," *Environmental Research Letters*, vol. 7, no. 3, article 034010, 2012.

[7] D. O. Fuller, T. C. Jessup, and A. Salim, "Loss of forest cover in Kalimantan, Indonesia, since the 1997-1998 El Niño," *Conservation Biology*, vol. 18, no. 1, pp. 249–254, 2004.

[8] R. Tsujino, T. Yumoto, S. Kitamura, I. Djamaluddin, and D. Darnaedi, "History of forest loss and degradation in Indonesia," *Land Use Policy*, vol. 57, Supplement C, pp. 335–347, 2016.

[9] World bank, *Indonesia and Climate Charge: Current Status and Policies*, 2007, September 2017, https://siteresources.worldbank.org/INTINDONESIA/Resources/Environment/ClimateChange_Full_EN.pdf.

[10] Republic of Indonesia, *Intended Nationally Contribution Republik of Indonesia*, 2015, April 2017, http://www4.unfccc.int/submissions/INDC/Published%20Documents/Indonesia/1/INDC_REPUBLIC%20OF%20INDONESIA.pdf.

[11] Republic of Indonesia, *Peraturan Menteri Lingkungan Hidup dan Kehutanan Republik Indonesia Nomor P.83/MenLHK/Setjen/Kum.1/10/2016 tentang Perhutanan Sosial*, 2016, July 2017, http://www.forda-mof.org/files/P.83_2016.pdf.

[12] L. Goers, J. Lawson, and E. Garen, "Economic drivers of tropical deforestation for agriculture," in *Managing Forest Carbon in a Changing Climate*, M. Ashton, M. Tyrrell, D. Spalding, and B. Gentry, Eds., pp. 305–320, Springer, Dordrecht, 2012.

[13] T. Tokola, "Remote sensing concepts and their applicability in REDD+ monitoring," *Current Forestry Reports*, vol. 1, no. 4, pp. 252–260, 2015.

[14] R. A. Houghton, "Aboveground forest biomass and the global carbon balance," *Global Change Biology*, vol. 11, no. 6, pp. 945–958, 2005.

[15] H. K. Gibbs, S. Brown, J. O. Niles, and J. A. Foley, "Monitoring and estimating tropical forest carbon stocks: making REDD a reality," *Environmental Research Letters*, vol. 2, no. 4, article 045023, 2007.

[16] K. Mokany, R. J. Raison, and A. S. Prokushkin, "Critical analysis of root: shoot ratios in terrestrial biomes," *Global Change Biology*, vol. 12, no. 1, pp. 84–96, 2006.

[17] D. M. A. Rozendaal, R. L. Chazdon, F. Arreola-Villa et al., "Demographic drivers of aboveground biomass dynamics during secondary succession in neotropical dry and wet forests," *Ecosystems*, vol. 20, no. 2, pp. 340–353, 2017.

[18] H. Gallaun, G. Zanchi, G.-J. Nabuurs, G. Hengeveld, M. Schardt, and P. J. Verkerk, "EU-wide maps of growing stock and above-ground biomass in forests based on remote sensing and field measurements," *Forest Ecology and Management*, vol. 260, no. 3, pp. 252–261, 2010.

[19] M. Wilnhammer, A. Rothe, W. Weis, and S. Wittkopf, "Estimating forest biomass supply from private forest owners: a case study from Southern Germany," *Biomass Bioenergy*, vol. 47, Supplement C, pp. 177–187, 2012.

[20] T. Santika, E. Meijaard, S. Budiharta et al., "Community forest management in Indonesia: avoided deforestation in the context of anthropogenic and climate complexities," *Global Environmental Change*, vol. 46, pp. 60–71, 2017.

[21] A. Wijaya, S. Kusnadi, R. Gloaguen, and H. Heilmeier, "Improved strategy for estimating stem volume and forest biomass using moderate resolution remote sensing data and GIS," *Journal of Forest Research*, vol. 21, no. 1, pp. 1–12, 2010.

[22] P. Wicaksono, "Mangrove above-ground carbon stock mapping of multi-resolution passive remote-sensing systems," *International Journal of Remote Sensing*, vol. 38, no. 6, pp. 1551–1578, 2017.

[23] C. J. Tucker, C. L. Vanpraet, M. J. Sharman, and G. Van Ittersum, "Satellite remote sensing of total herbaceous biomass production in the senegalese sahel: 1980–1984," *Remote Sensing of Environment*, vol. 17, no. 3, pp. 233–249, 1985.

[24] N. I. Gasparri, M. G. Parmuchi, J. Bono, H. Karszenbaum, and C. L. Montenegro, "Assessing multi-temporal Landsat 7 ETM + images for estimating above-ground biomass in subtropical dry forests of Argentina," *Journal of Arid Environments* vol. 74, no. 10, pp. 1262–1270, 2010.

[25] A. A. Wani, P. K. Joshi, and O. Singh, "Estimating biomass and carbon mitigation of temperate coniferous forests using spectral modeling and field inventory data," *Ecological Informatics*, vol. 25, Supplement C, pp. 63–70, 2015.

[26] C. Huang, X. Ye, C. Deng, Z. Zhang, and Z. Wan, "Mapping above-ground biomass by integrating optical and SAR imagery: a case study of Xixi National Wetland Park, China," *Remote Sensing*, vol. 8, no. 8, p. 647, 2016.

[27] A. R. Huete, H. Q. Liu, K. Batchily, and W. van Leeuwen, "A comparison of vegetation indices over a global set of TM images for EOS-MODIS," *Remote Sensing of Environment*, vol. 59, no. 3, pp. 440–451, 1997.

[28] O. Mutanga and A. K. Skidmore, "Narrow band vegetation indices overcome the saturation problem in biomass estimation," *International Journal of Remote Sensing*, vol. 25, no. 19, pp. 3999–4014, 2004.

[29] Sentinel-2_Team, *Sentinel-2 User Hand Book*, Europian Space Agency, 2015.

[30] W. J. Frampton, J. Dash, G. Watmough, and E. J. Milton, "Evaluating the capabilities of Sentinel-2 for quantitative estimation of biophysical variables in vegetation," *ISPRS Journal of Photogrammetry and Remote Sensing*, vol. 82, Supplement C, pp. 83–92, 2013.

[31] T. Majasalmi and M. Rautiainen, "The potential of Sentinel-2 data for estimating biophysical variables in a boreal forest: a simulation study," *Remote Sensing Letters*, vol. 7, no. 5, pp. 427–436, 2016.

[32] A. Hinrichs, D. Muhtaman, and N. Irianto, *Forest Certification on Community Land in Indonesia*, Deutsche Gesellschaft für Technische Zusammenarbeit (GTZ), Jakarta, Indonesia, 2008.

[33] J. W. Nibbering, "Tree planting on deforested farmlands, Sewu Hills, Java, Indonesia: impact of economic and institutional changes," *Agroforestry Systems*, vol. 46, no. 1, pp. 65–82, 1999.

[34] D. Rohadi et al., *Improving Economic Outcomes for Smallholders Growing Teak in Agroforestry Systems in Indonesia*, Australian Centre for International Agriculture Research, Canberra, 2012.

[35] R. H. Purwanto, R. Rohman, A. Maryudi, T. Yuwono, D. B. Permadi, and M. Sanjaya, "Potensi Biomasa dan Simpanan Karbon Jenis-jenis Tanaman Berkayu di Hutan Rakyat Desa Nglanggeran, Gunungkidul, Daerah Istimewa Yogyakarta," *Jurnal Ilmu Kehutanan*, vol. 6, no. 2, pp. 128–141, 2015.

[36] J. Louis, V. Debaecker, B. Pflug et al., "Sentinel-2 Sen2Cor: L2A processor for users," in *Living Planet Symposium*, p. 91, Prague, Czech Republic, May 2016.

[37] T. J. Wormald, *Mixed and Pure Forest Plantations in the Tropics and Sub Tropics*, FAO, Rome, 1992.

[38] Republic of Indonesia, *Statistik Lingkungan hidup dan kehutanan tahun 2016*, 2017, August 2018, http://www.menlhk.go.id/downlot.php?file=Statistik_KLHK_2016.pdf.

[39] S. Eckert, "Improved forest biomass and carbon estimations using texture measures from WorldView-2 satellite data," *Remote Sensing*, vol. 4, no. 4, pp. 810–829, 2012.

[40] C. J. Tucker, "Red and photographic infrared linear combinations for monitoring vegetation," *Remote Sensing of Environment*, vol. 8, no. 2, pp. 127–150, 1979.

[41] A. Huete, K. Didan, T. Miura, E. P. Rodriguez, X. Gao, and L. G. Ferreira, "Overview of the radiometric and biophysical performance of the MODIS vegetation indices," *Remote Sensing of Environment*, vol. 83, no. 1-2, pp. 195–213, 2002.

[42] J. M. Chen, "Evaluation of vegetation indices and a modified simple ratio for boreal applications," *Canadian Journal of Remote Sensing*, vol. 22, no. 3, pp. 229–242, 1996.

[43] C. F. Jordan, "Derivation of leaf-area index from quality of light on the Forest floor," *Ecology*, vol. 50, no. 4, pp. 663–666, 1969.

[44] J. Delegido, J. Verrelst, L. Alonso, and J. Moreno, "Evaluation of Sentinel-2 red-edge bands for empirical estimation of green LAI and chlorophyll content," *Sensors*, vol. 11, no. 7, pp. 7063–7081, 2011.

[45] A. A. Gitelson, Y. J. Kaufman, and M. N. Merzlyak, "Use of a green channel in remote sensing of global vegetation from EOS-MODIS," *Remote Sensing of Environment*, vol. 58, no. 3, pp. 289–298, 1996.

[46] J. T. Curtis and R. P. McIntosh, "The interrelations of certain analytic and synthetic phytosociological characters," *Ecology*, vol. 31, no. 3, pp. 434–455, 1950.

[47] S. Brown and A. E. Lugo, "The storage and production of organic matter in tropical forests and their role in the global carbon cycle," *Biotropica*, vol. 14, no. 3, pp. 161–187, 1982.

[48] P. S. Thenkabail, R. B. Smith, and E. De Pauw, "Hyperspectral vegetation indices and their relationships with agricultural crop characteristics," *Remote Sensing of Environment*, vol. 71, no. 2, pp. 158–182, 2000.

[49] B. Matsushita, W. Yang, J. Chen, Y. Onda, and G. Qiu, "Sensitivity of the enhanced vegetation index (EVI) and normalized difference vegetation index (NDVI) to topographic effects: a case study in high-density cypress forest," *Sensors*, vol. 7, no. 11, pp. 2636–2651, 2007.

[50] E. L. Garroutte, A. J. Hansen, and R. L. Lawrence, "Using NDVI and EVI to map spatiotemporal variation in the biomass and quality of forage for migratory elk in the greater yellowstone ecosystem," *Remote Sensing*, vol. 8, no. 5, p. 404, 2016.

[51] K. Ioki, S. Tsuyuki, Y. Hirata et al., "Estimating above-ground biomass of tropical rainforest of different degradation levels in Northern Borneo using airborne LiDAR," *Forest Ecology and Management*, vol. 328, Supplement C, pp. 335–341, 2014.

[52] N. R. A. Jachowski, M. S. Y. Quak, D. A. Friess, D. Duangnamon, E. L. Webb, and A. D. Ziegler, "Mangrove biomass estimation in Southwest Thailand using machine learning," *Applied Geography*, vol. 45, pp. 311–321, 2013.

[53] E. Santi, C. Tarantino, V. Amici et al., "Fine-scale spatial distribution of biomass using satellite images," *Journal of Ecology and The Natural Environment*, vol. 6, no. 2, pp. 75–86, 2014.

[54] K. C. Thumaty, R. Fararoda, S. Middinti, R. Gopalakrishnan, C. S. Jha, and V. K. Dadhwal, "Estimation of above ground biomass for central Indian deciduous forests using ALOS PALSAR L-band data," *Journal of the Indian Society of Remote Sensing*, vol. 44, no. 1, pp. 31–39, 2016.

[55] L. Ji, B. K. Wylie, D. R. N. Brown et al., "Spatially explicit estimation of aboveground boreal forest biomass in the Yukon River Basin, Alaska," *International Journal of Remote Sensing*, vol. 36, no. 4, pp. 939–953, 2015.

[56] Z. Shao and L. Zhang, "Estimating forest aboveground biomass by combining optical and SAR data: a case study in Genhe, Inner Mongolia, China," *Sensors*, vol. 16, no. 6, 2016.

[57] T. Dube and O. Mutanga, "The impact of integrating WorldView-2 sensor and environmental variables in estimating plantation forest species aboveground biomass and carbon stocks in uMgeni Catchment, South Africa," *ISPRS Journal of Photogrammetry and Remote Sensing*, vol. 119, pp. 415–425, 2016.

[58] E. W. Maina, P. A. Odera, and M. J. Kinyanjui, "Estimation of above ground biomass in forests using Alos Palsar data in Kericho and Aberdare ranges," *Open Journal of Forestry*, vol. 7, no. 2, pp. 79–96, 2017.

[59] S. Baig, W. A. Qazi, A. M. Akhtar et al., "Above ground biomass estimation of *Dalbergia sissoo* forest plantation from dual-polarized ALOS-2 PALSAR data," *Canadian Journal of Remote Sensing*, vol. 43, no. 3, pp. 297–308, 2017.

[60] D. Lu, Q. Chen, G. Wang, L. Liu, G. Li, and E. Moran, "A survey of remote sensing-based aboveground biomass estimation methods in forest ecosystems," *International Journal of Digital Earth*, vol. 9, no. 1, pp. 63–105, 2016.

15

Urban Expansion Assessment in Huaihe River Basin, China, from 1998 to 2013 using Remote Sensing Data

Yong Fan,[1,2] Xi Zhu,[3] Zongyi He,[2] Shengdi Zhang,[4] Junbo Gao,[1] Feiyan Chen,[2] Xiaolin Peng,[2] and Junli Li[5]

[1]*School of Geographic Sciences, Xinyang Normal University, No. 237 Nanhu Road, Xinyang 464000, China*
[2]*School of Resource and Environmental Sciences, Wuhan University, No. 129 Luoyu Road, Wuhan 430079, China*
[3]*Shanghai Municipal Institute of Surveying and Mapping, No. 419 Wuning Road, Shanghai 200063, China*
[4]*School of Geo-Sciences, The University of Edinburgh, Central Campus, Drummond Street, Edinburgh EH89XP, UK*
[5]*School of Resources and Environment, Anhui Agricultural University, No. 130 Changjiang West Road, Hefei 230036, China*

Correspondence should be addressed to Zongyi He; zongyihe@tom.com

Academic Editor: Eduard Llobet

Urbanization reflects the overall behavior of human society; thus, characterization of its associated spatial and temporal trends has been extensively researched. This study examines the process of urban expansion in the Huaihe River Basin (HRB) which is a key transition region within China's urban system. In order to grasp the urban expansion process in different temporal sequences objectively, rapidly, and accurately, we used remote sensing data to assess the urban expansion in time and space. Urban expansion rules were defined for the urban area, urbanization intensification, extended dynamic degree, and spatial pattern. The research findings show that the urban area expansion speed was at medium level throughout the entire HRB and within each province. Presently, the formation of a whole urban agglomeration or urban system is not complete in the HRB; urban expansion in the HRB displayed space-time disequilibrium tendencies during 1998–2013.

1. Introduction

Urbanization is an important spatial expression of the human social development process [1]. The similarities and differences between different types of urban expansion can be analyzed by reconstructing the urban expansion process, thereby revealing the inner and outer driving mechanisms in the emergence and development of urban land [2]. In 2008, for the first time in history, the world reached an important milestone: half of the world population was shown to live in urban areas. Moreover, most of the urban population growth has been occurring in developing countries. The space occupied by urban areas is increasing faster than the urban population itself. Between 2000 and 2030, the world's urban population is expected to increase by 72%, and the built-up areas of cities of 100,000 people or more could increase by 175% [3]. Dramatic urban expansion has occurred universally in developing countries in recent decades [4], and problems such as environmental contamination and urban flooding have consequently emerged [5]. The Huaihe River Basin (HRB) in China represents a transition region between the north and south areas of China's urban system and has a large population, rich resources, and relatively rudimentary development. HRB has experienced high population growth in the past two decades, particularly after the Chinese reform and opening. As a result, intense competition of land for residential, industrial, agricultural, and other developmental uses is increasing in the absences of a clear urban growth policy and adequate control. The Press Conference on New Urbanization Plan of China [6] stated that the Government has supreme authority in formulating the country's urbanization policy and can promote the balanced development of geographical space to create new regional economic growth poles. Therefore, this region holds great theoretical and practical significance for studies of sustainable development in regional towns. The focus of the present study is reconstruction of urban extension process of this area in different time periods by using remote sensing data.

Remote sensing images contain large amounts of information of continuous space and time characteristics, which enable cyclical simulation analyses of urban expansion at different scales [7]. This approach is advantageous for extracting urban information on individual cities or large metropolises to analyze the changes in urban land. This can be accomplished by using high resolution remote sensing images [8, 9] such as those provided by Landsat satellite data [10], SPOT (Satellite Pour l'Observation de la Terre) satellite data [11], and ZY-3 (Ziyuan-3) satellite data [12], as well as radar data [13, 14]. High precision information for urban extraction can be achieved by merging several data sets [15, 16]. However, owing to the large amounts of imagery data required, studies can be constrained by the high costs of obtaining coverage; thus, small coverage analyses with monoimages are often used to extract information on single cities [17]. Remote sensing images with low resolution can also be used, which are derived from DMSP/OLS (Defense Meteorological Satellite Program/Operational Linescan System) data and MODIS (Moderate Resolution Imaging Spectroradiometer) data; such data have been used to generate static city global maps [18]. In the early 1970s, researchers developed a linear scanning business system known as the OLS of the United States DMSP that can capture urban lighting data at night in cloudless cases; such data were found to be useful for studies of urbanization [19]. Currently, DMSP/OLS data are widely used by scholars at home and abroad as effective data sources for monitoring human activities and researching urbanization [20–25].

Two types of methods are used to extract urban information from DMSP/OLS data. The first category is selected for image segmentation according to the area threshold, and the second is characterized by image classification, which uses pixels as the basic unit. The latter includes mainly classification methods that employ support vector machines (SVMs) [26] and the method of linear decomposition of mixed pixels [27]. The first method is complex and time-consuming at both the regional and global scales, particularly when applied in research areas showing development differences among towns [28]. Compared with the threshold dichotomy, the method of image classification circumvents the threshold determination, which converts the process of urban information extraction into a classification problem whereby urban information is distinguished from noncity related information. Combining SPOT-vegetation (VGT) data, which reflect the amount of vegetation cover, with DMSP/OLS data can improve the accuracy of information extraction in urban areas. Some studies have demonstrated that remote sensing data can be used with multiple sources and low resolution for research involving urban information extraction [29, 30]. Hence, for research of large areas and long time series of the urban expansion process, the extent of urban and make maps for spatial analysis can be extracted on the basis of DMSP/OLS data and SPOT-VGT data along with the SVM classification method [31, 32].

OLS sensors neither perform data correction nor consider atmospheric changes. Interference from clouds and pollution must be considered as influencing factors of the continuity and truth of DMSP/OLS and SPOT-VGT data sets during collection [33], which limits the applications of the extracted urban information for the two types of data.

The present study has three main objectives. First, on the basis of the mutual correction of the light data and reconstruction of the quality of the Normalized Difference Vegetation Index (NDVI) derived from SPOT-VGT data, this study aims to extract urban spatial information in the experimental area by using SVM methods. Second, an accurate reconstruction of the spatial and temporal patterns of urban expansion in the HRB is attempted by using remote sensing data. Finally, on the basis of the reconstruction of the sequences of time and space involved in the process of urban expansion, this study discusses the urban expansion process detected in the HRB, thereby providing technical support and decision-making references for regional urban planning and development efforts.

2. Study Area

The HRB is located in the eastern part of China between the Yangtze River and the Yellow River basin and spans Henan, Hubei, Anhui, Jiangsu, and Shandong provinces with a total area of 2.7 × 105 km^2. The HRB connects the east and midwest regions of China and is the central economic zone between the Yangtze River Economic Zone and the Yellow River Economic Zone. These regions have multiple transitivity on nature and society including transitions in climate, agriculture, and culture. The HRB population is 170 million, accounting for 12.3% of the Chinese population. This high population density has an average of 611 people per square kilometer, which is 4.8 times the average Chinese population density [34]. Because the HRB is a natural geographic unit, it has not overlapped with various administrative unit boundaries. Therefore, this study includes only 33 urban areas of the HRB, which have complete administrative units (Figure 1).

3. Data and Methods

3.1. Data Processing Flow. In this paper, the HRB urban location information for 1998, 2001, 2004, 2007, 2010, and 2013 was extracted by using multisource remote sensing data, and expansion of the urban system was analyzed. The specific process employed is shown in Figure 2.

First, multiple sensor DMSP/OLS and SPOT-VGT data were used for mutual correction and data preprocessing. Second, we used the SVM classification method to extract urban information for the experimental area. Finally, we analyzed the extraction results regarding the index features and spatial characteristics from global and various regional perspectives to obtain the expansion rules of the HRB for 1998–2013.

3.2. Data and Preprocessing

(1) DMSP/OLS Data. The DMSP/OLS lighting data were used for urban information extraction and were downloaded from

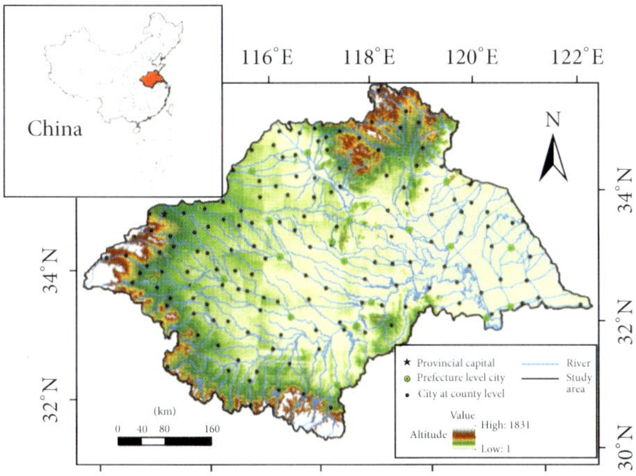

Figure 1: Study area.

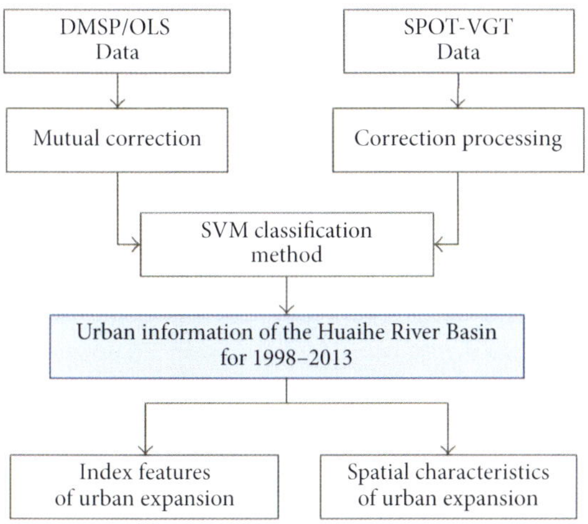

Figure 2: Data processing flow.

the United States National Geophysical Data Center (NGDC) [35]. To ensure that the DMSP/OLS data in the long time series have continuity and comparability, data correction was performed manually. According to the principle of accumulating high digital number (DN) values and good continuity, this study applied the data the F16 satellite recorded in 2007 for reference. Specifically, from this reference data, the lighting data at night for Jixi City were compared with our data set [24].

(2) SPOT-VGT Data. This work employed a data set of NDVI remote sensing data compiled from synthetic images that were composed over 10-day periods by the SPOT-VGT satellite; these data were obtained from the Belgium Flemish Institute for Technological Research [36]. This study used the data to improve the extraction process for urban information, and NDVI data were extracted for the long time series. Because NDVI data reflect changes in vegetation, those in the time series illustrate the rising, falling, and stable trends in vegetation in the HRB. In this study, quality reconstruction work for the NDVI data was conducted by using the Moving Weighted Harmonic Analysis (MWHA) method [33].

(3) Other Secondary Data. Auxiliary data that were used in this study included data on administrative divisions, statistical, social, and economic data [37].

3.3. Extraction Method for Urban Spatial Information

This study employed the SVM method to extract the urban information for the entire research extent during the different time periods.

The principle of SVM is that certain nonlinear mapping (kernel functions) of the input vector map is changed into a high-dimensional feature space, where it can build the optimal separation hyperplane. The algorithm establishes a decision function, and the data can be classified during testing as being the most accurate through the decision function. The decision function is calculated by formula (1) in the nonlinear case.

$$f(x) = \operatorname{sign}\left(\sum_{i=1}^{n} y_i a_i K(x_i, x) + b\right),$$
$$b = \frac{1}{N_{\mathrm{NSV}}} \sum_{x_i \in JN} \left(y_i - \sum_{x_i \in J} a_j y_j K(x_i, x_j)\right). \quad (1)$$

Among these terms, $K(x_i, x)$ is the kernel function, N_{NSV} is the standard supporting vector number, JN is the set of the standard supporting vector, J is the set of the supporting vector, and α_i is a Lagrange multiplier. The key link of the SVM algorithm is the choice of the kernel function, and the kernel function selected needs to be suitable for the Mercer's conditions [38]. In the process of classification, the pixels are divided into the urban category. The initial samples in that category are combined as training samples for the urban reclassification. This iterative process is repeated until no new pixels for urban category are present during the separation. All of the pixels for urban category at this time are the final classification results.

The DMSP/OLS and SPOT-VGT data were applied to the data merging steps as follows. First, the DMSP/OLS and SPOT-VGT data were pretreated to obtain strong continuity and consistency of the DMSP/OLS and NDVI results. The DMSP/OLS results are based on the cumulative DN value and the good continuity principle of the intercalibration data. Second, the principle for the selection of the training samples was determined. Here, the DN values greater than or equal to 40 pixels were used as urban area training samples, that is, as seeds, and DN values less than 40 and NDVI values greater than or equal to 0.4 pixels were used as nonurban area training seed samples. Finally, the two types of data were combined. We used the SVM classification method to extract the urban information.

3.4. Analysis Method for Urban Expansion

3.4.1. Average Annual Rate of Change in the Urban Expansion Area

The average annual change rate of the urban expansion

area is the average growth rate of the urban area per year, as shown in formula (2).

$$V_T = \frac{S_B - S_A}{T}, \tag{2}$$

which represents the growth rate of the urban area in the T time period. S_A and S_B represent the urban area for times A and B, respectively, where time B follows time A.

3.4.2. Intensity of Urban Expansion. The Urbanization Intensification Index (UII) can be used to describe the speed and intensity of urban expansion. Large UII values are indicative of faster rates of expansion in urban areas and thus suggest that a city has assumed a large amount of nonurban space in the process of development. Conversely, small UII values are indicative of lower rates of expansion in urban areas and suggest that fewer nonurban areas have been taken.

Compared with the average annual rate of change in urban expansion, the UII introduces the regional urban area to carry out standardized treatments, which makes the expanding speed of different periods of urban area utilization more comparable; it also illustrates the relationship between urban area and nonurban area during the expansion of urban area. The formula for calculating the UII is as shown in formula (3).

$$\text{UII} = \frac{U_{ib} - U_{ia}}{TLA} \times \frac{1}{T} \times 100. \tag{3}$$

Among these terms, UII is the period of the urban expansion intensity index, U_{ib} represents the end point for the urban area, U_{ia} denotes the early urban area, TLA is the urban area within the study area, and T is the length of the study period, where the units are in years.

3.4.3. Dynamic Degree of Urban Expansion. The annual average rate of change in urban expansion reveals the change in urban expanding area only in terms of the absolute volume. The dynamic degree of urban expansion introduces the initial urban area in the annual average rate of change, which means that the calculation results can reflect the relative dynamic changing rate of urban expansion. The formula for calculating the dynamic degree of urban expansion is shown in formula (4).

$$K = \frac{U_{ib} - U_{ia}}{U_{ia}} \times \frac{1}{T} \times 100\%, \tag{4}$$

wherein K is the dynamic degree of urban expansion in the study time period, U_{ib} represents the urban area at the end, U_{ia} is the initial urban area, and T is the interval length of time, where the units are in years.

3.4.4. Spatial Characteristics of Urban Expansion. The spatial feature structure changes during the processes of development and expansion of a city. Gao et al. [39] divided the urban expansion in China into the following four types: (1) filling type, (2) epitaxial type, (3) enclave type, and (4) corridor type. Urban extension axes are an indispensable part of the spatial structure of urban expansion and can be divided into two types based on the angle of the extension of the urban area: (1) extension axes along land transportation lines and (2) extension axes along the river system. In the course of the actual expansion of the urban area, along the stretching axes formed two main types of urban space: centralized types of urban space and multiple grouping types of urban space.

4. Results

In this study, urban information for the HRB was extracted from 1998 to 2013; the results are shown in Figure 3. The urban system space-time evolution laws of the HRB over the 16 years period were analyzed from the perspective of global and regional aspects based on the extraction results.

4.1. Average Annual Change Rate for the Urban Expansion Area. The gradually increasing urban area was the most significant and intuitive feature of the urban expansion. The changes in urban area within the HRB are shown in Figure 4.

Overall, the urban area in the HRB increased between 1998 and 2013. The growth rate from 1998 to 2001 was slow, although the urban area showed a marked increase after 2001. The urban expansion increased linearly from 2001 to 2013.

As shown in Figure 4, from 1998 to 2013, the urban area's annual growth rate in the HRB did not maintain a rising trend. Specifically, the growth slowed between 1998 and 2001 and increased suddenly in 2001. Thereafter, the fast growth stage continued, and the growth rate reached 665.33 km^2 per year from 2001 to 2004, which was the largest value detected. The growth slightly decreased from 2004 to 2007; afterward, the growth continued at a rate of 550–650 km^2 per year. On the whole, the average annual growth rate of the HRB urban remained in a relatively stable state.

4.2. Urban Expansion Intensity. In this study, 33 urban areas were analyzed according to the urban expansion intensity, as shown in Table 1.

Table 1 shows that in most of the HRB, the UII values were minimum for 1998 to 2001 and maximum between 2007 and 2013. In Henan province, the UII values of Zhengzhou were obviously higher than those in the other cities for each period, and the index values for Xinyang were relatively low. In Anhui province, the UII reached 0.8 from 2001 to 2007 and, then, plunged to 0.05 between 2007 and 2010; the urban expansion intensity of the other regions was relatively stable. In Jiangsu province, Taizhou's UII values maintained a high trend from 1998 to 2013, and a peak of 2.05 appeared between 2010 and 2013. As a whole, the urban expansion intensity of Jiangsu province was slightly higher than that of the other three provinces from 1998 to 2013.

4.3. Dynamic Degree of Urban Expansion. According to Li et al. [40] and the situation in HRB, urban expansion in the basin was divided into five types: slow speed expansion, at ($K \leq 5$); low speed expansion, at ($5 < K \leq 10$); medium speed expansion, at ($10 < K \leq 20$); rapid speed expansion, at ($20 < K \leq 25$); and high speed expansion, at ($K > 25$).

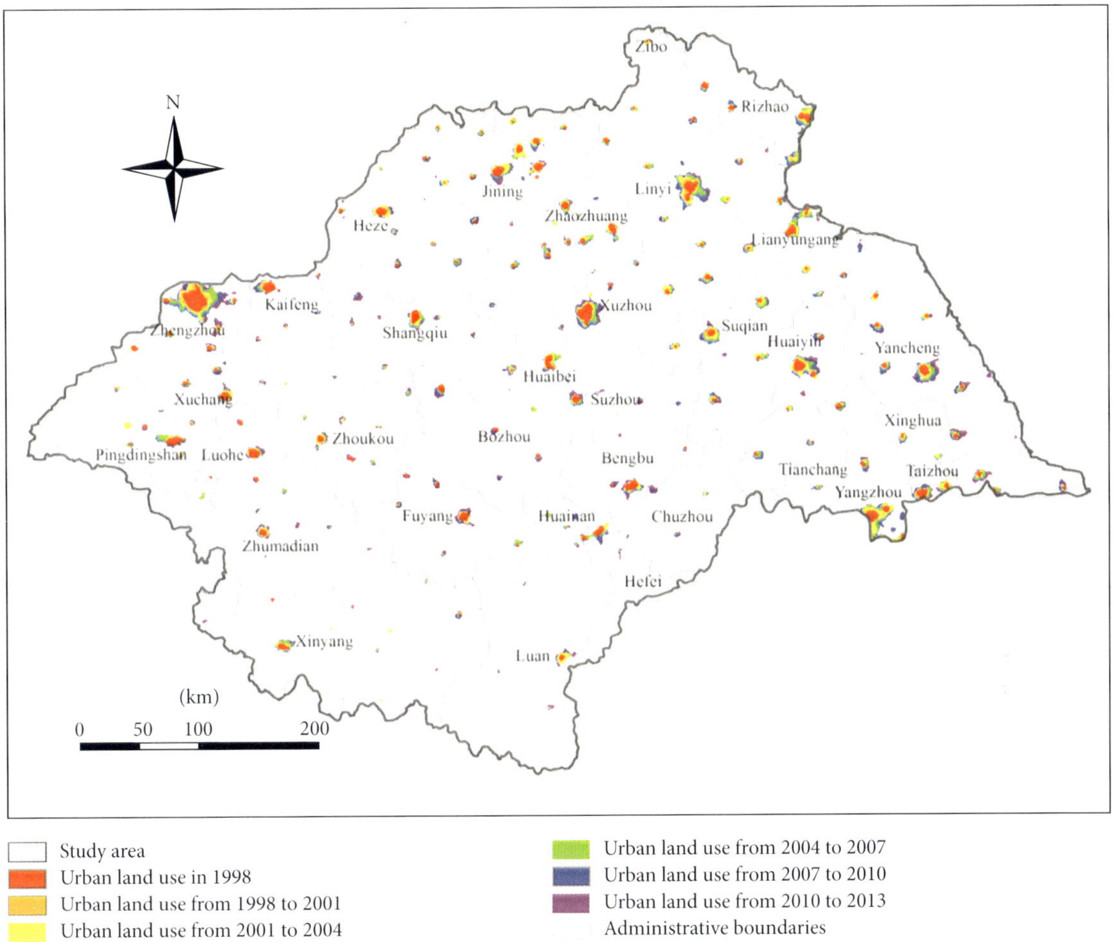

FIGURE 3: Urban information extraction results from 1998 to 2013 in HRB.

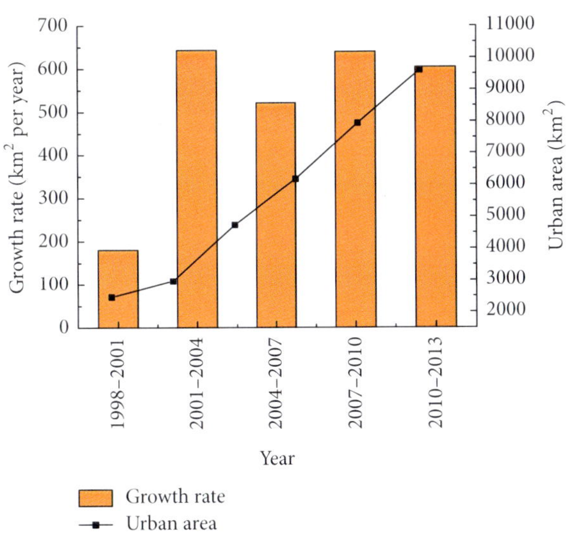

FIGURE 4: Changes in urban area between 1998 and 2013 in HRB.

As shown in Table 2, HRB's urban area was in slow expansion from 1998 to 2001, rapid expansion from 2001 to 2004, medium speed expansion from 2004 to 2007, and low speed expansion from 2007 to 2013, which may be related to reinforcement of the regional development planning regulation policy since 2000.

The dynamic degree of expansion in each city in different periods varied significantly. Henan was in low and medium speed expansion stages from 1998 to 2013. Anhui was in low and rapid expansion stages in 1998 to 2001 and 2001–2004, respectively; afterward, medium speed expansion was maintained. Jiangsu was in low and high speed expansion stages in 1998–2001 and 2001–2004; afterward, medium speed expansion was maintained. Shandong was in a medium speed expansion stage from 1998 to 2001; afterward, slow expansion was maintained.

Although the entire HRB and each of its provinces showed low and medium speed expansion, some cities were in the high expansion stage during particular periods, such as Tianchang from 1998 to 2001; Zhumadian, Xinyang, Pingdingshan, Liuan, Huainan, Yancheng, Suqian, Huaian, Yangzhou, Xinghua, Nantong, and Zaozhuang from 2001 to 2004; Chuzhou and Tianchang from 2004 to 2007; Hefei and Chuzhou from 2007 to 2010; and Shangqiu, Zhoukou, Zhumadian, Suzhou, Liuan, Hefei, and Chuzhou from 2010 to 2013. This phenomenon indicates that the overall growth of HRB was not high, although some cities experienced a high

Table 1: Urban expansion intensity index between 1998 and 2013 in HRB.

	1998–2001	2001–2004	2004–2007	2007–2010	2010–2013
Henan Province	*0.06*	*0.21*	*0.16*	*0.14*	*0.29*
Zhengzhou	0.64	1.21	1.00	0.94	1.08
Kaifeng	0.07	0.10	0.06	0.20	0.31
Shangqiu	0.01	0.16	0.19	0.08	0.48
Zhoukou	0.01	0.09	0.07	0.03	0.31
Zhumadian	0.01	0.08	0.04	0.04	0.22
Xinyang	0.01	0.09	0.06	0.03	0.07
Luohe	0.12	0.03	0.11	0.21	0.53
Xuchang	0.01	0.30	0.12	0.33	0.27
Pingdingshan	0.06	0.33	0.25	0.05	0.09
Anhui Province	*0.03*	*0.11*	*0.10*	*0.13*	*0.22*
Suzhou	0.02	0.07	0.06	0.12	0.33
Huaibei	0.18	0.23	0.22	0.18	0.15
Bozhou	0.01	0.02	0.08	0.16	0.20
Fuyang	0.04	0.10	0.17	0.13	0.29
Luan	0.01	0.12	0.02	0.03	0.15
Hefei	0.00	0.00	0.03	0.22	0.19
Chuzhou	0.00	0.02	0.02	0.13	0.16
Bengbu	0.07	0.20	0.07	0.28	0.29
Huainan	0.08	0.82	0.85	0.05	0.10
Tianchang	0.04	0.04	0.11	0.16	0.11
Jiangsu Province	*0.07*	*0.43*	*0.33*	*0.46*	*0.27*
Xuzhou	0.01	0.36	0.32	0.55	0.19
Lianyungang	0.20	0.36	0.44	0.29	0.02
Yancheng	0.01	0.28	0.19	0.46	0.46
Suqian	0.05	0.36	0.38	0.24	0.29
Huaian	0.05	0.38	0.13	0.67	0.12
Yangzhou	0.11	1.10	0.57	0.59	0.04
Xinhua	0.03	0.34	0.17	0.23	0.03
Taizhou	1.01	1.70	1.35	0.85	2.05
Nantong	0.04	0.32	0.53	0.32	0.90
Shandong Province	*0.15*	*0.31*	*0.28*	*0.38*	*0.11*
Heze	0.17	0.10	0.12	0.29	0.09
Jining	0.05	0.45	0.17	0.44	0.12
Zhaozhuang	0.09	0.57	0.49	0.06	0.10
Linyi	0.16	0.23	0.39	0.40	0.14
Rizhao	0.36	0.52	0.32	0.41	0.25

growth rate, rapidly developing urbanization, and emerging urban regions.

4.4. Spatial Characteristics of Urban Expansion. As shown in Figure 4, the expansion of most urban areas in the HRB in 1998 to 2013 was extensive type; that is, the urban expansion occurred at the original edge of the city. In addition, new urban regions were identified around expanding cities such as Zhengzhou, Linyi, Suqian, Yancheng, and Xuzhou, which reveal the features of an enclave. In Pingdingshan and Lianyungang, urban expansion as appeared to be corridor type; that is, linear extension occurred along a particular direction, usually for traffic lines or river systems. Within the extent of the HRB, the spatial form of most expanding cities appeared to be block-shaped. In addition, the extension shaft of Lianyungang and Pingdingshan was long and extended along a particular direction with the entire city assuming a strip form. Linyi urban expansion was multidirectional and appeared to be star-like. In Huaian and Yangzhou, urban expansion in 1998 and 2001 translated into massive urban areas in 2001 and 2004. Jining had multiple independent urban areas, revealing the trend of connecting urban outer contour with each increasing urban range.

TABLE 2: Dynamic degree of urban expansion between 1998 and 2013 in HRB.

	1998–2001	2001–2004	2004–2007	2007–2010	2010–2013
Henan Province	*6.70*	*17.83*	*9.15*	*6.15*	*10.95*
Zhengzhou	10.30	14.89	8.53	6.37	6.13
Kaifeng	5.88	7.50	3.40	10.49	12.68
Shangqiu	1.09	23.28	17.76	4.88	25.71
Zhoukou	3.18	21.74	10.97	3.30	31.23
Zhumadian	3.45	33.33	8.33	6.67	29.17
Xinyang	6.86	40.65	12.45	5.07	8.33
Luohe	4.35	0.86	3.33	6.06	12.82
Xuchang	0.46	20.27	5.04	11.92	7.17
Pingdingshan	7.88	36.77	13.29	2.00	3.30
Anhui Province	*7.98*	*24.29*	*12.42*	*11.76*	*15.00*
Suzhou	10.00	24.36	12.59	18.82	32.65
Huaibei	8.33	8.57	6.44	4.44	3.36
Bozhou	3.45	5.21	18.02	23.98	17.69
Fuyang	10.00	18.59	20.99	9.85	16.76
Luan	7.69	112.50	3.81	5.13	25.19
Hefei	0.00	0.00	0.00	222.22	26.09
Chuzhou	0.00	0.00	33.33	93.33	29.83
Bengbu	8.16	18.58	4.21	14.33	10.68
Huainan	5.75	48.04	20.48	0.75	1.46
Tianchang	33.33	16.67	33.33	25.00	9.52
Jiangsu Province	*5.05*	*27.27*	*11.40*	*11.91*	*5.27*
Xuzhou	0.31	18.04	10.42	13.38	3.29
Lianyungang	15.94	19.36	15.04	6.84	0.36
Yancheng	1.33	31.80	10.63	20.20	12.45
Suqian	7.78	42.34	20.04	7.68	7.75
Huaian	4.02	26.41	5.01	22.39	2.31
Yangzhou	4.62	41.44	9.54	7.65	0.38
Xinhua	7.41	72.73	11.43	11.35	1.06
Taizhou	12.94	15.77	8.52	4.26	9.11
Nantong	3.33	26.26	24.86	8.41	19.12
Shandong Province	*11.85*	*18.38*	*10.63*	*10.36*	*2.92*
Heze	24.89	8.65	7.88	16.01	3.42
Jining	2.94	23.97	5.36	11.70	2.42
Zhaozhuang	5.02	28.97	13.38	1.21	1.90
Linyi	14.84	14.53	17.36	11.84	3.06
Rizhao	23.89	20.07	7.68	7.88	3.98

From the perspective of a different province, the expansion degree in Zhengzhou was the most rapid in Henan province; that in Anhui province was average. The expansion degree in Xuzhou was the most rapid in Jiangsu province, and that in Jining and Linyi was higher than the expansion degree in other cities in Shandong Province.

5. Discussions

During the study period, built-up areas increased rapidly, and the distribution urban expansion was not balanced both temporally and spatially. For example, the value at southeast coast was higher than that in other areas and peaked in 2013. These results are in agreement with the basic trend of China's urban expansion. Considering the dynamic expansion, the overall expansion of the HRB and the provinces were low and medium speeds. Overall, in addition to that in Zhengzhou, Linyi, and Xuzhou, urban expansion between 1998 and 2013 in the rest of the urban areas of HRB was not huge. After development, the urban distribution was more dispersed; the urban density was not high, and the study area did not form a complete urban agglomeration.

Table 3: Elastic coefficient of residential expansion and expanding type of city.

Year	Built-up area	Growth rate of nonagricultural population	Elasticity coefficient	Expansion type
1998–2001	0.90%	7.25%	0.12	Insufficient expansion
2001–2004	54.12%	23.74%	2.28	Rapid expansion
2004–2007	31.25%	12.55%	2.49	Rapid expansion
2007–2010	40.13%	19.32%	2.08	Rapid expansion
2010–2013	9.87%	75.61%	0.13	Insufficient expansion

The expansion of Xuzhou from 1998 to 2013 can be divided into three stages; 1998–2004 was a low speed development stage. Although the urban areas expand annually, the elasticity coefficient of the urban areas increased but still did not meet the nonagricultural population growth. 2004–2010 was a rapid development stage. The elasticity coefficient of urban growth jumped to 2.49 (Table 3), which is classified as rapid expansion. This indicates that the speed of construction land expansion was too fast in this period, which opened the national real estate market and created a relationship with economic construction. However, the uncoordinated growth of construction land and nonagricultural population led to an unreasonable land use pattern, which to some extent increased the carrying capacity of land to result in agricultural land loss. In 2010–2013, the relative relationship between the rate of urban expansion and the growth rate of the nonagricultural population was more harmonious, although it was still slightly higher than the population growth rate.

This paper analyzed the correlation of urban expansion from the aspects of population and economy. The urban nonagricultural population and the GDP (Gross Domestic Product) data characterize the urban population and economic conditions. SPSS (Statistical Product and Service Solutions) was used to analyze the correlation among built-up areas, populations, and the GDP. The results show that the correlation coefficient between the built-up areas and the GDP was 0.89 with significance over 95%, which indicates that Xuzhou had a strong positive correlation between GDP and built-up areas. Population growth is an important driving force for urban expansion. An increase in urban population has strongly promoted the construction of commercial and residential buildings, the construction of public transport facilities and urban public infrastructure, and the development of urban secondary and tertiary industries, thus accelerating the expansion of built-up areas and promoting the process of urban expansion. On the contrary, the expansion of the built-up areas attracts additional labor to engage in nonagricultural activities, resulting in a corresponding increase in the number of nonagricultural population. The correlation coefficient between built-up areas and nonagricultural population in Xuzhou was 0.88 with significance more than 95%; that is, the nonagricultural population in Xuzhou strongly correlated with the built-up areas. These results show that the built-up areas in Xuzhou between 1998 and 2013 correlated strongly with the nonagricultural population and the GDP, which shows that the expansion of population, economy, and built-up areas is mutually influenced and promoted. Economic development and the increase in nonagricultural population has greatly promoted the expansion of built-up area.

It should be noted that, owing to the complexity of the surface features, some nonurban pixel light noises remained in the nighttime light data. Furthermore, geometric correction was not adopted; therefore, error can be introduced in the location and false positives can be generated. Accordingly, the geometric correction of light data should be studied to improve the accuracy of information extraction in the future. The HRB includes characteristics such as a high population density and an obvious contradiction between the water supply and demand. Considering these features in disaster-prone areas, follow-up research can explore various natural factors, the relationship between economic and political factors, and the development of urban areas in the HRB to analyze the driving forces on urban expansion.

6. Conclusions

This study used remote sensing data and processed night light data and NDVI data in a long time series. The SVM classification method was used to extract urban information, and urban expansion rules were analyzed in the HRB. The main conclusions are given in the following points:

(1) DMSP/OLS night light data and SPOT-VGT data were selected to examine urban expansion. The SVM classification method was used for extracting urban information.

(2) The urban expansion processes between 1998 and 2013 in the HRB were reconstructed, and quantitative analysis was performed from the aspects of urban area size, expansion intensity, expansion dynamic degree, and expansion pattern.

Conflicts of Interest

The authors declare that there are no conflicts of interest regarding the publication of this paper.

Acknowledgments

This work is supported by the Human and Social Sciences of Chinese Ministry of Education (Grant no. 14YJCZH028) and the National Natural Science Foundation of China (Grant no. 41571400\41401171\41671405).

References

[1] B. Catalán, D. Saurí, and P. Serra, "Urban sprawl in the Mediterranean?: patterns of growth and change in the Barcelona Metropolitan Region 1993–2000," *Landscape and Urban Planning*, vol. 85, no. 3-4, pp. 174–184, 2008.

[2] G. Mills, "Cities as agents of global change," *International Journal of Climatology*, vol. 27, no. 14, pp. 1849–1857, 2007.

[3] S. Angel, S. C. Sheppard, and D. Civco, *The Dynamics of Global Urban Expansion*, The World Bank Transportation and Urban Development Department, Washington, DC, USA, 2005.

[4] L. Tu, Z. Qin, W. Li et al., "Surface urban heat island effect and its relationship with urban expansion in Nanjing, China," *Journal of Applied Remote Sensing*, vol. 10, no. 2, Article ID 026037, 2016.

[5] C. L. Fang, "The urbanization and urban development in china after the reform and opening-up," *Economic Geograph*, vol. 29, pp. 19–25, 2009 (Chinese).

[6] China's State Council, "Chinese new urbanization plan (2014–2020)," http://www.gov.cn/zhengce/2014-03/16/content_2640075.htm.

[7] P. Griffiths, P. Hostert, O. Gruebner, and S. V. der Linden, "Mapping megacity growth with multi-sensor data," *Remote Sensing of Environment*, vol. 114, no. 2, pp. 426–439, 2010.

[8] B. S. Sokhi, N. D. Sharma, and P. S. Uttarwar, "Satellite remote sensing in urban sprawl mapping & monitoring a case study of Delhi," *Journal of the Indian Society of Remote Sensing*, vol. 17, no. 3, pp. 57–69, 1989.

[9] F. M. Henderson and Z.-G. Xia, "SAR applications in human settlement detection, population estimation and urban land use pattern analysis: a status report," *IEEE Transactions on Geoscience and Remote Sensing*, vol. 35, no. 1, pp. 79–85, 1997.

[10] M. T. Rahman, "Detection of land use/land cover changes and urban sprawl in Al-Khobar, Saudi Arabia: an analysis of multi-temporal remote sensing data," *ISPRS International Journal of Geo-Information*, vol. 5, no. 2, article 15, 2016.

[11] C. Weber and A. Puissant, "Urbanization pressure and modeling of urban growth: example of the Tunis Metropolitan Area," *Remote Sensing of Environment*, vol. 86, no. 3, pp. 341–352, 2003.

[12] H. Luo, L. Li, H. Zhu, X. Kuai, Z. Zhang, and Y. Liu, "Land cover extraction from high resolution ZY-3 satellite imagery using ontology-based method," *ISPRS International Journal of Geo-Information*, vol. 5, no. 3, article 31, 2016.

[13] B. Rodríguez-Cuenca, S. García-Cortés, C. Ordóñez, and M. A. Alonso, "Morphological operations to extract urban curbs in 3D MLS point clouds," *ISPRS International Journal of Geo-Information*, vol. 5, no. 6, p. 93, 2016.

[14] Z. Zhang, P. Tang, H. Y. Li, and Z. Feng, "Refined domain model for multisource data synergized quantitative remote sensing production system," *Journal of Remote Sensing*, vol. 20, no. 2, pp. 184–196, 2016 (Chinese).

[15] M. K. Jat, P. K. Garg, and D. Khare, "Monitoring and modelling of urban sprawl using remote sensing and GIS techniques," *International Journal of Applied Earth Observation and Geoinformation*, vol. 10, no. 1, pp. 26–43, 2008.

[16] H. Taubenböck, T. Esch, A. Felbier, M. Wiesner, A. Roth, and S. Dech, "Monitoring urbanization in mega cities from space," *Remote Sensing of Environment*, vol. 117, pp. 162–176, 2012.

[17] X. H. Wang, *Urban Expansion in China over the Past 30 Years Detected Using DMSP/OLS Nighttime Light Data*, Nanjing University, Nanjing, China, 2013 (Chinese).

[18] I. Ciesin, "CIAT Global Rural-Urban Mapping Project (GRUMP)," Alpha Version: Population grids, 2004.

[19] D. Li and X. Li, "An overview on data mining of nighttime light remote sensing," *Acta Geodaetica et Cartographica Sinica*, vol. 44, no. 6, pp. 591–601, 2015 (Chinese).

[20] T. A. Croft, "Nighttime images of the earth from space," *Scientific American*, vol. 8, pp. 86–98, 1978.

[21] M. L. Imhoff, W. T. Lawrence, D. C. Stutzer, and C. D. Elvidge, "A technique for using composite DMSP/OLS 'city lights' satellite data to map urban area," *Remote Sensing of Environment*, vol. 61, no. 3, pp. 361–370, 1997.

[22] C. Milesi, C. D. Elvidge, R. R. Nemani, and S. W. Running, "Assessing the environmental impacts of human settlements using satellite data," *Management of Environmental Quality: An International Journal*, vol. 14, no. 1, pp. 99–107, 2003.

[23] C. Y. He, J. G. Li, J. Chen et al., "the urbanization model and process in Bohai surrounding area in the 1990s by using DMSP/OLS data," *Acta Geographica Sinica*, vol. 60, pp. 409–417, 2005 (Chinese).

[24] Z. Liu, C. He, Q. Zhang, Q. Huang, and Y. Yang, "Extracting the dynamics of urban expansion in China using DMSP-OLS nighttime light data from 1992 to 2008," *Landscape and Urban Planning*, vol. 106, no. 1, pp. 62–72, 2012.

[25] B. Pandey, P. K. Joshi, and K. C. Seto, "Monitoring urbanization dynamics in india using DMSP/OLS night time lights and SPOT-VGT data," *International Journal of Applied Earth Observation and Geoinformation*, vol. 23, no. 1, pp. 49–61, 2013.

[26] Y. Megahed, P. Cabral, J. Silva, and M. Caetano, "Land cover mapping analysis and urban growth modelling using remote sensing techniques in greater Cairo region—Egypt," *ISPRS International Journal of Geo-Information*, vol. 4, no. 3, pp. 1750–1769, 2015.

[27] Y. Yang, C. Y. He, Y. Zhao, T. Li, and Y. Qiao, "Research on the layered threshold method for extracting urban land using the DMSP/OLS stable night time light data," *Journal of Image and Graphics*, vol. 16, no. 4, pp. 666–673, 2011 (Chinese).

[28] X. Cao, J. Chen, H. Imura, and O. Higashi, "A SVM-based method to extract urban areas from DMSP-OLS and SPOT VGT data," *Remote Sensing of Environment*, vol. 113, no. 10, pp. 2205–2209, 2009.

[29] Y. Zhou, L. Liu, Y. Lu, Z. Ma, F. Xia, and Z. Shi, "Regional scale mapping of soil organic matter using remote sensing and visible-near infrared spectroscopy," *Journal of Remote Sensing*, vol. 19, no. 6, pp. 998–1006, 2015 (Chinese).

[30] A. Schneider, M. A. Friedl, and C. E. Woodcock, "Mapping urban areas by fusing multiple sources of coarse resolution remotely sensed data," *Photogrammetric Engineering & Remote Sensing*, vol. 4, pp. 2623–2625, 2003.

[31] D. Lu, H. Tian, G. Zhou, and H. Ge, "Regional mapping of human settlements in southeastern China with multisensor remotely sensed data," *Remote Sensing of Environment*, vol. 112, no. 9, pp. 3668–3679, 2008.

[32] C. D. Elvidge, D. Ziskin, K. E. Baugh et al., "A fifteen year record of global natural gas flaring derived from satellite data," *Energies*, vol. 2, no. 3, pp. 595–622, 2009.

[33] G. Yang, H. F. Shen, L. P. Zhang, Z. He, and X. Li, "A moving weighted harmonic analysis method for reconstructing high-quality SPOT VEGETATION NDVI time-series data," *IEEE Transactions on Geoscience and Remote Sensing*, vol. 53, no. 11, pp. 6008–6021, 2015.

[34] The Survey and Research Group with the Central Committee of China Zhi Gong Party and The Central Committee of China Zhi Gong Party, "Research report on economic development

and environmental protection of Huaihe River Basin," *China Development*, vol. 15, pp. 1–7, 2015 (Chinese).

[35] NCGIA (National Center for Geographic Information & Analysis), https://ngdc.noaa.gov/eog/download.html.

[36] Flemish Institute for Technological Research, http://free.vgt.vito.be.

[37] Geospatial Data Cloud, (Chinese) http://www.gscloud.cn.

[38] V. N. Vapnik, *The Nature of Statistical Learning Theory*, Springer, 1995.

[39] J. L. Gao, J. J. Chen, and X. Su, "Urban expansion and its driving mechanism in China: from three main shools' perspectives," *Progress in Geography*, vol. 32, pp. 743–754, 2013 (Chinese).

[40] X. W. Li, J. Y. Fang, and S. L. Piao, "The intensity and modes of urban landuse growth in Shanghai," *Journal of Natural Resources*, vol. 18, pp. 412–422, 2003 (Chinese).

16

Application of Fractional Differential Calculation in Pretreatment of Saline Soil Hyperspectral Reflectance Data

Anhong Tian,[1,2] Junsan Zhao,[1] Heigang Xiong,[3,4] Shu Gan,[1] and Chengbiao Fu[1,2]

[1]*Faculty of Land Resource Engineering, Kunming University of Science and Technology, Kunming, Yunnan 650093, China*
[2]*College of Information Engineering, Qujing Normal University, Qujing, Yunnan 655011, China*
[3]*College of Applied Arts and Science, Beijing Union University, Beijing 100083, China*
[4]*College of Resource and Environment Sciences, Xinjiang University, Urumqi, Xinjiang 830046, China*

Correspondence should be addressed to Junsan Zhao; jszht@kmust.edu.cn

Guest Editor: Christine Pohl

Pretreatment of spectrum data is a necessary and effective method for improving the accuracy of hyperspectral model building. Traditional differential calculation pretreatment only considers the integer-order differential, such as the 1st order and 2nd order, and overlooks important spectrum information located at fractional order. Since fractional differential (FD) has rarely been applied to spectrum field measurement, there are few reports on the spectrum features of saline soils under different degrees of human interference. We used FD to analyze field spectrum data of saline soil collected from Xinjiang's Fukang City. Order interval of 0.2 was adopted to divide 0–2 orders into 11-order differentials. MATLAB programming was used to convert the raw spectral reflectance and its four common types of mathematics and to conduct FD calculation. Spectrum data for area A (no human interference area) and area B (human interference area) was separately processed. From the statistical analysis of soil salinization characteristics, the salinization degree and type for area B were more diverse and complicated than area A. MATLAB simulation results showed that FD calculation could depict the minute differences between different FD order spectra under different human interference areas. The overall differential value trend appeared to move towards 0 reflectance value. After differential processing, the trend of bands that passed the 0.05 significance test of correlation coefficient (CC) showed an increase first then decrease later. The maximum CC absolute value for five spectrum transformations all appeared in the fractional order. FD calculation could significantly increase the correlation between spectral reflectance and soil salt content both for full-band and single-band spectra. Results of this study can serve as a reference for the application of FD in field hyperspectral monitoring of soil salinization for different human interference areas.

1. Introduction

Pretreatment of spectrum data is a very effective way of increasing the precision of hyperspectral model building. Spectrum differential (derivative) technology is a very valuable analytical method in hyperspectral data pretreatment that is often applied to increasing the strength of spectrum signals [1], which can quickly determine the inflection point and maximum and minimum reflectivity band positions. Studies have shown that the first-order differential can eliminate the influence of a partially linear or near-linear noise spectrum on the target spectrum [2]. The second-order differential can effectively eliminate the baseline drift and background signal and improve the pretreatment accuracy [3]. However, in the study of monitoring soil salinization by hyperspectral techniques, most of the spectral data are preprocessed based on the integer-order (1st order or 2nd order) differential algorithm to enhance the correlation between partial bands' reflectivity and surface parameters. But some studies have pointed out that the spectral reflectance curve of the first-order and second-order differential transformation differs greatly from the original spectral reflectance curve [4]. Only considering the integer-order differential transformation ignores the important spectral information located in the fractional order, resulting in a problem of reduced modeling accuracy. At the same time, many systems in reality belong to the fractional order. When the model is expressed by integer differential, the accuracy is not ideal.

At present, FD has a small amount of research in the field of spectral analysis. FD was first used in diffuse reflectance spectroscopy. In recent years, FD has been used in public datasets such as diesel, corn, and wheat, as well as to analyze the spectral pretreatment effects of Xinjiang's saline soil in laboratory environments. For example, Schmitt [5] used fractional differentiation to process the diffuse-reflectance spectra; simulations revealed that fractional derivative spectroscopy separated the overlapping peaks between spectra and deleted baseline variations. Also, due to the features of memory and nonlocality of FD, it was allowed to control the sensitivity to the slope and curvature of spectrum curve according to the weight of different differential orders; this offered more range for band selection in regression modeling. Kharintsev and Salakhov [6] adopted the FD method to acquire the spectrum parameters for overlapping wavelength bands, in order to overcome the high variance features of spectrum parameters for the tradition ordinary least squares method. Also, a statistical regularization method is used to determine the stable and unbiased estimation for fractional differentiation and the overlapping wavelength bands identified by a Tsallis distribution model. Results showed that fractional derivatives could be effectively used in the field of spectral analysis. Zheng et al. [7] proposed a fractional-order Savitzky-Golay differentiation (FOSGD) method on the basis of Riemann-Liouville (R-L) fractional calculus theory; it was used to preprocess the near-infrared (NIR) spectrum datasets for corn data, diesel data, and wheat data. FOSGD could increase the model performance for least squares regression, and the root mean square error (RMSE) was smaller than integral differential, especially for the non-concentration indexes such as viscosity, density, and hardness. In the field of soil spectral analysis, Zhang et al. [8] first used the FD algorithm to preprocess soil spectral data in the laboratory. Zhang et al. collected the saline soils in Yutian Oasis of Xinjiang, China, and analyzed the correlation between soil salinity and spectra by a Grünwald-Letnikov (G-L) FD algorithm. Simulation showed that the G-L algorithm enriched the preprocessing method of hyperspectral data from the perspective of fractional order. The depth of the deep mining of the potential information in the spectral data could greatly raise the correlation between salt content and the spectra and the CC obtained at the fractional order instead of the traditional integer order (1st order, 2nd order). Also, this novel pretreatment method for saline soils could prevent the loss of important information caused by only considering the traditional integer order. Xia et al. [9] took the 43 saline soil samples of Ebinur Lake in Xinjiang, China, as data sources, and analyzed the correlation between original spectral reflectance and electrical conductivity. Results showed that fractional derivate could improve the CC for some single bands, such as 505 nm, 2239 nm, and 2443 nm. Also, it could detail the change trends between integer and fractional curve.

In addition, the type of human activities, level, and the duration has an important impact on soil feature and heterogeneity [10]. The changes of regional salt are not only deeply affected by global warming but also by the interaction of regional environmental elements and human activities [11]. The impact of human activities on the soil is conscious and purposeful. The influence of human activities on soil development can be in two ways, that is, the soil can be developed in a direction of benign circulation through rational use of soil and soil degradation can be caused by irrational use. Therefore, different degrees of human activities will inevitably lead to differences in regional soil morphological characteristics (structure, texture, tightness, porosity, dry humidity, etc.), triggering changes in salt transfer channels and salt transmit rates and thus increasing the complexity of salt spatial variability of soil salinity. As a result, the time-space dynamics of the oasis salt and the changes in the ecological environment are also very different. Therefore, the relationship between land use change and soil salinization is the focus of current attention. Soil salinization is an important environmental problem in the arid and semiarid areas of Xinjiang [12, 13]. Soil salinization causes decreased fertility, pH imbalance, and degradation of the land, which severely restricts the development of China's agricultural economy [13, 14]. Because hyperspectral remote sensing technology has many spectrum reflectance bands, has high spectral resolution, and does not damage the testing sample [15–18], it has been widely used in hyperspectral data study on saline soil under human interference [19, 20]. However, hyperspectral remote sensing has been rarely used in study on correlation between saline soil features under different levels of human interference and their spectrum features.

However, these aforementioned soil studies are focused on soils under human interference and the soil spectrum reflectance value was obtained in ideal indoor environment; the solar light source was simulated by a halogen lamp, without considering the influence of actual environmental factors in the field, and the inversion model established by indoor spectra may not be directly applicable to field remote sensing inversion. Therefore, the construction of soil physical and chemical properties' inversion model based on field spectroscopy is necessary. Field spectral measurements use the sun as a source of light to ensure equivalence with remote sensing image acquisition [21]. Currently FD has rarely been reported to field spectrum measurement. Thus, we used saline soil from areas with and without human interference in Xinjiang's Fukang City as the subject to explore the pretreatment effect of FD for field spectrum measurements. MATLAB R2015a software was used for FD calculation. Spectrum data from area A (no human interference) and area B (with human interference) were processed. For the order interval, 0.2 was used. The raw spectrum and the four types of transformation were put through $0–2^{nd}$-order differential processing with a total of 11 orders. The application prospects of the FD algorithm in the pretreatment of field measured spectra under different human interference levels are discussed, which provides a new idea for studying the temporal and spatial variations of salt in oasis in the arid regions.

2. Study Area and Data

2.1. Study Area. The study area is located between the southern edge of Dzungarian Basin and the Northern Foothill of

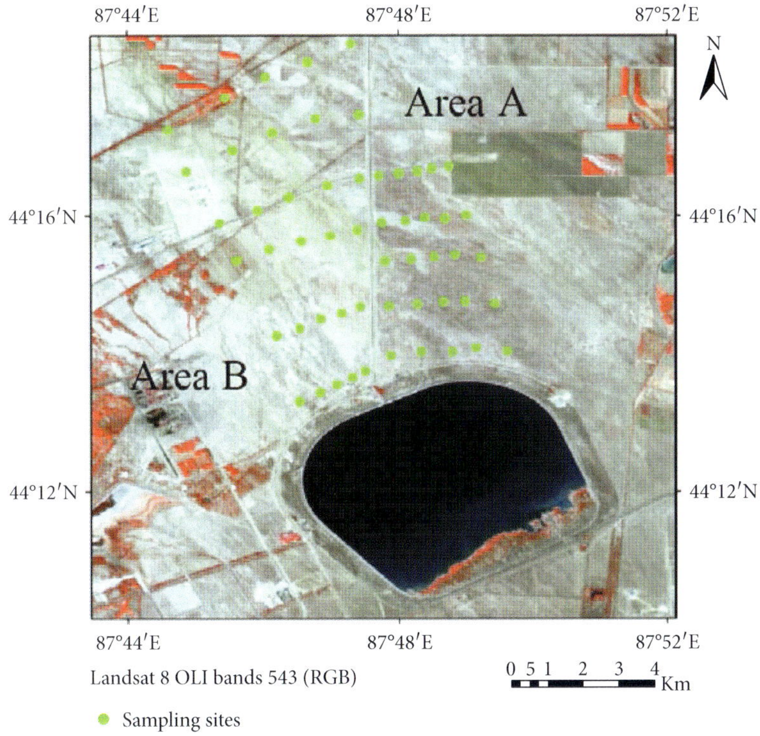

FIGURE 1: Location of the study area and the distribution of soil samples.

(a) (b)

FIGURE 2: Sampling environment for area A: (a) one, (b) two.

Xinjiang's Tianshan Mountain (87°44′–88°46′E, 43°29′–45°45′N). This area has temperate continental climate. Thus, this area has hot summers and cold winters, and the day and night temperatures differ significantly. This area has ample light and heat, and the annual average temperature is 6.7°C. However, this area has little rainfall, and the annual rainfall is around 164 mm. Five east-west running sampling lines with a 600–800-meter gap in between were established in area A (starting from the northern part of area A to the southern part of area A). Six sampling lines with a 800–1000-meter gap were established in area B. Five representative sampling points were selected for each sampling line. The intervals between the sampling points were 300–500 meters. GPS positioning was implemented for the 55 sampling points, as shown in Figure 1. Overall, 25 sampling points were used for area A and 30 sampling points were used for area B.

2.2. Division of Areas with Different Interference Levels. The sampling area is the alluvial fan edge located southeast of Xinjiang's Fukang City 102 Army Corp. There is a north-south travelling canal laid with impermeable membrane in the center of this area that is 24 m wide on top, 6 m wide on the bottom, 5 m deep, and 15.30 km long. Thus, we used this canal as the boundary to divide the research area into areas A and B for sampling. Area A is located further away from human habitat and is segregated by the canal; therefore, it is not affected by human activity. The soil is an alkali soil with sparse Haloxylon ammodendron, Tamarix ramosissima, and Ceratocarpus arenarius growth, as shown in Figure 2. Testing showed that the soil salinization was more severe and the pH value was approximately 7.76–8.98. The salt content of the top layer [10] was 5.34–44.45 g/kg.

Area B is near the 102 Army Corp and experiences strong human activities. The land has mostly been developed, and

(a) (b)

FIGURE 3: Sampling environment for area B: (a) one, (b) two.

the development method is primarily artificial Haloxylon ammodendron woodland, seedling cultivation land, and elm woods, as shown in Figure 3. Because area B is regular woodland, it has only been tilled but does not have fertilizer added. Therefore, the research area has been divided into two categories based on the threat level of human activities on the oasis ecological environment, area A with no human interference and area B with human interference.

2.3. Soil Sample Collection and Field Spectrum Measurement. The sample collection time was in the middle of May in 2017. A portable FieldSpec® 3 Hi-Res spectrum spectrometer produced by American ASD Corporation was used to obtain spectrum data from the soil test. The spectrum effective range for this instrument is 350–2500 nm. The spectrum sampling interval is 1.4 nm for the 350–1000 nm range and 2 nm for the 1000–2500 nm range. The resampling interval is 1 nm. ASD ViewSpec Pro was used as the analysis software. Because weather conditions can affect spectrum testing, the spectrum test for this study was conducted between 11:00 and 15:00 hours local time to reduce data error caused by weather factors. The spectrum test was conducted during times when the sky was clear with no cloud and no wind. Before each spectrum test, the spectrum spectrometer was conducted on the whiteboard calibration to remove effects from dark current. The probe of the spectrum spectrometer was placed 15 cm directly above the sampling surface to collect the spectrum. To prevent surface cracks and surrounding plants from affecting the soil spectrum test, each spectrum test sampling point was chosen to be away from items that can interfere with the soil spectrum. Spectrum from five locations within 1 meter of the sampling point with a similar soil background value was collected. Each point was tested 10 times to obtain 10 spectrum curves, with a result of 50 curves for each sampling location. The mean of these spectrum tests was used as the actual tested spectrum value for each sampling point.

While conducting the spectrum field measurement, 0–20 cm of the soil sample was collected from each sampling point and placed in a numbered bag. Once taken back to the laboratory, the samples were air dried naturally and had impurities removed. After sieving through 1 mm holes, the samples were sent to the Xinjiang Institute of Ecology and Geography (Chinese Academy of Sciences) to test the soil salt content by professional personnel.

2.4. Spectrum Data Analysis. At first, to reduce the effect from noise, peripheral bands from 350 to 399 nm and 2401 to 2500 nm with lower signal noise were removed. Bands in the moisture absorption segment (1355–1410 nm and 1820–1942 nm) were also removed. Moreover, Savitzky-Golay filtering was then used to conduct spectrum smoothing to remove noise. At last, to reduce the effects of atmospheric background noise, four different types of mathematical transformations are used for the raw spectral reflectance (R), that is, root mean square ($\sqrt{R}$) transformation, reciprocal ($1/R$) transformation, logarithmic reciprocal ($1/\log R$) transformation, and logarithm ($\log R$) transformation.

3. Methodology

3.1. Comparison of Fractional Differential and Integer Differential. According to the theoretical knowledge of integer-order calculus, the function $f(x)$ has n-order continuous derivative characteristics and h is an increment. The first-order, second-order, and third-order derivatives of function $f(x)$ can be expressed as

$$f'(x) = \frac{df}{dx} = \lim_{h\to 0}\frac{f(x) - f(x-h)}{h},$$
$$f''(x) = \left(\frac{df}{dx}\right)' = \lim_{h\to 0}\frac{f(x) - 2f(x-h) + f(x-2h)}{h^2},$$
$$f'''(x) = \frac{d^3 f}{dx^3} = \lim_{h\to 0}\frac{f(x) - 3f(x-h) + 3f(x-2h) - 3f(x-3h)}{h^3}. \qquad (1)$$

FD has been successfully applied to system model building, signal filter, pattern recognition, and fractal theory [22–24]. This is mainly because in these applications, the fractional-order model has more accurate results than the original integer differential model. FD mainly has three forms of expressions [8, 9]: Riemann-Liouville (R-L), Grünwald-

Table 1: Salinization degree and soil sample classification statistics.

Salinization degree	Salt content (g/kg)	Area A			Area B		
		Samples	Proportion (%)	Mean (g/kg)	Samples	Proportion (%)	Mean (g/kg)
Nonsaline soil	<3.0	0	0.000%	0	1	3.333%	2.700
Mild saline soil	3.0~5.0	0	0.000%	0	2	6.667%	3.775
Moderate saline soil	5.0~10.0	1	4.000%	7.150	6	20.000%	7.358
Severe saline soil	10.0~20.0	21	84.000%	15.220	13	43.333%	15.519
Saline soil	>20.0	3	12.000%	26.475	8	26.667%	25.678

Table 2: Classification criteria of the soil salinization type and soil sample classification statistics.

$C(Cl^-)/2C(SO_4^{2-})$	Salinization type	Zone A		Zone B	
		Samples	Proportion (%)	Samples	Proportion (%)
<0.20	SO_4^{2-} salt type	22	88.000%	14	46.667%
0.20–1.00	Cl^--SO_4 salt type	3	12.000%	12	40.000%
1.00–2.00	SO_4-Cl^- salt type	0	0	3	10.000%
>2.00	Cl^- salt type	0	0	1	3.333%

Note: $C(Cl^-)$ and $C(SO_4^{2-})$, respectively, represent the millimole amount of Cl^- and SO_4^{2-} contained in 100 g soil.

Letnikov (G-L), and Caputo. Of which, the generally G-L FD can be defined as

$$d^{\alpha} f(x) = \lim_{h \to 0} \frac{1}{h^{\alpha}} \sum_{m-0}^{(t-a/h)} (-1)^m \cdot \frac{\Gamma(\alpha+1)}{m!\Gamma(\alpha-m+1)} f(x-mh), \tag{2}$$

where α is the order, h is the differential step size, and t and a are the upper and lower limits of the differential, respectively. Gamma function $\Gamma(z) = \int_0^{\infty} e^{-u} u^{z-1} du = (z-1)!$.

Set one variable function $f(x)$ to define domain $x \in [a, t]$ so that $h = 1$ (this is consistent with the 1 nm sampling interval of the spectrum instrument used for this study). $n = [t - a]$, then (2) can be used to derive the FD difference expression of function $f(x)$:

$$\frac{d^{v} f(x)}{dx^{v}} \approx f(x) + (-v) f(x-1) + \frac{(-v)(-v+1)}{2} f(x-2) + \cdots + \frac{\Gamma(-v+1)}{n!\Gamma(-v+n+1)} f(x-n), \tag{3}$$

where v is the order, the 0 order of function $f(x)$ is $f(x)$ itself. When $v = 1$ and 2, (3) is consistent with the 1st-order differential and 2nd-order differential formula when differential window size $\omega = 1$. In this study, the 0-order differential mathematical transformation refers to the transformation itself and does not undergo differential processing.

It can be seen from (3) that the FD has the features of memory and nonlocality; it is allowed to control the sensitivity to the slope and curvature of spectrum curve according to the weight of different differential orders, and this offers more range for band selection.

3.2. Correlation Coefficient Model Validation Index. CC is adopted to measure the linear correlation between two variables X and Y; its calculation formula can be described as [25–27]

$$CC_{XY} = \frac{\sum_{i=1}^{n} (X_i - \bar{X})(Y_i - \bar{Y})}{\sqrt{\sum_{i=1}^{n} (X_i - \bar{X})^2 \cdot \sum_{i=1}^{n} (Y_i - \bar{Y})^2}}, \tag{4}$$

where $\bar{X}$ and $\bar{Y}$ are, respectively, the average of X_i, and Y_i. CC_{XY} is in the range of −1 to +1. If the value of $|CC_{XY}|$ is closer to 1, it means that the linear relationship between X and Y is higher. In general, when CC_{XY} is higher than 0.8, X variable and Y variable can be regarded as highly relevant. When CC_{XY} is between 0.5 and 0.8, X variable and Y variable can be seen as moderately related. When CC_{XY} is between 0.3 and 0.5, these two variables can be considered as low correlation. When CC_{XY} is lower than 0.3, it means that the correlation between two variables is very weak and can be regarded as nonlinear correlation.

4. Results and Discussion

4.1. Analysis of Soil Salinity Characteristics for Different Interference Areas. According to the soil salt content [28], the degree of salinization was divided into nonsaline soil, mild saline soil, moderate saline soil, severe saline soil, and saline soil (Table 1). Area A was basically heavy saline soil, accounting for 84% of the total sample. Area B was affected by human activities, and the degree of salinization was more diverse, mainly heavy saline soil, saline soil, and moderate saline soil, accounting for 43.333%, 26.667%, and 20% of the total samples, respectively.

In view of the classification of the soil salinization type [28], the salinization type was divided into SO_4^{2-} salt type, Cl^--SO_4 salt type, SO_4-Cl^- salt type, and Cl^- salt type (Table 2). Zone A contained only the SO_4^{2-} and Cl^--

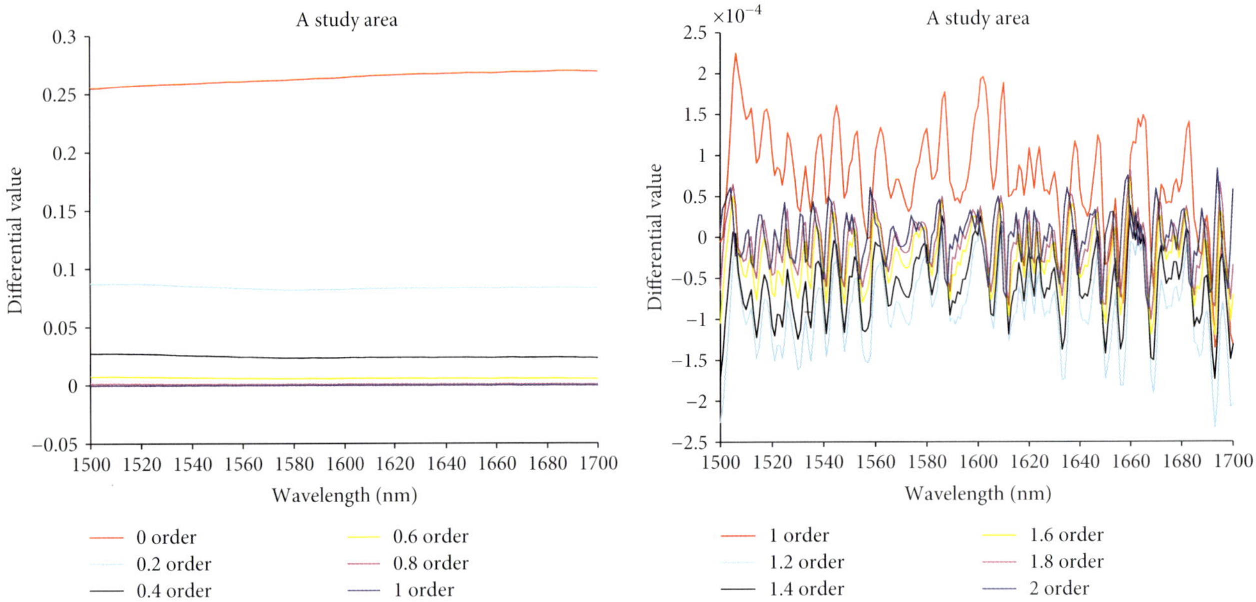

(a) (b)

Figure 4: Wavelength 1500–1700 reflectance FD value in area A. (a) 0–1 orders. (b) 1–2 orders.

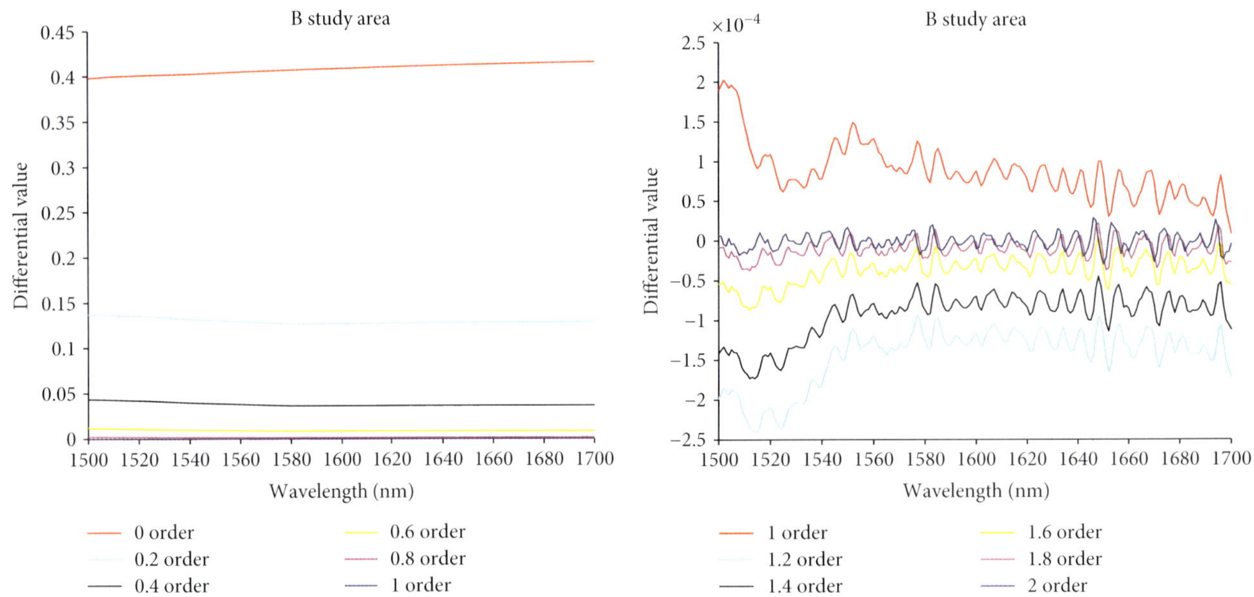

(a) (b)

Figure 5: Wavelength 1500–1700 reflectance FD value in area B. (a) 0–1 orders. (b) 1–2 orders.

SO_4 salt types. The soil in zone B contained all four types of salinization.

4.2. FD Simulation Results for Different Interference Areas. Based on the aforementioned formula derivation, the MATLAB R2015a was used for FD calculation and to process the spectrum data from area A and area B. The raw spectrum and the four types of transformation were put through 0–2nd-order differential processing with a total of 11 orders. Using the mean of the raw spectrum from areas A and B as an example, 11-order differential result was shown in Figures 4 and 5. The *X* axis was the wavelength, and the *Y* axis showed the reflectance value of each order after differential processing. Because there were too many bands, simulation result from 1500 to 1700 band was used for comparison purposes. Figures 4 and 5 showed that the overall differential value trend moved towards the reflectance value of 0. The change from 0 order to 1st order was the most obvious. The reflectance value of area A decreased from 0.25 at 0 order to 0.02 at 1st order. The reflectance value of area B decreased from 0.4 at 0 order to 0.02 at 1st order. This showed that the change in differential curve from 0 order to 1st order

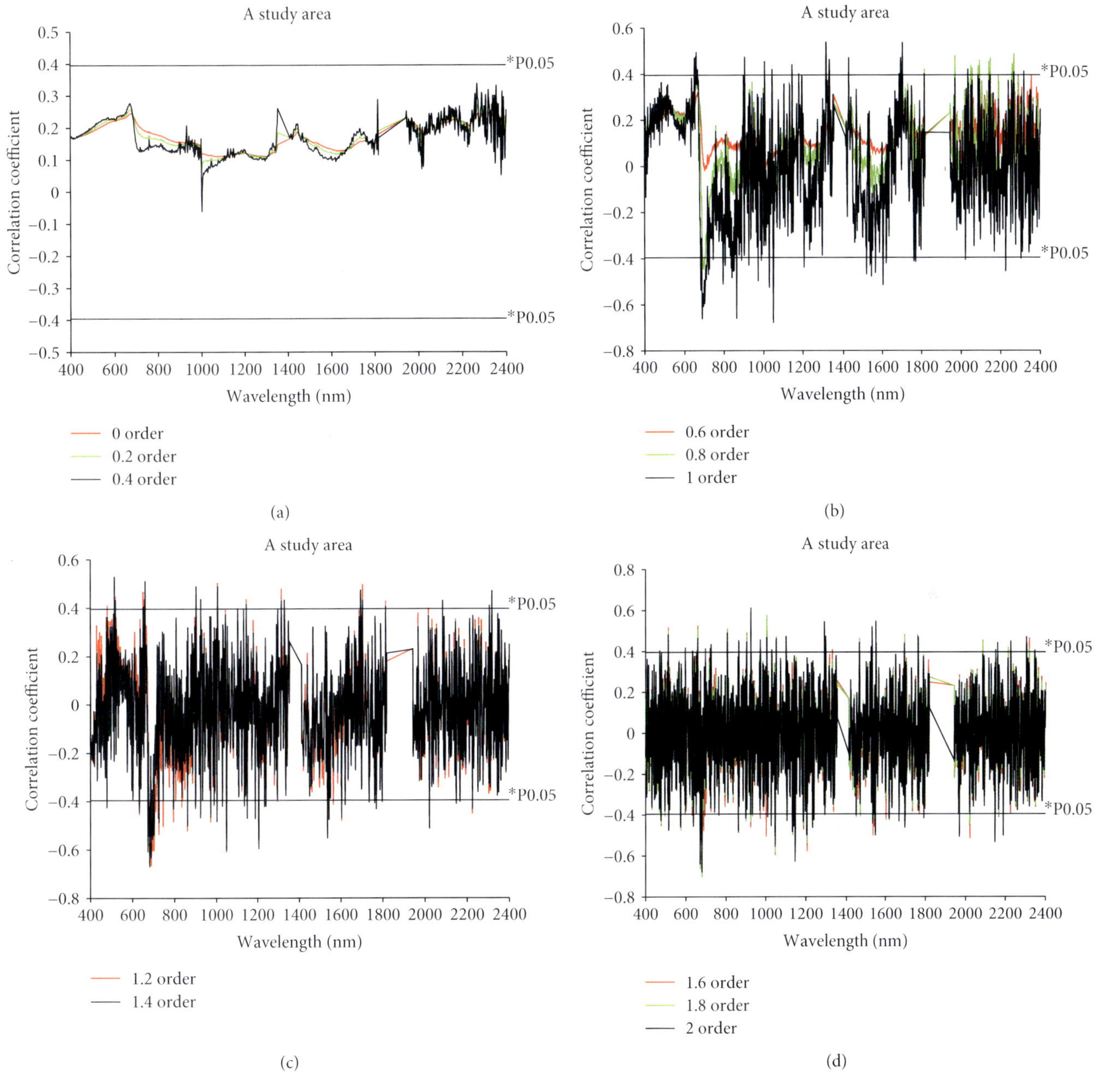

FIGURE 6: CC between salt content and raw spectral reflectance for area A. (a) 0–0.4-order differential. (b) 0.6–1st-order differential. (c) 1.2–1.4-order differential. (d) 1.6–2nd-order differential.

slowly moved towards the 1st-order differential curve. The difference between 1st-order and 2nd-differential values was very small. But the change trend still showed that in the process from 1st order to 2nd order, the differential curve slowly moved towards the 2nd-order differential curve. This verified that FD was sensitive and that FD calculation could render small differences between spectra under different human interference levels.

4.3. Effect of FD on the Correlation for the Full-Band Spectrum. Before quantitative analysis is conducted on the spectrum features of saline soil under different human interferences, we should first study the correlation between spectra and salt content and find the sensitive bands, which can increase the accuracy of the inversion model. In this study, we used raw spectral reflectance (*R*) as an example. Spectral reflectance was processed with MATLAB R2015 software using 0.2 as the order interval to analyze the correlation between salt content and *R*'s 11-order differential (from 0 to 2nd order). The 0.05 test was used to determine the significance level. The ∗P0.05 for area A and area B was equal to plus or minus 0.396 and 0.361, respectively. Simulation result was shown in Figures 6 and 7.

When area A's raw spectral reflectance was between 0 and 0.5 orders, no band passed the 0.05 significance test. However, starting from 0.6 order, the bands passed the 0.05 significance test. Raw spectral reflectance bands in area B passed the 0.05 significance test starting from 0 order.

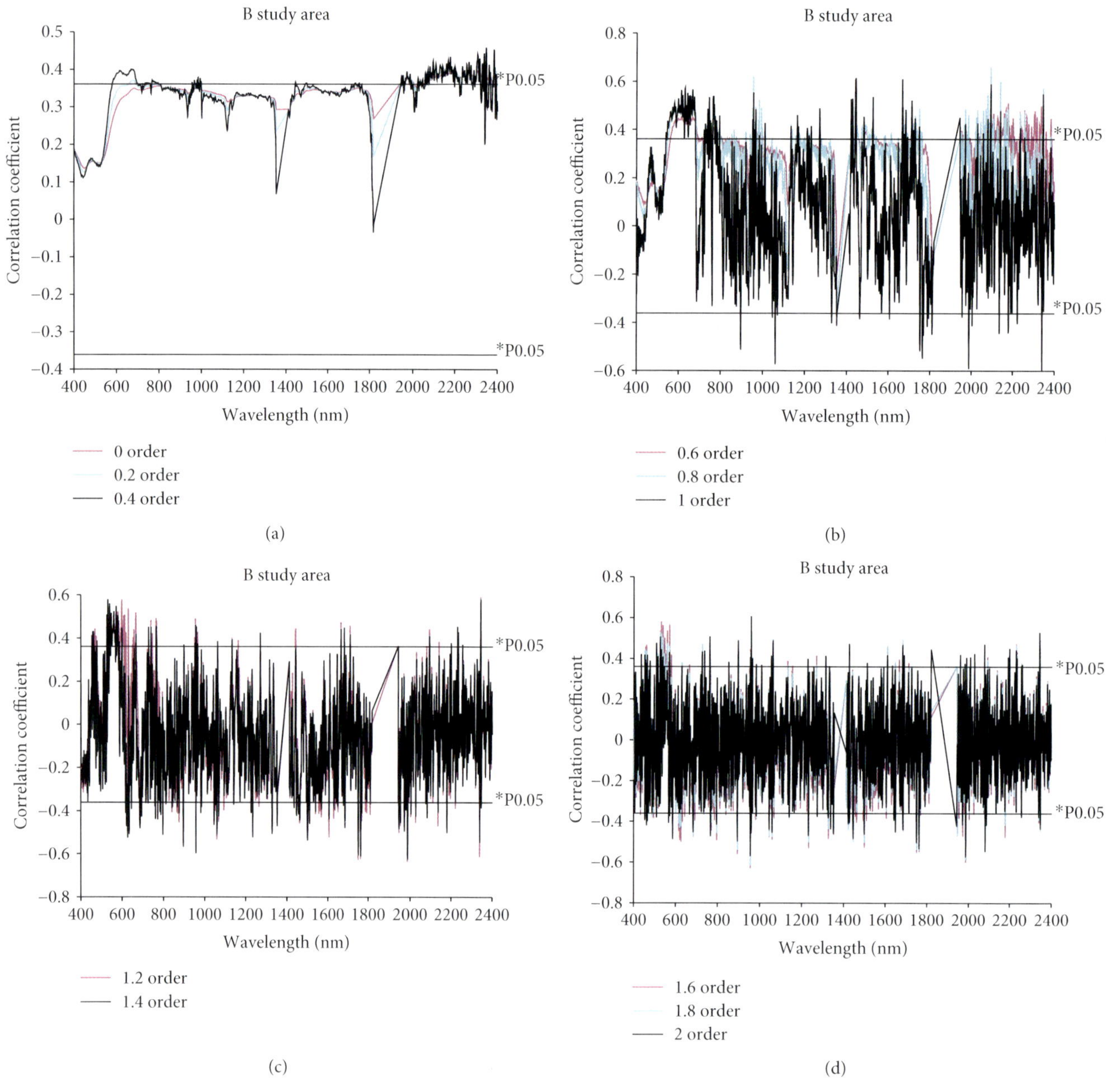

FIGURE 7: CC between salt content and raw spectral reflectance for area B. (a) 0–0.4-order differential. (b) 0.6–1st-order differential. (c) 1.2–1.4-order differential. (d) 1.6–2nd order differential.

Simulation result showed that FD calculation refined the curve change trend and reduced the loss of differential information. From 0 order to 0.7 order, the CC curve gradually showed a gradual trend and the gradual trend between each order's differential curve was presented. From 0.8 order to 2nd order, the CC curve of a gradual trend became less obvious and the curve fluctuation became greater.

4.4. Number of Bands Passed the 0.05 Significance Test. From Figures 6 and 7, it could be seen that as the order increases, so did the number of bands from areas A and B that passed the 0.05 significance test. However, the specific number of increase still required further statistical analysis. Thus, the number of bands from different order differentials for the five spectrum transformations that passed the 0.05 significance test was shown in Figure 8.

In area with no human interference, the sequence of the number of bands that passed the 0.05 significance test was as follows (from high to low): $1/R > \text{Lg}R > \sqrt{R} > R > 1/\text{Lg}R$. In area with human interference, the sequence of the number of bands that passed the 0.05 significance test was as follows (from high to low): $1/\text{Lg}R > R > \sqrt{R} > \text{Lg}R > 1/R$. In addition, the trend of bands that passed the 0.05 significance test showed an increase first then decrease later.

4.5. Bands with the Largest CC Absolute Value. In the five spectrum transformations, there were corresponding band information to the largest CC absolute value of 11 orders,

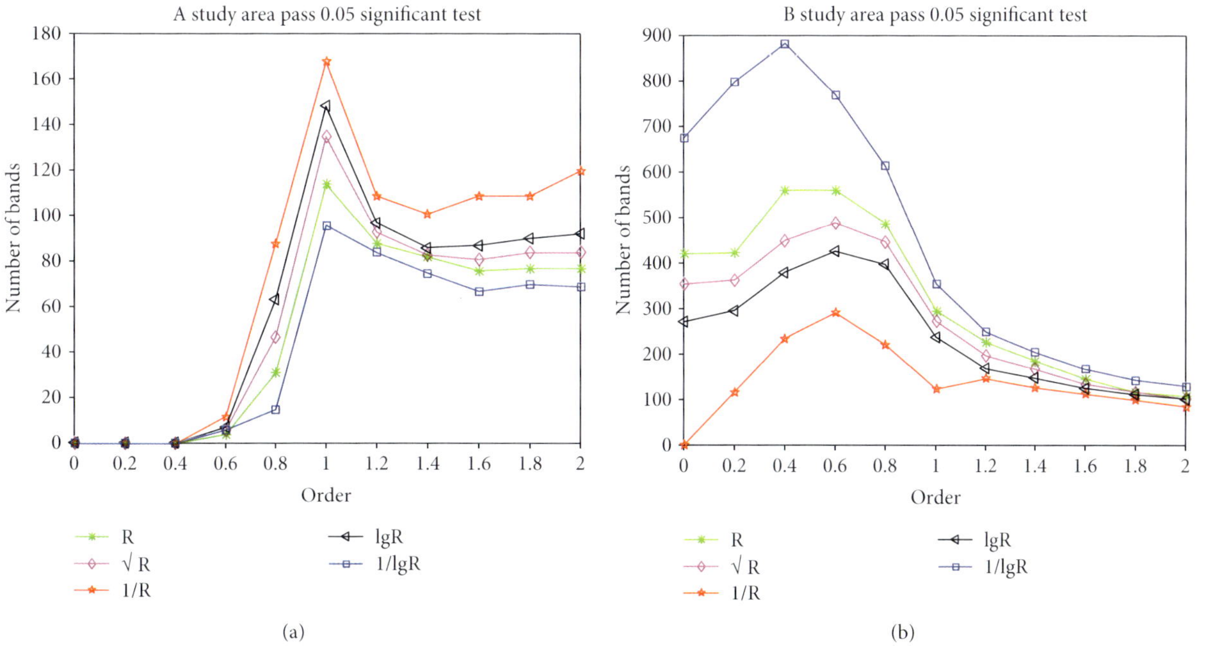

Figure 8: Number of bands passed the 0.05 significance test. (a) Area with no human interference. (b) Area with human interference.

Table 3: The bands with the largest CC absolute value in the no human interference area.

Order	R		$\sqrt{R}$		$1/R$ 676		LgR 676		1/LgR 1946	
	Absolute	Band	Absolute	Band	Absolute	Band	Absolute	Band	Absolute	Band
0	0.247549	2355	0.258711	676	0.303965	676	0.273546	676	0.24171	1946
0.2	0.272918	2355	0.279275	2355	0.325149	672	0.289407	672	0.263505	2355
0.4	0.337521	2268	0.343669	2268	0.362546	2268	0.349886	2268	0.330819	2355
0.6	0.452251	2267	0.454245	2268	0.478337	2268	0.462786	2268	0.448415	2267
0.8	0.488129	2267	0.558134	693	0.651624	690	0.635344	693	0.487335	2267
1	0.680359	1049	0.694649	1049	0.693701	1049	0.700773	1049	0.649393	862
1.2	0.670618	689	0.674993	690	0.664982	690	0.678021	690	0.654813	689
1.4	0.672355	682	0.678587	682	0.666791	682	0.678755	682	0.656751	682
1.6	0.704214	682	0.70968	682	0.699427	682	0.710109	682	0.69164	682
1.8	0.708322	682	0.714082	682	0.709584	682	0.715906	682	0.697884	682
2	0.680375	681	0.686301	681	0.700033	1148	0.68948	681	0.671469	681

Table 4: The bands with the largest CC absolute value in the human interference area.

Order	R		$\sqrt{R}$		$1/R$ 676		LgR 676		1/LgR 1946	
	Absolute	Band	Absolute	Band	Absolute	Band	Absolute	Band	Absolute	Band
0	0.39693	2384	0.390787	2384	0.361249	2384	0.382675	2384	0.407767	2310
0.2	0.41892	2384	0.412882	2384	0.384573	2383	0.405559	2383	0.4258	2384
0.4	0.45916	2346	0.456893	2383	0.457625	614	0.455213	2383	0.489956	2346
0.6	0.57097	2346	0.545842	2346	0.491625	668	0.515645	2346	0.608853	2346
0.8	0.66103	2095	0.640903	2095	0.593338	955	0.612877	955	0.67799	2095
1	0.61547	1443	0.615054	1988	0.598832	1988	0.625937	1988	0.64202	2095
1.2	0.63752	1988	0.64851	1988	0.631102	1764	0.652144	1988	0.629525	630
1.4	0.62432	1988	0.635681	1988	0.638534	1988	0.643132	1988	0.62093	531
1.6	0.63035	957	0.621285	957	0.589642	1987	0.610791	1987	0.64168	957
1.8	0.61592	957	0.605922	957	0.563155	957	0.593538	957	0.630378	957
2	0.60516	958	0.595053	958	0.553827	958	0.582938	958	0.621153	958

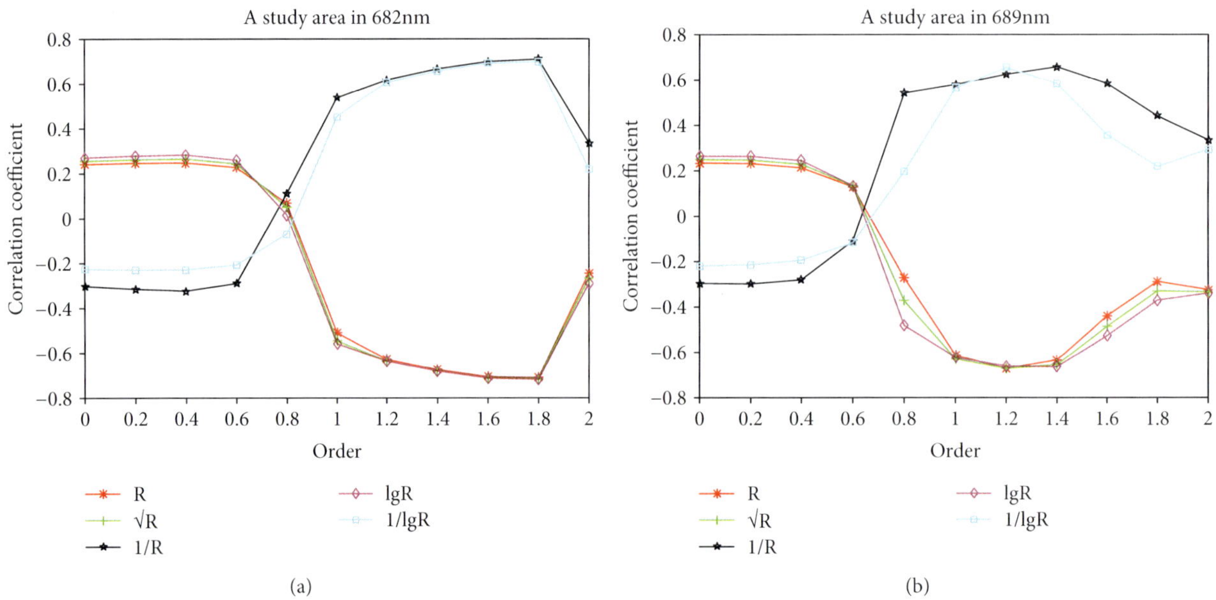

Figure 9: Changes in CC for area A. (a) 682 nm. (b) 689 nm.

as shown in Tables 3 and 4. In area A, the corresponding bands of the largest CC absolute value from various order differentials in the five transformations were in the 682 nm and they were all at 1.8 order. Also, the corresponding bands were in the 2095 nm and 1998 nm for area B, and they were located at 0.8 order or 1.2 order or 1.4 order.

Moreover, in the no human interference area, the largest CC of R is 0.708322, which was raised to the maximum CC of 0.715906 after LgR transformation at 1.8 order. Also, in the human interference area, R's biggest CC was 0.66103, it was promoted to the maximum CC of 0.67799 after 1/LgR transformation at 0.8 order. Statistical results showed that the largest CC from the five spectrum transformations in the no human interference area and human interference area appeared in the fractional order.

In addition, because the area A was affected by human activities, the soil characteristics were more complicated, so the maximum CC for area A was greater than area B. Moreover, the maximum CC absolute value was greater than 0.6, thus verified that the spectral reflectance and salt content were moderately related, close to the height correlation especially for area A, indicating that our research was valuable and the salt content could be predicted by spectra. So, we can use the pretreatment method for fractional differential to further verify the model in subsequent studies.

4.6. Impact of FD on the Correlation for Some Single-Band Spectrum. To further clarify the effect of FD calculation on some band's correlation, we chose the four bands in Tables 3 and 4 for our study on changes in CC as fractional orders change. For area A, we chose 682 nm, and 689 nm. For area B, we chose 957 nm and 1988 nm, as shown in Figures 9 and 10. FD calculation processing of the five spectral reflectance transformations significantly increased the absolute value of the CC; the maximum CC all appeared in the fractional order, and it was not in the integer differential (such as 1st or 2nd order) as in traditional analysis methods. We used 682 nm and 957 nm as examples to discuss the single-band correlation.

Figure 9(a) showed that the maximum CC absolute value of 682 nm for R, $\sqrt{R}$, $1/R$, LgR, and 1/LgR was 0.70832, 0.71408, 0.70958, 0.71591, and 0.6978, respectively, which were at 1.8 order. Compared with 1st order, the maximum CC absolute value of five transformations were increased by 0.20091, 0.17003, 0.17307, 0.15902, and 0.24417, respectively. And they were improved by 0.46656, 0.44956, 0.37373, 0.42777, and 0.47757, respectively, when compared with 2nd order. Among them, the 1.8-order 1/LgR transformation had the largest improvement compared with the traditional integer order 1st order or 2nd order.

Figure 10(a) revealed that the largest CC absolute value of 957 nm for R, $\sqrt{R}$, $1/R$, LgR, and 1/LgR was 0.63035, 0.62128, 0.57891, 0.60951, and 0.64168, respectively, all located at 1.6 order. Compared with the first order, the largest CC absolute value of five transformations were increased by 0.32788, 0.29408, 0.20445, 0.26160, and 0.39874, respectively. Compared with the second order, they were improved by 0.51078, 0.50334, 0.48060, 0.49566, and 0.53342, respectively. Among them, the 1.6-order 1/LgR transformation had the biggest improvement compared with the traditional integer order 1st order or 2nd order.

Furthermore, from Figures 9 and 10, it could be seen that the maximum CC for area A was also greater than area B for some single-band spectrum; it was also caused by human activities in area B to influence the soil features.

The main reason for the aforementioned results is that FD is an extension of the integer differential concept. Integer differential is only one of the special cases. Those can be explained in two ways.

On the one hand, due to fractional differential's memory and nonlocality, integer differential is only related to the band reflectance within the differential window. Equation

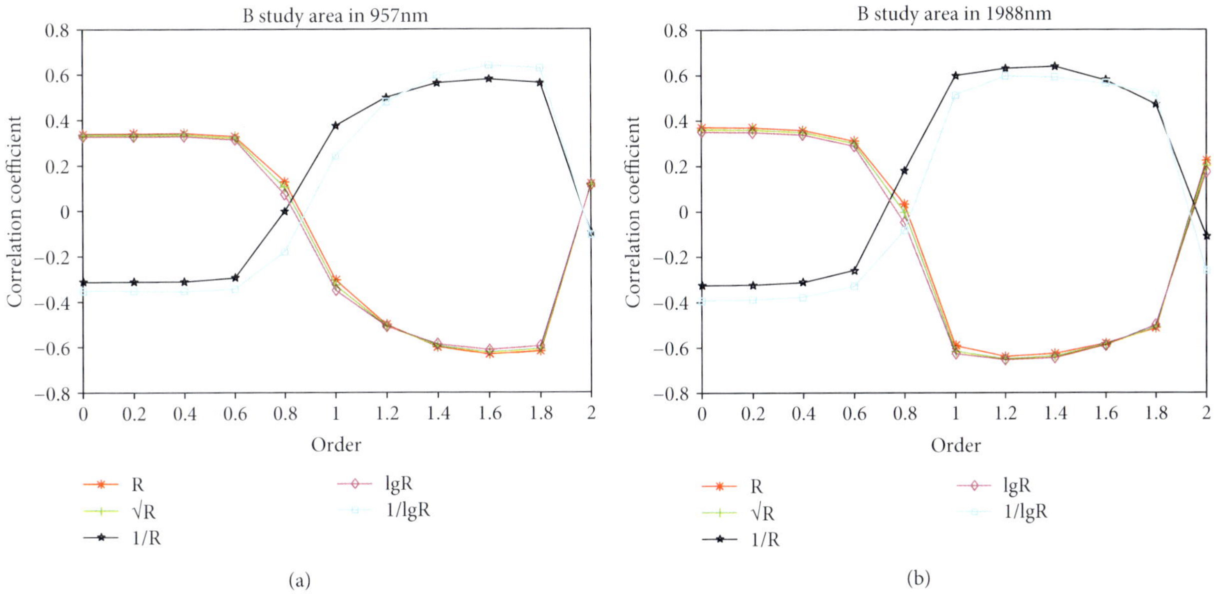

FIGURE 10: Changes in CC for area B. (a) 957 nm. (b) 1988 nm.

(3) shows that the FD value of certain bands is not only related to that band's reflectance but is also related to all the previous reflectance of that band [8]. Also, each point's reflectance is given a different weight according to the differential order. Greater weight is given to closer points. Thus, that band's FD impact is also greater. In addition, this is the one of the greatest difference between FD and integer differential.

On the other hand, some studies show that many systems belong to fractional order [8, 9, 29, 30], and we traditionally use the integer-order differential model for spectral processing; this leads to a greater difference between the model and actual results, and low accuracy of system simulation and prediction. As a result, this also neglects the system's realism.

5. Conclusion

In this study, we introduced FD into hyperspectral data pretreatment. Saline soil spectral reflectance data from field spectrum measurements taking in Xinjiang's Fukang City was adopted as a basis. FD processing was used on raw spectral reflectance and its common mathematical transformation to explore the effect of FD calculation on the CC between saline soil hyperspectral reflectance and soil salt content in fields with different levels of human interference. The main conclusion was that the soil properties of zone B were more complex than zone A, because zone B was affected by human activities. Under different human interference levels, FD calculation could describe small difference between spectra. The overall differential value trend was towards a reflectance value of 0. After differential processing, the number of bands that passed the 0.05 significance test first increased then decreased; the changes in trends were depicted through fractional differential. For different human interference areas, the largest CC absolute value of the five spectrum transformations all appeared in the fractional order. For full-band or some single bands (689 nm, 682 nm, 1988 nm, and 957 nm), the differential transformation could significantly increase the correlation between spectral reflectance and soil salt content. The extreme value could be obtained in the fractional order. Therefore, FD enriched the hyperspectral data pretreatment. We explored latent spectrum data information from a FD perspective and used hyperspectral data to provide a new perspective for field hyperspectral monitoring of soil salinization in areas with different levels of human interference.

Conflicts of Interest

The authors declare that they have no conflict of interest.

Acknowledgments

This research was supported by the financial support of the National Natural Science Foundation of China (41761081, 41671198, and 41861054) and Yunnan Province Science and Technology Department and Education Department Project of China (2017FH001-067, 2017FH001-117, and 2016ZDX127).

References

[1] X. X. Qiao, C. Wang, M. C. Feng et al., "Hyperspectral estimation of soil organic matter based on different spectral preprocessing techniques," *Spectroscopy Letters*, vol. 50, no. 3, pp. 156–163, 2017.

[2] Y. M. Zheng, T. Q. Zhang, J. Zhang, X. D. Chen, and X. G. Shen, "Influence of smooth, 1st derivative and baseline correction on the near-infrared spectrum analysis with PLS,"

Spectroscopy and Spectral Analysis, vol. 24, no. 12, pp. 1546–1548, 2004.

[3] F. Zhang, J. L. Ding, Tashpolat, Tiyip, and Q. S. He, "Spectral data analysis of salinity soils with ground objects in the delta oasis of Weigan and Kuqa Rivers," *Spectroscopy and Spectral Analysis*, vol. 28, no. 12, pp. 2921–2926, 2008.

[4] J. Wang, J. Ding, A. Abulimiti, and L. Cai, "Quantitative estimation of soil salinity by means of different modeling methods and visible-near infrared (VIS-NIR) spectroscopy, Ebinur Lake Wetland, Northwest China," *PeerJ*, vol. 6, article e4703, 2018.

[5] J. M. Schmitt, "Fractional derivative analysis of diffuse reflectance spectra," *Applied Spectroscopy*, vol. 52, no. 6, pp. 840–846, 1998.

[6] S. S. Kharintsev and M. K. Salakhov, "A simple method to extract spectral parameters using fractional derivative spectrometry," *Spectrochimica Acta Part A: Molecular and Biomolecular Spectroscopy*, vol. 60, no. 8-9, pp. 2125–2133, 2004.

[7] K.-Y. Zheng, X. Zhang, P.-J. Tong, Y. Yao, and Y.-P. Du, "Pretreating near infrared spectra with fractional order Savitzky–Golay differentiation (FOSGD)," *Chinese Chemical Letters*, vol. 26, no. 3, pp. 293–296, 2015.

[8] Z. Dong, T. Tiyip, Z. Fei, and A. Kelimu, "Application of fractional differential in preprocessing hyperspectral data of saline soil," *Transactions of the Chinese Society of Agricultural Engineering*, vol. 30, no. 24, pp. 151–160, 2014.

[9] N. Xia, T. Tiyip, A. Kelimu et al., "Influence of fractional differential on correlation coefficient between $EC_{1:5}$ and reflectance spectra of saline soil," *Journal of Spectroscopy*, vol. 2017, Article ID 1236329, 11 pages, 2017.

[10] P. C. Duan, H. G. Xiong, R. R. Li, and L. Zhang, "A quantitative analysis of the reflectance of the saline soil under different disturbance extent," *Spectroscopy and Spectral Analysis*, vol. 37, no. 2, pp. 571–576, 2017.

[11] C. R. Bern, M. L. Clark, T. S. Schmidt, J. A. M. Holloway, and R. R. Mcdougal, "Soil disturbance as a driver of increased stream salinity in a semiarid watershed undergoing energy development," *Journal of Hydrology*, vol. 524, pp. 123–136, 2015.

[12] H. Triki Fourati, M. Bouaziz, M. Benzina, and S. Bouaziz, "Modeling of soil salinity within a semi-arid region using spectral analysis," *Arabian Journal of Geosciences*, vol. 8, no. 12, pp. 11175–11182, 2015.

[13] M. Elhag, "Evaluation of different soil salinity mapping using remote sensing techniques in arid ecosystems, Saudi Arabia," *Journal of Sensors*, vol. 2016, Article ID 7596175, 8 pages, 2016.

[14] A. Abbas, S. Khan, N. Hussain, M. A. Hanjra, and S. Akbar, "Characterizing soil salinity in irrigated agriculture using a remote sensing approach," *Physics and Chemistry of the Earth, Parts A/B/C*, vol. 55-57, pp. 43–52, 2013.

[15] J. Ding and D. Yu, "Monitoring and evaluating spatial variability of soil salinity in dry and wet seasons in the Werigan–Kuqa Oasis, China, using remote sensing and electromagnetic induction instruments," *Geoderma*, vol. 235-236, pp. 316–322, 2014.

[16] Z. Miao and W. Shi, "A new methodology for spectral-spatial classification of hyperspectral images," *Journal of Sensors*, vol. 2016, Article ID 1538973, 12 pages, 2016.

[17] T. Shi, L. Cui, J. Wang, T. Fei, Y. Chen, and G. Wu, "Comparison of multivariate methods for estimating soil total nitrogen with visible/near-infrared spectroscopy," *Plant and Soil*, vol. 366, no. 1-2, pp. 363–375, 2013.

[18] P. Pornchaloempong, P. Sirisomboon, and S. Pongkuan, "Rapid evaluation of the salt content of canned sardines in brine using near-infrared diffuse reflectance spectroscopy," *Spectroscopy Letters*, vol. 49, no. 10, pp. 613–618, 2016.

[19] J. L. Ding, Y. Yao, and F. Wang, "Detecting soil salinization in arid regions using spectral feature space derived from remote sensing data," *Acta Ecologica Sinica*, vol. 34, no. 16, 2014.

[20] S. Chen, X. Bin, Y. X. Jin et al., "Remote sensing monitoring and spatial-temporal features analysis of soil salinization in agricultural area of Northern Xinjiang," *Scientia Geographica Sinica*, vol. 35, no. 12, pp. 1607–1615, 2015.

[21] D. An, G. Zhao, C. Chang et al., "Hyperspectral field estimation and remote-sensing inversion of salt content in coastal saline soils of the Yellow River Delta," *International Journal of Remote Sensing*, vol. 37, no. 2, pp. 455–470, 2016.

[22] R. Sahadevan and P. Prakash, "Exact solutions and maximal dimension of invariant subspaces of time fractional coupled nonlinear partial differential equations," *Communications in Nonlinear Science and Numerical Simulation*, vol. 42, pp. 158–177, 2017.

[23] B. Safarinejadian, M. Asad, and M. S. Sadeghi, "Simultaneous state estimation and parameter identification in linear fractional order systems using coloured measurement noise," *International Journal of Control*, vol. 89, no. 11, pp. 2277–2296, 2016.

[24] S. M. Shah, R. Samar, N. M. Khan, and M. A. Z. Raja, "Fractional-order adaptive signal processing strategies for active noise control systems," *Nonlinear Dynamics*, vol. 85, no. 3, pp. 1363–1376, 2016.

[25] A. C. Iznaga, M. R. Orozco, E. A. Alcantara et al., "Vis/NIR spectroscopic measurement of selected soil fertility parameters of Cuban agricultural Cambisols," *Biosystems Engineering*, vol. 125, no. 125, pp. 105–121, 2014.

[26] F. Zhang, T. Tiyip, J. Ding et al., "Studies on the reflectance spectral features of saline soil along the middle reaches of Tarim River: a case study in Xinjiang Autonomous Region, China," *Environmental Earth Sciences*, vol. 69, no. 8, pp. 2743–2761, 2013.

[27] M. C. Sarathjith, B. S. Das, S. P. Wani, and K. L. Sahrawat, "Dependency measures for assessing the covariation of spectrally active and inactive soil properties in diffuse reflectance spectroscopy," *Soil Science Society of America Journal*, vol. 78, no. 5, 2014.

[28] Z. Dong, *Application Research of Fractional Derivative on the Remote Sensing Monitoring of Soil Salinization [D]*, Xinjiang University, Urumqi, 2017.

[29] T. J. Freeborn, B. Maundy, and A. S. Elwakil, "Measurement of supercapacitor fractional-order model parameters from voltage-excited step response," *IEEE Journal on Emerging and Selected Topics in Circuits and Systems*, vol. 3, no. 3, pp. 367–376, 2013.

[30] G. S. M. Ngueuteu and P. Woafo, "Dynamics and synchronization analysis of coupled fractional-order nonlinear electromechanical systems," *Mechanics Research Communications*, vol. 46, pp. 20–25, 2012.

PERMISSIONS

All chapters in this book were first published in JS, by Hindawi Publishing Corporation; hereby published with permission under the Creative Commons Attribution License or equivalent. Every chapter published in this book has been scrutinized by our experts. Their significance has been extensively debated. The topics covered herein carry significant findings which will fuel the growth of the discipline. They may even be implemented as practical applications or may be referred to as a beginning point for another development.

The contributors of this book come from diverse backgrounds, making this book a truly international effort. This book will bring forth new frontiers with its revolutionizing research information and detailed analysis of the nascent developments around the world.

We would like to thank all the contributing authors for lending their expertise to make the book truly unique. They have played a crucial role in the development of this book. Without their invaluable contributions this book wouldn't have been possible. They have made vital efforts to compile up to date information on the varied aspects of this subject to make this book a valuable addition to the collection of many professionals and students.

This book was conceptualized with the vision of imparting up-to-date information and advanced data in this field. To ensure the same, a matchless editorial board was set up. Every individual on the board went through rigorous rounds of assessment to prove their worth. After which they invested a large part of their time researching and compiling the most relevant data for our readers.

The editorial board has been involved in producing this book since its inception. They have spent rigorous hours researching and exploring the diverse topics which have resulted in the successful publishing of this book. They have passed on their knowledge of decades through this book. To expedite this challenging task, the publisher supported the team at every step. A small team of assistant editors was also appointed to further simplify the editing procedure and attain best results for the readers.

Apart from the editorial board, the designing team has also invested a significant amount of their time in understanding the subject and creating the most relevant covers. They scrutinized every image to scout for the most suitable representation of the subject and create an appropriate cover for the book.

The publishing team has been an ardent support to the editorial, designing and production team. Their endless efforts to recruit the best for this project, has resulted in the accomplishment of this book. They are a veteran in the field of academics and their pool of knowledge is as vast as their experience in printing. Their expertise and guidance has proved useful at every step. Their uncompromising quality standards have made this book an exceptional effort. Their encouragement from time to time has been an inspiration for everyone.

The publisher and the editorial board hope that this book will prove to be a valuable piece of knowledge for researchers, students, practitioners and scholars across the globe.

LIST OF CONTRIBUTORS

Jinru Xue and Baofeng Su
College of Mechanical and Electronic Engineering, Northwest A&F University, Yangling, Shaanxi 712100, China

Pei Liu, Shoujun Jia, Ruimei Han and Hanwei Zhang
Key Laboratory of Mine Spatial Information Technologies of SBSM, Henan Polytechnic University, Jiaozuo 454003, China
School of Surveying and Mapping Land Information Engineering, Henan Polytechnic University, Jiaozuo 454003, China

Shumin Wang, Ling Ding, Zihan Chen and Aixia Dou
Institute of Earthquake Forecasting, China Earthquake Administration, Beijing 100036, China

Xiangyang Zhou
School of Instrumentation Science and Opto-electronics Engineering, Beihang University (BUAA), Beijing 100191, China
State Key Laboratory for Manufacturing Systems Engineering, Xi'an Jiaotong University, Xi'an 710049, China

Libo Zhao
State Key Laboratory for Manufacturing Systems Engineering, Xi'an Jiaotong University, Xi'an 710049, China

Chao Yang, Beilei Zhao and Zhuangsheng Zhu
School of Instrumentation Science and Opto-electronics Engineering, Beihang University (BUAA), Beijing 100191, China

Chun Liu, Mengchi Ai, Zhuo Chen and Hangbin Wu
College of Surveying and Geo-Informatics, Tongji University, Shanghai 200092, China

Yuan Zhou
Tongfang Surveying Engineering and Technology Co. Ltd., Shanghai 200092, China

Youkyung Han
School of Convergence & Fusion System Engineering, Kyungpook National University, Sangju 37224, Republic of Korea

Anjin Chang
School of Engineering and Computing Sciences, Texas A&M University-Corpus Christi, 6300 Ocean Dr., Corpus Christi, TX 78412, USA

Seokkeun Choi, Honglyun Park and Jaewan Choi
School of Civil Engineering, Chungbuk National University, 1, Chungdae-ro, Seowon-gu, Cheongju, Chungbuk 28644, Republic of Korea

Hongmin Gao, Shuo Lin, Yao Yang and Chenming Li
College of Computer and Information Engineering, Hohai University, Nanjing 211100, China

Mingxiang Yang
State Key Laboratory of Simulation and Regulation of Water Cycle in River Basin, China Institute of Water Resources and mHydropower Research, Beijing 100038, China

Pengyan Zhang and Yunzeng Chen
College of Environment and Planning, Henan University, Kaifeng 475004, China

Wenlong Jing
Guangzhou Institute of Geography, Guangzhou 510070, China
Key Laboratory of Guangdong for Utilization of Remote Sensing and Geographical Information System, Guangzhou 510070, China
Guangdong Open Laboratory of Geospatial Information Technology and Application, Guangzhou 510070, China

Xiaochen Wang, Wei Tian, Kun Li and Long Liu
Laboratory of Target Microwave Properties, Deqing Academy of Satellite Applications, Huzhou 313200, China
Institute of Remote Sensing and Digital Earth, Chinese Academy of Sciences, 1 Beichen Road, Beijing 100101, China
University of Chinese Academy of Sciences, 19A Yuquan Road, Beijing 100049, China

Yun Shao
Laboratory of Target Microwave Properties, Deqing Academy of Satellite Applications, Huzhou 313200, China

Institute of Remote Sensing and Digital Earth, Chinese Academy of Sciences, 1 Beichen Road, Beijing 100101, China
University of Chinese Academy of Sciences, 19A Yuquan Road, Beijing 100049, China
State Key Laboratory of Remote Sensing Science, Institute of Remote Sensing and Digital Earth, Chinese Academy of Sciences, 1 Beichen Road, Beijing 100101, China

Lu She
Institute of Remote Sensing and Digital Earth, Chinese Academy of Sciences, 1 Beichen Road, Beijing 100101, China
University of Chinese Academy of Sciences, 19A Yuquan Road, Beijing 100049, China

Na-Yeon Park, Jae-Dong Jang, Eun-Ha Sohn and Jun-Dong Park
National Meteorological Satellite Center, Korea Meteorological Administration, Jincheon-gun 27803, Republic of Korea

Youngmi Kim
National Institute of Meteorological Sciences, Korea Meteorological Administration, Seogwipo- si 63568, Republic of Korea

Mi-Lim Ou
Climate Change Monitoring Bureau, Korea Meteorological Administration, Seoul 07062, Republic of Korea

Minji Seo, Kyeong-Sang Lee and Kyung-Soo Han
Department of Spatial Information Engineering, Pukyong National University, Busan 48513, Republic of Korea

Yun-Jae Choung
Research Institute of Spatial Information Technology, Geo C&I Co. Ltd., 435 Hwarang-ro, Dong gu, Daegu 41165, Republic of Korea

Myung-Hee Jo
Department of Aero-Satellite Geo-Informatics Engineering, School of Convergence and Fusion System Engineering, College of Science and Technology, Kyungpook National University, 2559 Gyeongsang-daero, Sangju 37224, Republic of Korea

Yan Gao, Zeyu Liang, Biao Wang, Yanlan Wu and Penghai Wu
School of Resources and Environmental Engineering, Anhui University, Hefei, Anhui 230601, China

Hai Lan
Department of Computer Science and Engineering, Tandon School of Engineering, New York University, Brooklyn, NY 11201, USA

Xinshi Zheng
Center for Urban Science + Progress, New York University, Brooklyn, NY 11201, USA

Paul M. Torrens
Department of Computer Science and Engineering, Tandon School of Engineering, New York University, Brooklyn, NY 11201, USA
Center for Urban Science + Progress, New York University, Brooklyn, NY 11201, USA

Askar, Narissara Nuthammachot and Worradorn Phairuang
Faculty of Environmental Management, Prince of Songkla University (PSU), 90110 Songkhla, Thailand

Pramaditya Wicaksono
Cartography and Remote Sensing, Faculty of Geography, Gadjah Mada University, 55281 Yogyakarta, Indonesia

Tri Sayektiningsih
Research and Development Institute for Natural Resource Conservation Technology, Ministry of Environment and Forestry of Indonesia, 76112 East Kalimantan, Indonesia

Yong Fan
School of Geographic Sciences, Xinyang NormalUniversity, No. 237 Nanhu Road, Xinyang 464000, China
School of Resource and Environmental Sciences, Wuhan University, No. 129 Luoyu Road, Wuhan 430079, China

Xi Zhu
ShanghaiMunicipal Institute of Surveying andMapping, No. 419Wuning Road, Shanghai 200063, China

Zongyi He, Feiyan Chen and Xiaolin Peng
School of Resource and Environmental Sciences, Wuhan University, No. 129 Luoyu Road, Wuhan 430079, China

Shengdi Zhang
School of Geo-Sciences, The University of Edinburgh, Central Campus, Drummond Street, Edinburgh EH89XP, UK

Junbo Gao
School of Geographic Sciences, Xinyang NormalUniversity, No. 237 Nanhu Road, Xinyang 464000, China

Junli Li
School of Resources and Environment, Anhui Agricultural University, No. 130 ChangjiangWest Road, Hefei 230036, China

Anhong Tian and Chengbiao Fu
Faculty of Land Resource Engineering, Kunming University of Science and Technology, Kunming, Yunnan 650093, China
College of Information Engineering, Qujing Normal University, Qujing, Yunnan 655011, China

Junsan Zhao and Shu Gan
Faculty of Land Resource Engineering, Kunming University of Science and Technology, Kunming, Yunnan 650093, China

Heigang Xiong
College of Applied Arts and Science, Beijing Union University, Beijing 100083, China
College of Resource and Environment Sciences, Xinjiang University, Urumqi, Xinjiang 830046, China

Index

S

T

U

V

Y